FOOD MICROBIOLOGY AND HYGIENE

Second Edition

FOOD MICROBIOLOGY AND HYGIENE

Second Edition

P. R. HAYES

Department of Microbiology, University of Leeds, UK

ELSEVIER APPLIED SCIENCE
LONDON and NEW YORK

ELSEVIER SCIENCE PUBLISHERS LTD
Crown House, Linton Road, Barking, Essex IG11 8JU, England

WITH 39 TABLES AND 97 ILLUSTRATIONS

First edition 1985
Second edition 1992

© 1992 ELSEVIER SCIENCE PUBLISHERS LTD

British Library Cataloguing in Publication Data

A catalogue record for this book is
available from the British Library

ISBN 1–85166-873–X

Library of Congress Cataloging-In-Publication Data

Hayes, P. R.
    Food microbiology and hygiene / P. R. Hayes. — 2nd ed.
    p.   cm.
    Includes bibliographical references and index.
    ISBN 1–85166–873-X
    1. Food Handling.   2. Food—Microbiology.   3. Food service—
Sanitation.   I. Title.
TX537.H35   1992
664'.0028'9—dc20   92–10943
                                                                  CIP

Printed by Galliard (Printers) Ltd, Great Yarmouth, England

PREFACE TO THE SECOND EDITION

The aims of this book remain the same, that is, that it should be of interest to all those people concerned with, or about, food hygiene in the broadest sense.

There was clearly a need for a book of this sort and its success has necessitated a second edition. It will, I hope, answer criticisms that were justifiably made about certain omissions and shortcomings levelled at the earlier edition. The whole book has been thoroughly revised with the introduction of several new sections to various chapters.

During the time that has elapsed since the earlier edition appeared there has been much publicity about newer forms of 'food poisoning'. Thus listeriosis is discussed in some detail whilst the problems of salmonellas in eggs and BSE are also considered. Interest in irradiated foods has waxed and waned but it is rightly included in the relevant chapter. There has been much progress in methodology with the advent of advanced molecular techniques such as gene probes and that of PCR; these are discussed briefly.

I have included sections on HACCP which has come into great prominence in recent years thus answering a specific criticism made of the earlier edition. The chapter on water and waste disposal contains material on Legionnaires' disease and cryptosporidiosis, infections of much concern at the present time. Finally, the chapter on legislation has undergone a major revision with far greater emphasis being placed on EC food hygiene legislation.

I have again received much help from experts in their field whom I acknowledge here; without them this book could not have been written.

v

Finally, I reiterate that I take full responsibility, as before, for any errors or omissions.

Additional acknowledgements for the second edition

I should again like to acknowledge the invaluable help given to me by many individuals and organisations.

I wish to express my gratitude to: Miss Dorothy Flowerdew and her colleagues at the Leatherhead Food Research Association for once again supplying vital information and advice on national food legislation and microbiological standards on those countries included in this book; Professor L. Wassermann, Ulmer Spatz, Neu-Ulm, Germany, for further help on German food legislation; David Timperley, Campden Food and Drink Research Association, for again devoting so much time in advising me on the intricacies of food equipment design; Dr Tony Baird-Parker, Unilever, for help on HACCP; and Dr Richard Gilbert, Public Health Laboratory Service, for supplying me with food poisoning statistics.

It is a pleasure to be able to thank the companies and research establishments listed below for helpful discussions and/or supplying new photographic material or drawings:

APV, Crawley, West Sussex
Campden Food and Drink Research Association,
    Chipping Campden, Gloucestershire
Fristam (Pumps) Ltd, Harrow, Middlesex
KEW Industry Ltd, Penrith, Cumbria
Northern Foods plc, Nottingham
Rotowash UK Ltd, Isleworth, Middlesex
Don Whitley Scientific Ltd, Shipley, West Yorkshire

I am indebted to the following publishers and authors for permission to produce extracts, tables or figures from their works: Academic Press for material from *Journal of Infection* (author, Barbara A. Bannister); Blackwell Scientific Publications, *Microorganisms in Foods — Vol. 4*; Butterworth–Heinemann, *Safe Processing of Food* (D. A. Shapton and N. F. Shapton), and *Food Control* (A. C. Baird-Parker); Cambridge University Press, *Epidemiology and Infection*, **103**, 425–33 (A. Stevens *et al.*); Institution of Water and Environmental Management for Water,

Pollution Control (M. Shore *et al.*), and the Public Health Laboratory Service for data from PHLS Reports.

Thanks are also due to my departmental colleagues including Kevin Kerr, Jerry Knapp and Richard Lacey who, in their different ways, made contributions to this book. Particular thanks are due to Freddie Webster, not only for her skills on the word processor, but also for her extreme patience with me over a long period of hard work. Finally, I should like to thank my wife, Rita, for her stoicism in carrying out, so conscientiously, the onerous duties of proof-reading.

P. R. Hayes

PREFACE TO THE FIRST EDITION

Books on food hygiene and its various ramifications are scarce and when I was approached by the publishers to write this book I felt that it might answer a need. That need was for a book to be written in a language that was intelligible to all those involved in food hygiene and also, I hoped, to the layman.

My interest in food hygiene goes back many years to when I worked in the food processing industry as a food bacteriologist. I realized then that so many different disciplines with their often divergent viewpoints were involved in this subject and that it was necessary to reconcile these differences to achieve the best results. This is not always as easy as it seems: engineers and microbiologists, architects and chemists, production managers and quality controllers do not talk the same language but they must be made to where food hygiene is concerned.

Being a microbiologist I must confess to some bias. I hope, however, that over the years I have been associated with the food processing industry I have come to appreciate food hygiene objectively and can make suggestions that resolve rather than exacerbate conflicting interests.

The first four chapters deal almost exclusively with microbiological aspects. The basic principles are presented initially so that the following three chapters dealing, respectively, with food poisoning, food spoilage and microbiological examining methods can be more readily understood by the uninitiated. The chapters following consider factory design, construction and more detailed factory layout, and the design of food processing equipment. The chapter on quality assurance and production control includes a review of critical points in selected food processes;

this reflects a recurrent theme of this book, that of practicability—there are many constraints applied in the industrial world which are often not fully appreciated by the theoretician but, so often, the simple modification can be made which improves product quality.

The vitally important area of cleaning and disinfection is covered next as methods and their practical application. The most sophisticated and expensive cleaning methods are vitiated if equipment design, factory layout, etc. are inadequate; these chapters are therefore linked with those preceding. The final three chapters cover the disparate subjects of waste disposal, training and legislation. The last-named topic is reviewed particularly in relation to food hygiene legislation. This clearly cannot be presented in detail but it is useful to compare the philosophies of selected countries on this topic.

The book should have wide interest and is directed towards the myriads of people involved in food hygiene. Included are factory architects, builders, designers of food processing machinery, engineers, senior management in the food industry, microbiologists, hygiene officers, Environmental Health Officers, medical officers, chemists, caterers, educationalists and members of the legal profession.

The material covered is diverse and this book could not have been completed without the generous help of many experts more knowledgeable than I in their particular subjects. However, I am the sole author and take full responsibility for it; any errors or omissions are entirely due to my shortcomings and should not be levelled at those acknowledged elsewhere.

Acknowledgements for the first edition

It is a pleasure to acknowledge the help given to me by colleagues and other individuals, research establishments and firms during the preparation of this book.

Thanks are due to my departmental colleagues for undertaking many of my duties during my absence; I should also like to note the help given by colleagues, Mr J. Lamb and Dr H. G. Muller, in the Procter Department of Food Science.

I should like to thank Mr D. A. Timperley, Unilever Research, Sharnbrook, for the many helpful suggestions he made on Chapter 7; similar thanks are due to Messrs D. Littler and P. W. E. Bird, and

Dr N. Hughes, Lever Industrial, Liverpool, for comments on Chapter 9; Miss D. Flowerdew and her colleagues at the British Food Manufacturing Research Association, Leatherhead, for advice on international food legislation and microbiological standards included in Chapter 13; Professor L. Wassermann, Ulmer Spatz, Neu-Ulm, Germany, for more detailed help on German food legislation; Dr R. H. G. Charles, Department of Health and Social Security, on EEC food legislation; and Mr P. Mepham, Leeds City Council, on UK food legislation.

I am grateful to those companies and research establishments, listed below, that either allowed me unrestricted access into their factories or assisted me with helpful discussions, or supplied photographs or drawings:

Alfa-Laval Co. Ltd, Brentford, Middlesex
Alveston Kitchens, Leamington Spa, Warwickshire
APV International Ltd, Crawley, West Sussex
Batchelors Foods Ltd, Sheffield, Yorkshire
Birds Eye Walls Ltd, Walton-on-Thames, Surrey
Campden Food Preservation Research Association, Chipping Campden, Gloucestershire
Findus (UK) Ltd, Newcastle upon Tyne, Tyne and Wear
A. Johnson and Co. (London) Ltd, Reading, Berkshire
KEW Industry Ltd, Penrith, Cumbria
Marks and Spencer plc, London
Pork Farms Ltd, Nottingham, Nottinghamshire
Joshua Tetley and Sons Ltd, Leeds, Yorkshire
Toftejorg Ltd, High Wycombe, Buckinghamshire
Walls Meat Co., Evesham, Warwickshire

I am indebted to the many authors and publishers who allowed me to reproduce specified extracts from their works. Here I should acknowledge that: the extracts from the four DHSS Reports 'On the State of Public Health for the Years 1968, 72, 74 and 76' are reproduced with the permission of the Controller of Her Majesty's Stationery Office; the extract on canned foods (Put *et al.* 1972) is reproduced by permission of the editor of the *Journal of Applied Bacteriology*; and the extract from Kotschevar & Terrell's *Food Service Planning* is reproduced by permission of John Wiley and Sons. I am also indebted to Dr R. J. Gilbert and co-authors for allowing me to reproduce material from the *Journal of Hygiene, Cambridge*.

I wish to make a special note of the invaluable contribution made by my ex-colleague, Dr A. Neville Barker, in reading the first draft and in making the countless constructive suggestions and emendations that he did. Special thanks are also due to Mr Andrew West for preparation of the graphs and diagrams, and to Mrs Brenda Fisher for her care in typing the text.

Finally, I should like to thank my wife, Rita, for her practical help in proof-reading, and for her understanding and forbearance during the difficult periods that inevitably arise when one is writing a book.

P. R. Hayes

CONTENTS

✳

Contents

Chapter 1

FUNDAMENTAL PRINCIPLES OF MICROBIOLOGY

1.1. INTRODUCTION

Microbiology is the study of microscopic organisms and their activities; within this orbit is the study of the distribution of the organisms, their characteristics, and their beneficial and harmful effects, especially in relation to the changes they make in their environment of which food is one particular example. These small living organisms or microorganisms as they are called are capable of individual existence and are not aggregated into tissues or organs as in higher plants and animals.

Six major groups of microorganisms are generally recognized, namely bacteria, fungi, viruses, algae, protozoa and rickettsias. The bacteria are the most important in relation to food but fungi also have a significant role. Of less importance are the viruses whilst the remaining groups, algae (simple plants), protozoa (single-celled animals) and rickettsias (specialized insect parasites) are beyond the scope of this book and will not be considered further except in relation to food-borne illnesses.

Microorganisms, particularly bacteria, are the most ubiquitous of living organisms. Bacteria occur in the air in varying numbers and up to considerable heights. Large numbers are found in the soil (1 g of typical garden soil will contain many millions of bacteria) whilst considerably smaller numbers are found in fresh water and sea water. Bacteria, as a whole, are extremely adaptable and in view of this flourish in a wide range of environments. One such specialized group, the barophiles, has been isolated from the depths of the sea and from oil wells where pressures in excess of 100 atm may be experienced. Some bacterial species are adapted to live at

1

very high temperatures whilst others favour low temperatures. At one extreme bacteria, isolated from hot springs in New Zealand, are capable of growth at 90°C; at the other extreme bacteria, isolated from Antarctic regions, may be incapable of growth above 10°C. Certain bacteria are capable of surviving concentrations of injurious chemicals which would be fatal to other forms of life. Bacteria are found on the surfaces and within the alimentary and respiratory tracts of larger animals although tissues of healthy animals and plants are generally free from bacteria.

This adaptability of bacteria, enabling them to live in extreme environmental conditions, can often cause unexpected problems in the food industry.

1.2. CHARACTERISTICS OF BACTERIA

1.2.1. Shape and Size

Three principal shapes of bacteria exist: spheres (cocci; singular, coccus), rods (bacilli; singular, bacillus) and curved or twisted rods (spirilla; singular, spirillum). Typical bacteria measure 1–3 μm in length and 0.4–1 μm in width although others larger and smaller than this do occur.

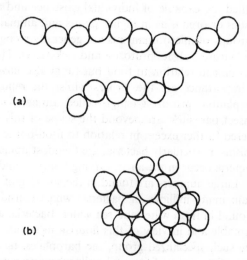

Fig. 1.1. Diagram showing: (a) chains of streptococci, and (b) a cluster of staphylococci.

1.2.2. Reproduction

Bacteria reproduce or multiply by a process known as binary fission which involves the simple division of the cell into two new daughter cells. Characteristic arrangements of cells are produced depending upon the plane of division. Thus cocci that divide in more than two planes produce bunches of cells (e.g. staphylococci) whilst cocci that divide in one plane only produce chains of cells (streptococci). These two arrangements are illustrated in Fig. 1.1. By contrast, rod-shaped bacteria divide at right angles to their long axis and so do not arrange themselves in the variety of ways possible with cocci. In fact rods normally separate after division and thus usually appear as single cells; similarly spirilla usually separate after division.

1.2.3. Bacterial Structure

Only a limited number of cell components are regarded as essential to bacteria, that is they are found in all bacteria (see Fig. 1.2); other inessential components are found in some, but by no means all, bacteria. Thus all but a few specialized forms possess a cell wall which is responsible for the rigidity of the bacterium. This wall surrounds the cytoplasm which is itself surrounded by a delicate cytoplasmic membrane. Membranes, as such, have a vitally important role in the life processes of the organism. They are involved in the transport of molecules, energy production, and in the maintenance of permeability barriers. The cytoplasm is a complex substance containing the nuclear body, which is rich in deoxyribonucleic acid (DNA), ribosomes which consist of ribonucleic acid (RNA) combined with protein, and a fluid portion which contains various dissolved nutrients. The nuclear body is involved in reproduction whilst ribosomes are involved in protein synthesis.

Other cell components of importance but not found in all bacteria include flagella, capsules and endospores (Fig. 1.2). Flagella (singular, flagellum) are the organs of locomotion in most motile bacteria. They are usually several times the length of the cell body but are extremely thin. Flagella may be polar (located only at one or both ends of the cell) or peritrichous (distributed more evenly around the cell). Capsules are formed by the excretion of polysaccharides from the cell proper and are viscous in character; they act as a protective covering and may vary considerably in thickness.

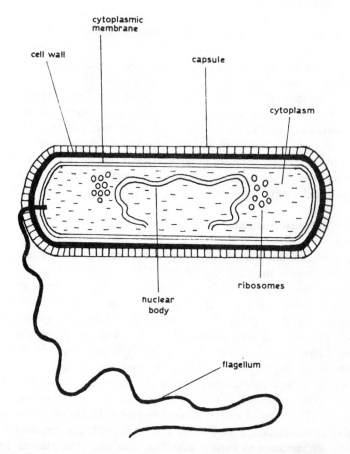

Fig. 1.2. Section of a typical bacterial cell showing essential structures. Also included are two inessential structures, the flagellum and the capsule.

Some bacteria produce resting structures known as endospores; this process is known as sporulation. These highly resistant bodies (see Fig. 1.3) commonly known as spores are produced within the cell, one spore being formed within a single bacterial cell. The spore has an unusual and characteristic chemical structure and is noted for its marked resistance to heat, UV light, chemicals and desiccation. When conditions are favourable the spores will germinate and produce fresh vegetative cells.

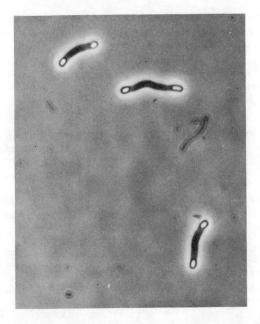

Fig. 1.3. Photomicrograph showing bacterial spores formed at the ends of paired cells before division (× 2000). (By courtesy of K. Johnstone.)

1.2.4. Gram Reaction

One of the most important distinguishing features of a bacterium is its reaction to a specific staining procedure named, after its Danish discoverer, the Gram stain. By the use of this technique bacteria can be divided into two types, Gram positive and Gram negative. The heat-fixed smear of bacteria, prepared on a glass slide, is stained with crystal violet (or other suitable basic dye). The smear is next treated with an iodine solution and then decolorized with alcohol. Gram positive bacteria retain the crystal violet whilst Gram negative bacteria are decolorized. By counter-staining the smear with a contrasting dye (e.g. carbol fuchsin, red) it is possible to distinguish readily the violet Gram positive forms from the red Gram negative forms. Although this staining technique was developed in 1884 the chemical basis of this differential reaction is still not fully understood but the division of bacteria into these two groups correlates surprisingly well with many other characteristics.

Table 1.1
Principal Genera of Bacteria Associated with Foods

Group description	Genus
Gram negative Microaerophilic spiral or curved rods	*Campylobacter*
Aerobic rods to cocci	*Acetobacter*
	Acinetobacter
	Alcaligenes
	Alteromonas
	Brucella
	Flavobacterium
	Halobacterium
	Halococcus
	Moraxella
	Pseudomonas
Facultatively anaerobic rods	*Aeromonas*
	Enterobacter
	Erwinia
	Escherichia
	Proteus
	Salmonella
	Serratia
	Shigella
	Vibrio
	Yersinia
Gram positive Cocci	*Leuconostoc*
	Micrococcus
	Pediococcus
	Staphylococcus
	Streptococcus
Endospore-forming rods	*Bacillus*
	Clostridium
	Desulfotomaculum
Regular non-sporing rods	*Brochothrix*
	Kurthia
	Lactobacillus
	Listeria
Irregular non-sporing rods	*Corynebacterium*
Mycobacteria	*Mycobacterium*

1.3. TYPES OF BACTERIA IMPORTANT IN FOODS

In the classification of living organisms those closely related (i.e. highly similar) can be associated together and regarded as a species. When different species have many characteristics in common they can be linked or grouped together into assemblages known as genera (singular, genus). All recognized bacterial species are listed and classified in *Bergey's Manual of Determinative Bacteriology*, the latest (9th) edition of which was published some years ago, in two volumes of interest (Krieg & Holt, 1984; Sneath *et al.*, 1986). Many of the genera and species listed have been isolated from foods but only the more important genera are discussed; they are also listed in Table 1.1.

1.3.1. Gram Negative Bacteria

1.3.1.1.. Microaerophilic, Spiral and Curved Bacteria
Only one genus, *Campylobacter*, is significant in foods being important as a cause of gastroenteritis. The genus is characterized by the formation of helically curved rods with less than one turn to many turns per cell, the rods being motile with a single polar flagellum at one or both ends of the cell.

1.3.1.2. Aerobic Rods and Cocci
By far the most important genus in this group is *Pseudomonas*. Almost all species are strictly aerobic and motile by means of one or more polar flagella. Many species grow at low temperatures causing food spoilage. Amongst these can be distinguished species producing diffusible green water-soluble pigments (fluorescent pseudomonads) but non-fluorescent pseudomonads can be equally important in food spoilage. Several species produce insoluble yellow, orange or blue pigments but these are not important in foods. Certain *Pseudomonas* species have been re-classified as *Alteromonas* species on the basis of DNA analysis; alteromonads are sometimes involved in the spoilage of chickens and fish, and produce strong ammoniacal odours.

The genera *Acinetobacter* and *Moraxella* are comprised of organisms that frequently occur as cocci or very short and plump rods; they have a tendency to resist decolorization in the Gram staining procedure and thus may appear somewhat Gram positive. Confusion still exists in differentiating between these two genera and they are often bracketed together. However, moraxellas are primarily animal parasites and if strains are isolated from foods they are probably atypical and are certainly much less common than acinetobacters. Acinetobacters are rather important spoilage organisms in poultry and shellfish but they can be isolated from a wide variety of uncooked foods.

The genus *Flavobacterium* comprises yellow pigmented species. It is regularly present on fresh meats and fish but species tend to be overgrown by other bacteria during spoilage of these foods; it has been implicated in the spoilage of milk and milk products.

The genera *Halobacterium* and *Halococcus* are distinguished by their requirement for a high salt concentration (>12%) for growth. They are thus involved in the spoilage of highly salted fish on which they produce characteristic red or pink patches. Halobacteria have many characters in common with pseudomonads whilst the halococci have been compared with micrococci (see below); however, the latter pair differ in many respects in addition to the Gram reaction.

The remaining three genera of interest are *Acetobacter*, *Alcaligenes* and *Brucella*. Acetobacters are found in alcoholic beverages where they convert ethanol to acetic acid causing souring and they can also cause rot in apples and pears. Cells are often ellipsoidal and motile by peritrichous flagella. *Alcaligenes* species are again peritrichous but coccal forms as well as rods are common. They occur particularly in dairy products and eggs causing spoilage problems. *Brucella* species, small non-motile ovoid rods ‚or cocco-bacilli, are mammalian parasites and pathogens. Food-borne illnesses in man (e.g. brucellosis) are caused by contact with the animals or, typically, by the consumption of unpasteurized milk.

1.3.1.3. Facultatively Anaerobic Rods
These Gram negative rod-shaped bacteria grow under either aerobic or anaerobic conditions (see Section 1.7.4). Two distinct families are recognized. The first, Enterobacteriaceae, contains eight genera of interest, viz. *Escherichia, Salmonella, Shigella, Enterobacter, Serratia, Proteus, Yersinia and Erwinia*. The second family, Vibrionaceae, contains only two genera of interest, *Vibrio* and *Aeromonas*.

All organisms in the family Enterobacteriaceae are either motile with peritrichous flagella or non-motile. There is only one species, *E. coli*, in the genus *Escherichia* of interest. This species occurs almost exclusively and in huge numbers in human and animal faeces, and is consequently described as a faecal coliform. It is important as an indicator of faecal pollution but some strains can cause food poisoning. Certain bacterial genera such as *Enterobacter* are described as coliforms because they have specific properties in common with *E. coli*. These genera are found in a variety of habitats, sometimes including faeces, and when testing for faecal pollution it is important to distinguish the coliforms from the true faecal coliforms.

The genus *Salmonella* contains nearly 2000 different serotypes or serovars (although there are only relatively few distinct species) which are very important causative organisms of food poisoning. *Shigella* species are fairly closely related to salmonellas except that the former are non-motile and the latter predominantly motile. Both genera are primarily associated with man and vertebrates, and shigellas are again pathogenic to man causing food-borne infections. *Yersinia* species are of no great significance in foods but one species, *Y. enterocolitica*, is now recognized as a cause of food poisoning in man. It is perhaps of interest to note that a closely related species, *Y. pestis*, is the causative organism of plague in man and rodents.

The remaining genera, *Serratia, Proteus* and *Erwinia*, are sometimes implicated in food spoilage. *Serratia* species are typically pigmented bright red and cause the characteristic red rot spoilage of eggs. *Proteus* species are also important in the spoilage of eggs and raw meats held at ambient temperatures, whilst *Erwinia* species are involved in the spoilage of vegetables.

The two genera in the family Vibrionaceae contain species which are typically motile with polar flagella. *Vibrio* species can often be presumptively identified by their appearance as short, curved rods under the microscope whereas *Aeromonas* species have the more conventional straight rod form. Vibrios are important in foods, different strains causing food poisoning, food-borne infections and food spoilage; aeromonads are sometimes involved in food spoilage and of some interest, but not discussed further, is *A. salmonicida*, an important fish pathogen.

1.3.2. Gram Positive Bacteria

1.3.2.1. Cocci

The family Micrococcaceae includes two genera of significance, *Micrococcus* and *Staphylococcus*. Micrococci are non-motile aerobes growing best between 25 and 37°C whereas staphylococci are facultatively anaerobic and grow best between 35 and 40°C; many species can grow at 45°C. Representatives of both genera can be isolated from a wide range of foods but whereas micrococci are principally spoilers of salted foods (e.g. bacon) the genus *Staphylococcus* contains one species, *S. aureus*, which is an important food poisoning organism.

Other genera within this group are facultative anaerobes and form non-motile cocci that occur typically in chains or tetrads depending upon the method of cell division. Three genera, *Streptococcus, Leuconostoc* and *Pediococcus*, comprise the 'lactic acid bacteria' together with *Lactobacillus* (see Section 1.3.2.3); they are so named due to their production of lactic acid by the fermentation of glucose. Both streptococci and leuconostocs produce chains of cells but only leuconostocs form ovoid rather than spherical cells on occasions. Pediococci produce spheres in pairs or tetrads due to cell division being in two planes. All three genera are involved in food spoilage and foods involved include bacon, vacuum-packed meats and milk but *Streptococcus* species are used as starter organisms in the preparation of cheeses and yoghurt.

1.3.2.2. Endospore Forming Rods

The genus *Bacillus* consists of species that are aerobic or facultatively anaerobic. Although they are Gram positive there is a tendency for many species to give a Gram negative staining reaction in older cultures. The position, size and shape of the spores vary and these characters help to distinguish the different species. *Bacillus* species are very commonly isolated from both raw and cooked foods. In fact, due to the extreme heat resistance of the spores of many species, they are often associated with the spoilage of under-processed canned foods but many other foods may be spoilt by members of this ubiquitous genus; one species, *B. cereus*, causes food poisoning in man.

The other spore forming genera, *Clostridium* and *Desulfotomaculum*, are anaerobic. The principal distinguishing feature of these genera is the ability of the latter to reduce sulphate to sulphide but until recently all anaerobic spore formers were accommodated in *Clostridium*. Species of both genera tend to give a Gram negative staining reaction in older cultures and both are involved in the spoilage of canned foods. Two species, *C. botulinum* and *C. perfringens*, cause food poisoning.

1.3.2.3. Regular Non-Sporing Rods

Lactobacillus is the remaining genus comprising the 'lactic acid bacteria'. Lactobacilli are non-motile rods that often occur in chains and they are anaerobic, facultatively anaerobic or microaerophilic (see Section 1.7.4). Lactobacilli cause spoilage of a variety of foods (e.g. bacon, vacuum-packed meats and alcoholic beverages) but, like streptococci, they are used as starter organisms in the food industry.

The genus *Listeria*, and *L. monocytogenes* in particular, has aroused much interest in the past few years having been implicated as an important cause of food poisoning in a variety of foods. Listeria are short rods, being motile with peritrichous flagella at 25°C but non-motile at 37°C; species are aerobic or facultatively anaerobic. Also of some significance is the genus *Brochothrix*, containing only one species, *B. thermosphacta*, a non-motile often elongated rod which can be involved in the spoilage of various meats. The aerobic *Kurthia* species, particularly *K. zopfii*, can also be isolated from spoiling meats but are never more than a minor part of the spoilage flora.

1.3.2.4. Irregular Non-Sporing Rods
Included in this diverse collection is the large genus *Corynebacterium* which contains a number of human and animal pathogens (e.g. *C. diphtheriae*). Corynebacteria can be isolated from many raw foods but they are of no real significance.

1.3.2.5. Mycobacteria
Bacteria comprising the genus *Mycobacterium* are aerobic, non-motile often curved rods which are not easily stained by Gram's method due to the presence of waxy materials in the cell walls. One species, *M. tuberculosis*, should be mentioned as at one time milk-borne tuberculosis was prevalent in the U.K. and many other countries.

1.4. CHARACTERISTICS OF FUNGI

Fungi are members of the plant kingdom which are not differentiated into the usual roots, stems and leaves; they do not possess the green photosynthetic pigment chlorophyll. Fungi exhibit a wide range of different forms including mushrooms, moulds and yeasts although only the two latter groups are of interest here.

1.4.1. Moulds

The term mould is used to describe certain multicellular fungi which form a filamentous branching growth known as a mycelium. The mycelium is composed of individual filaments called hyphae (singular, hypha). They may grow submerged within the food or on the surface of the food; in the latter case the growth is usually characterized by its fuzzy or cottony appearance. Hyphae may be classified as vegetative when concerned primarily with the uptake of nutrients or fertile when bearing reproductive structures on aerial stalks.

Reproduction of moulds is chiefly by means of asexual spores but many moulds also form sexual spores. Asexual spores, the functions of which are to propagate the species, are produced in large numbers and are small, light and resistant to drying. They are thus easily dispersed in the air and when they alight on suitable nutrient materials are able, under favourable conditions, to produce fresh mould growth. The

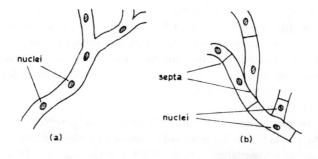

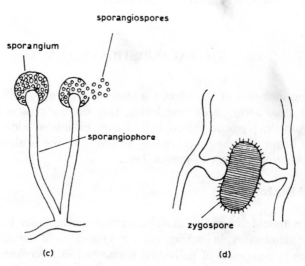

Fig. 1.4. Diagram showing: (a) non-septate hyphae, (b) septate hyphae, (c) asexually reproduced sporangiospores, and (d) sexually reproduced zygospore.

moulds which produce sexual spores are classified on the basis of the manner of formation of these spores and the types of spore produced.

1.4.1.1. Phycomycetes
These moulds produce hyphae which are not divided into typical uninucleate cells and thus have no cross-walls (septa); these non-septate hyphae have nuclei scattered throughout their length (Fig. 1.4). Asexual reproduction is accomplished by the production of spores borne within a structure called a sporangium; the stalk on which this fruiting body develops is termed a sporangiophore (Fig. 1.4). Phycomycetes, which are commonly found on foods, are members of the subclass Zygomycetes; they possess hyphae that, although apparently alike, are able to conjugate (i.e. form a sexual fusion) and form a zygospore (Fig. 1.4). These zygospores are surrounded by a tough wall and in consequence are very resistant to drying. Included in this group is the genus *Rhizopus*, the so called bread mould.

1.4.1.2. Ascomycetes
These moulds have septate hyphae and multiply asexually by 'pinching off' the tips of the fertile hyphae (conidiophores) to produce spores known as conidia (Fig. 1.5); these conidia are formed either singly, in chains or in irregular clusters on the conidiophores, and these character-

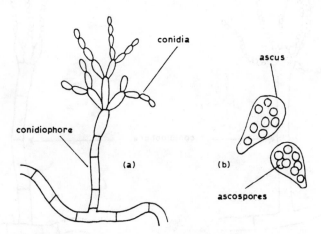

Fig. 1.5. Diagram of a typical ascomycete showing: (a) asexual conidia borne on a conidiophore, and (b) sexually reproduced ascospores within asci.

istic arrangements help in the identification of the various species. In the
Ascomycetes the sexual spores are termed ascospores. They are formed
following the union of two cells from the same mycelium or from two
different mycelia. A number of ascospores, usually eight, are subsequently formed within a sac known as the ascus (Fig. 1.5).

1.4.1.3. Fungi Imperfecti

Many moulds which are believed to be Ascomycetes are assigned to the
Fungi Imperfecti because of the apparent absence of a sexual stage, i.e.
they fail to produce ascospores. Like the Ascomycetes the Fungi
Imperfecti produce characteristic conidiophores and conidia. Thus since
it is possible for the same mould to be identified on the basis of its sexual (if present) or asexual structures, a mould may be placed in either
the Ascomycetes or the Fungi Imperfecti; in such a situation the same
species could be given either of two different names! *Aspergillus* and
Penicillium species are commonly found examples of imperfect fungi
and are illustrated in Fig. 1.6.

Table 1.2 lists the principal genera of fungi associated with foods.

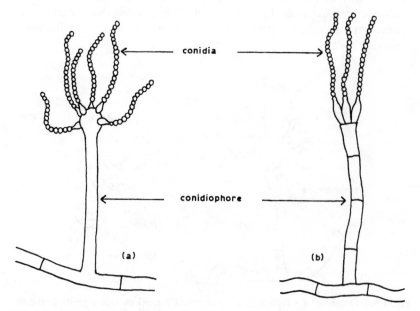

Fig. 1.6.   Diagram showing: (a) a simple *Aspergillus* and (b) a simple *Penicillium*.

Table 1.2
Principal Genera of Fungi Associated with Foods

	Genus
Moulds	
Class	
Phycomycetes	
(subclass: Zygomycetes)	*Absidia*
	Mucor
	Rhizopus
	Thamnidium
Ascomycetes	*Byssochlamys*
	Claviceps
	Monascus
	Neurospora
	Sclerotinia
Fungi Imperfecti	*Alternaria*
	Aspergillus
	Botrytis
	Cladosporium
	Fusarium
	Geotrichum
	Penicillium
	Phoma
	Scopulariopsis
	Sporendonema
	Sporotrichum
	Trichoderma
	Trichothecium
Yeasts	
Class	
Ascomycetes	*Debaryomyces*
	Pichia
	Saccharomyces
Fungi Imperfecti	*Candida*
	Rhodotorula
	Torulopsis

1.4.2. Yeasts

Most yeasts are microscopic unicellular fungi, which typically do not form a mycelium and therefore exist as single cells. Yeast cells may be round, egg-shaped or elongated, the shape being relatively constant for

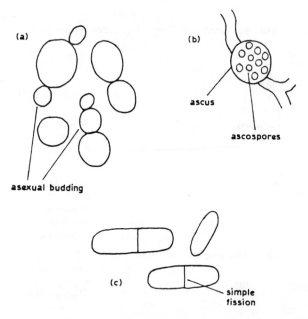

Fig 1.7. Diagram of yeasts showing different forms of reproduction: (a) asexual budding, (b) formation of ascospores by 'true' yeasts, and (c) multiplication by simple fission

a given species. The majority of yeasts reproduce asexually by budding, but a few species reproduce by simple fission as with the bacteria. Budding is a process in which some of the protoplasm bulges outward and the bud thus formed grows in size eventually reaching maturity and separating from the original cell (Fig. 1.7). The generation time for most yeast cells is about 20–30 min under ideal conditions.

Yeasts able to reproduce sexually are known as 'true' yeasts. The process involves the formation of ascospores, the yeast cell serving as the ascus (Fig. 1.7). Hence true yeasts are classified as Ascomycetes; conversely 'false' yeasts, producing no ascospores, belong to the Fungi Imperfecti.

1.5. CHARACTERISTICS OF VIRUSES

The precise definition of a virus is the subject of much controversy even today but a number of their properties can be described.

Viruses are very much smaller than bacteria. The larger ones, such as the cowpox virus which is used in vaccination against smallpox, are about 0·3 μm; the smaller ones, such as the foot-and-mouth disease virus, are about 0·01 μm. Because of their small size, viruses pass through bacteriological filters and are invisible under the light microscope. Hence viruses are often described as ultramicroscopic filterable agents. Shapes of virus particles, as seen using the electron microscope, include spheres, rods, ovals and tadpole-like structures.

Chemically, the simplest viruses consist of protein surrounding a nucleic acid core comprising either DNA or RNA but not both. More complex viruses contain carbohydrates, lipids and other materials. Biologically, viruses are strictly parasitic within the living cells of animals, plants or bacteria and are fundamentally different from all cellular forms of life. The high degree of specificity exhibited by viruses for certain hosts and even specific cells within the prescribed host is a characteristic which no doubt plays a major role in limiting their spread.

Bacterial viruses are termed bacteriophages, or often the diminutive, phages; they infect bacteria and 'dissolve' or lyse actively growing cells.

1.6. THE GROWTH CURVE OF BACTERIA

If bacteria are added to a substrate in which they can grow, and most food materials constitute such substrates, they will multiply by the process of binary fission as stated previously. This multiplication takes place by division of the cells thus: 1 into 2, 2 into 4, 4 into 8, 8 into 16 and so on. The rate of division depends upon a number of factors which will be considered later but, under favourable conditions, divisions could occur every 20 min or so (even less with some bacterial species); in this way a single cell is capable of producing over 16 million cells within 8 h and nearly 70 000 million cells after 12 h continuous growth. Such growth is known as logarithmic or exponential (see later).

However, this rapid growth of bacteria may not be established immediately. When bacteria are added to a fresh medium there is a period during which little, if any, growth occurs. This is called the lag phase and represents the period when the inoculated cells are adjusting to the new environment. During this period the cells increase in size and produce new materials but active division does not take place. The length of the lag phase is extremely variable but may be something like 2 h.

Food Microbiology and Hygiene

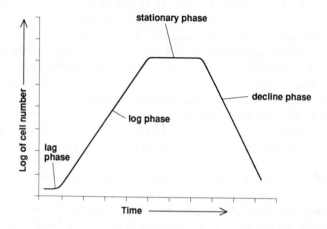

Fig. 1.8.   Growth curve of bacteria showing the four phases of growth, viz. lag, logarithmic or exponential, stationary, and decline or death.

The next phase of growth, termed the logarithmic or exponential phase, is characterized by a constant specific growth rate, as indicated earlier. In this phase the logarithms of the number of bacteria per unit volume plotted against time yield a straight line (Fig. 1.8). During this phase each cell in the population is duplicating itself in the same time and the cells formed are all viable and of constant size. In time the environment will change due to the growth of the bacteria causing a depletion in nutrients and an accumulation of waste products. The environmental change results in a decrease in the growth rate until the population enters the stationary phase. During this period the number of viable bacteria in the population remains constant; this may be due to the complete cessation of growth or to the fact that any growth is offset by the death of an equal number of bacteria. The final phase in the growth curve is known as the decline or death phase and represents a period when the death rate exceeds the rate of multiplication, if indeed any multiplication is occurring at all at this late stage. Death of the bacterial cell is often accompanied by lysis so that if old cultures are viewed under the microscope a large amount of cell debris may be seen with few intact cells in evidence.

1.7. FACTORS INFLUENCING BACTERIAL GROWTH

The principal environmental factors influencing bacterial growth are food, temperature, moisture, oxygen availability, hydrogen ion concentration and the presence of inhibitory substances. Whilst each of these factors is important and may limit growth, it is generally their combined effects which determine whether bacterial growth will occur and, given suitable growth conditions, which bacteria will grow and how quickly.

1.7.1. Food

Food is required by bacteria, not only as a source of energy but also for manufacturing protoplasm and structural materials. Bacteria differ greatly in their nutritional requirements but certain elements in food are essential and growth is impossible in their absence. The most important elements are carbon, hydrogen, oxygen, nitrogen, sulphur and phosphorus; smaller quantities of iron, magnesium, potassium and calcium are also required whilst several other elements are usually needed in minute amounts.

Bacteria exhibit great diversity with respect to both the kind and number of organic compounds they can utilize. Carbohydrates and amino acids are commonly used as carbon and energy sources; nitrogen and sulphur requirements are often satisfied by organic compounds containing these two elements, e.g. certain amino acids, peptides (compounds consisting of two or more amino acids) or proteins (compounds containing large numbers of amino acids).

An organism may require one or more organic compounds for cellular material that it is unable to synthesize from simpler compounds. Such essential nutrients are only required in small amounts and organic nutrients of this type are known as growth factors. They fall into three groups: (1) amino acids, required for protein synthesis; (2) purines and pyrimidines, required for nucleic acid synthesis, i.e. DNA and RNA; and (3) vitamins such as thiamine, riboflavine and nicotinic acid which are required for enzyme synthesis. Some bacteria have a complex list of requirements including a battery of amino acids, purines and pyrimidines, and perhaps eight or nine vitamins indicating an almost complete loss of the ability to synthesize these essential compounds for themselves. Use is made of these fastidious bacteria in measuring the quantities of vitamins such as riboflavine present in foods. *Lactobacillus* species are commonly used for these biological assays. Many species of this genus require minute amounts of many vitamins so that by removing an essential one from the medium growth will be inhibited. By adding graded amounts of the food containing the essential vitamins to the 'depleted' growth medium and measuring the growth response of the test organism an extremely accurate measure of the concentration of the vitamin in the food can be established.

1.7.2. Temperature

This is a factor of profound importance since temperature has a marked influence on the rates of all chemical reactions linked to the processes of growth. Thus the temperature of a growth medium or food determines the rate of growth of any bacteria associated therewith. For any organism the temperature at which growth is most rapid is known as the optimum temperature. The maximum temperature is the highest temperature at which the organism will grow; this is usually only a few degrees above the optimum temperature. The minimum temperature is the lowest temperature at which growth of the organism occurs and is usually substantially below the optimum temperature.

Three physiological groups of bacteria may be distinguished on the basis of their temperature relationships (Table 1.3). The definition of the temperature ranges of these three groups is still the subject of much controversy and it should be emphasized that the temperatures quoted are somewhat arbitrary and that some overlapping of the groups does occur. Thus, for example, a variety of mesophiles with optima of 30-35°C are able to grow, albeit slowly, at 5°C or less; such organisms are sometimes called psychrotrophic mesophiles. Growth rates of a typical mesophile and psychrotroph are illustrated in Table 1.4 and show the close proximity of optimum and maximum growth temperatures, and the relatively slow decline in growth rates below the optimum. It should

Table 1.3
Relationship Between Growth Rates of Bacteria and Incubation Temperatures

Group	Minimum temperature (°C)	Optimum temperature (°C)	Maximum temperature (°C)
Thermophiles	35-45	45-70	60-80
Mesophiles	5-20	30-45	40-50
Psychrotrophs	0-5	20-35	25-40

Table 1.4
Growth Rates of a Mesophile (*E. coli*) and a Psychrotroph (*Pseudomonas* sp.) at Different Temperatures[a]

E. coli		Pseudomonas	
Temperature (°C)	Generation time (min)	Temperature (°C)	Generation time (min)
47	NG[b]	35	NG[b]
46	32	34	180
44	22	32	34
40	21	28	45
38	22	24	51
34	28	20	77
30	33	16	100
26	56	12	130
22	96	8	240
18	260	4	440
14	400	0	1200
10	1200		

[a]From Ingraham (1958). [b]NG = no growth.

also be noted that the growth rate of the mesophile is much faster at its optimum than is that of the psychrotroph.

A fourth group of specialized organisms, the psychrophiles, is also recognized. They are characterized by an extremely low optimum growth temperature (5–20°C) and the ability to grow well at 0°C. However, in nature and particularly in foods these organisms are rare and hence of relatively little importance.

1.7.3. Moisture

Water accounts for some 80-90% of the total weight of living cells and all organisms require it for growth. Organisms vary tremendously in their water requirements but, generally, bacteria need more than fungi. It is the amount of *available* water and not the total water present which determines whether growth will occur and, if so, at what rate. That is, by adding a solute such as sodium chloride to a medium less water is made available to the microorganisms. The available moisture, expressed as water activity (a_w), is the vapour pressure of the solution (solutes in water in most foods), divided by the vapour pressure of pure water. The

a_w of pure water is 1·00, that of many bacteriological media is in excess of 0·995, that of fresh meat is 0·99, whilst a saturated solution of salt (sodium chloride, NaCl) has an a_w of only 0·75.

Most bacteria grow well in media with water activities of 0·990–0·998; growth in pure water ($a_w = 1·0$) is, of course, impossible. Many bacteria fail to grow at a_w levels below 0·95 although the commonly encountered micrococci and staphylococci will tolerate levels down to 0·86 (i.e. they withstand higher concentrations of solutes than most other bacteria). Salt is a commonly used preservative, and increases in the salt concentration suppress the growth of progressively more and more micro-organisms. At intermediate salt concentrations, say 5–7% NaCl, micro-

Table 1.5
Minimal a_w for Growth of Microorganisms Associated with Foods[a]

a_w	Bacteria	Yeasts	Moulds
0·99	*Moraxella/Acinetobacter*		
0·97	*Clostridium botulinum* Type E *Clostridium perfringens*		
0·96–0·94	Minimum a_w range for growth of most bacteria (e.g. *Escherichia coli* *Salmonella* *Pseudomonas* *Clost. botulinum* Types A & B)		
0·92	*Bacillus cereus*		
0·90		Minimum a_w for most yeasts (e.g. *Pichia* *Rhodotorula*	*Fusarium* *Cladosporium* *Trichothecium*
0·87–0·76		*Saccharomyces*	Minimum a_w range for most moulds (e.g. *Aspergillus flavus* *Penicillium* *Byssochlamys nivea*
0·86	*Staphylococcus*		
0·75	*Halobacterium* *Halococcus*	*Debaryomyces*	
0·62		Osmophilic yeasts	

[a]Data from Beuchat (1983), Sperber (1983) and Pitt & Hocking (1985). Based on growth in NaCl.

cocci and staphylococci may survive although they have no absolute requirement for NaCl. However, there is a specialized group of bacteria, namely the halophiles, which do require salt for growth and this group includes the so-called 'extreme halophiles' which grow best in 25-30% NaCl.

Moulds will generally tolerate lower a_w levels than bacteria (Table 1.5); many types will grow down to 0·75 or 0·70. Yeasts fall between moulds and bacteria in terms of their water requirements, an a_w of *ca* 0·90 being the limit for the majority. Mention should be made of the osmophiles, a group of yeasts growing in high sugar concentrations, which flourish in jams, syrups and jellies.

1.7.4. Oxygen

Bacteria have a range of activities with regard to their oxygen requirements, as do microorganisms in general. Bacteria that are dependent on free oxygen present in air are termed obligate or strict aerobes. Few bacteria are strictly aerobic but many such as pseudomonads and the moulds growing on the surface of foods are generally regarded as aerobic. At the other end of the scale are the obligate anaerobes which only grow in the absence of free oxygen; indeed, even trace amounts of oxygen are toxic to them. Again, strictly speaking, there are few obligate anaerobes but many bacteria are so called because they only tolerate extremely low levels of oxygen, e.g. certain members of the genus *Clostridium.*

The majority of bacteria fall between these extremes of oxygen requirements and can grow in either the absence or the presence of free oxygen. In fact most of these bacteria display a preference for aerobic conditions and organisms in this group are called facultative anaerobes. A fourth group of bacteria is sometimes recognized and termed microaerophiles; these bacteria, including many lactobacilli, have a requirement for oxygen but at concentrations considerably below those present in air.

There is a relationship between the oxygen tension, that is the concentration of oxygen in the environment and the oxidation-reduction (OR) potential. The OR potential is essentially a measure of the oxidizing or reducing capacity of the medium. Thus if a strong reducing agent is present it will lower the OR potential with a consequential shift towards the growth of anaerobes. Conversely, since oxygen is an oxidizing agent its presence ensures relatively high OR potentials and encourages the growth of the more aerobic organisms. The OR potential may also be increased by increasing the concentrations of other oxidizing agents thus allowing 'aerobic' bacteria to develop in conditions devoid of oxygen itself.

1.7.5. Hydrogen Ion Concentration (pH)

The hydrogen ion concentration has a marked effect on the growth of bacteria. This concentration is normally expressed in terms of pH which is defined as the logarithm of the reciprocal of the hydrogen ion concentration. It should be remembered that substances which produce an excess of hydrogen ions (H^+) in solution are acidic, the greater the excess the more strongly acidic they will be. Substances producing an excess of hydroxyl ions (OH^-) are termed bases and will be alkaline. For example, water ionizes to a small extent and 1 litre of 'pure' water contains 10^{-7} moles each of H^+ and O^-. Thus there is a balance of H^+ and OH^- and the solution is neutral (pH 7). This is therefore the mid-point of the pH scale which extends from 0 (strongly acidic) to 14 (strongly alkaline). Again, vinegar which is a dilute solution of acetic acid has a H^+ concentration of 10^{-4} per litre (and consequently an OH^- concentration of 10^{-10}) and thus has a pH of 4.

All microorganisms have an optimum pH at which they grow best, a minimum pH which is the most acid reaction permitting growth and a maximum pH which is the most alkaline reaction that allows their growth. Most bacteria favour a pH near neutrality or slightly on the alkaline side (6·8–7·5). Some prefer a lower pH (4–6) usually creating these conditions by themselves and producing acids from carbohydrates. A few bacteria (e.g. certain *Vibrio* species) are known which prefer definitely alkaline conditions of pH 8·5–9·0. Yeasts and moulds grow well in an acid environment of pH 3·5–4·5; moulds, whilst favouring acid conditions, usually grow over a wide pH range (3·5–8·0).

1.7.6. Inhibitory Substances

Foods may contain a variety of substances which may affect microbial growth; these may occur naturally, be produced by microbial growth or added artificially. Examples of naturally occurring substances present in foods which suppress microbial growth are the lysozyme and conalbumin of eggs, the lactenins of milk and the oils present in various plants. Inhibitory substances which accumulate as by-products of microbial growth include acids; lactobacilli break down carbohydrates with the production of lactic acid causing a marked pH drop which ultimately inhibits the growth of most bacteria. Finally, chemicals may be added deliberately to foods during processing to control the growth of undesirable organisms; thus sorbates are added to bread to control mould growth and sulphur dioxide is widely used to inhibit microbial growth in wines, beers and fruit juices as well as in comminuted meat products.

BIBLIOGRAPHY

Alexopoulos, C. J. & Mims, C. W. (1979). *Introductory Mycology*, 3rd edn, New York, Wiley.

Hawksworth, D. L., Sutton, B. C. & Ainsworth, G. C. (1983). *Dictionary of the Fungi*, 7th edn, Kew, Commonwealth Mycological Institute.

Pelczar, M. J., Chan, E. C. S. & Krieg, N. R. (1986). *Microbiology*, 5th edn, New York, McGraw-Hill.

Samson, R. A. & van Reenen-Hoekstra, E. S. (1988). *Introduction to Food-Borne Fungi*, 3rd edn, Baarn, Centraallbureau voor Schimmelcultures.

Stanier, R. Y., Ingraham, J. L., Wheelis, M. L. & Painter, P. R. (1986). *The Microbial World*, 5th edn, London, Macmillan.

REFERENCES

Beuchat, L. R. (1983). Influence of water activity on growth, metabolic activities and survival of yeasts and moulds. *Journal of Food Protection*, **46**, 135-41.

Ingraham, J. L. (1958). Growth of psychrophilic bacteria. *Journal of Bacteriology*, **76**, 75-80.

Krieg, N. R. & Holt, J. G. (1984). *Bergey's Manual of Systematic Bacteriology*, Vol. 1. Baltimore, Williams & Wilkins.

Pitt, J. I. & Hocking, A. D. (1985). *Fungi and Food Spoilage*, Sydney, Academic Press.

Sneath, P. H. A., Mair, N. S., Sharpe, M. E. & Holt, J. G. (1986). *Bergey's Manual of Systematic Bacteriology*, Vol. 2. Baltimore, Williams & Wilkins.

Sperber, W. H. (1983). Influence of water activity on foodborne bacteria – a review. *Journal of Food Protection*, **46**, 142-50.

Chapter 2

FOOD POISONING AND OTHER FOOD-BORNE HAZARDS

2.1. INTRODUCTION

The term 'food poisoning' is commonly used to cover a wide variety of illnesses or clinical conditions affecting the gastrointestinal tract. The very large majority of such illnesses found in developed countries result from the consumption of contaminated food or drink, and because they are caused by infection with or the presence of bacteria, these organisms will receive the greatest attention here. However, it is necessary to consider, albeit more briefly, other forms of food poisoning and food-borne hazards since these may sometimes be of concern and pose serious health hazards in other parts of the world.

The types of food poisoning may conveniently be grouped on the basis of the causative agent as follows:

1. Bacteria
2. Fungi
3. Viruses
4. Animals
5. Plants
6. Chemicals

Other food-induced illnesses such as indigestion and sensitivity to specific foods (e.g. allergies) will not be discussed; although distressing to the individuals concerned, they can in no way be regarded as forms of food poisoning.

2.2. INCIDENCE OF FOOD POISONING

In England food poisoning statistics have been published annually since the 1940s. It is well recognized that only a very small proportion, perhaps as low as 1%, of the total number of food poisoning cases figure in the annual statistics. This is due either to faulty reporting or, more usually, to the patient failing to consult his general practitioner. Figures are available from several sources including the reports of the Chief Medical Officer of Health, the Office of Population Censuses and Surveys and the Public Health Laboratory Service. The statistics which are presented in slightly different ways are never in total agreement but they suffice to show (Table 2.1) that over the five year period from 1984 to 1988 the number of cases ranged between 12 000 and 26 000. In the late 1980s numbers have risen dramatically to over 30 000 per annum primarily due to increases in salmonellosis. Table 2.1 refers to terms such as 'general outbreaks' and 'cases' and these terms have been defined by the Department of Health and Social Security (DHSS) as:

'Food poisoning'   — any acute illness attributable to the recent consumption of food.

'A case'   — a person affected with food poisoning.

'Sporadic case'   — an affected person whose illness is not connected with a similar illness of any other person.

'Household outbreak'   — an outbreak affecting two or more persons in the same private household, not connected with any other cases or outbreaks.

'General outbreak'   — an outbreak affecting two or more persons which was not confined to one private household.

2.2.1. The Bacteria Responsible

As can be seen from Table 2.1, *Salmonella* spp. are by far the most common cause of food poisoning in England and Wales. On average about 90% of the total number of cases can be attributed to salmonellas with *Clostridium perfringens* being responsible for a further 5–10% of cases; in 1988 salmonellas were reponsible for 93% of cases.

Food Microbiology and Hygiene

Table 2.1

Bacterial Food Poisoning in England and Wales (1984–1988)[a]

Causative organism	1984	1985	1986	1987	1988
Salmonella spp	10 603 (13 201)	9 966 (11 172)	12 775 (14 177)	– (18 096)	– (24 123)
Clostridium perfringens	68 (1 716)	64 (1 466)	61 (906)	51 (1 266)	57 (1 312)
Staphylococcus aureus	10 (181)	13 (118)	11 (76)	12 (178)	9 (111)
Bacillus cereus[b]	28 (214)	22 (81)	26 (65)	21 (137)	20 (418)
Total	10 709 (15 312)	10 065 (12 837)	12 873 (15 224)	– (19 677)	– (25 964)

[a] Data from PHLS Communicable Disease Reports. Annual figures represent the total number of incidents, including general and household outbreaks, and sporadic cases. Total number of cases in parentheses.
[b] Other *Bacillus* spp. may be included.

Staphylococcus aureus and *Bacillus cereus* each normally account for under 1% of cases whilst *Vibrio parahaemolyticus*, which formally accounted for a further 1%, is apparently no longer a significant problem in this country. Different types of *Escherichia coli* have been occasionally implicated in food poisoning outbreaks; the number of cases fluctuates widely but typically does not exceed 100 per annum in the UK.

Botulism, caused by *Clostridium botulinum*, is fortunately very rare in the UK as it is the most lethal type of bacterial food poisoning; a small number of cases were recorded in 1978 and 1987 and there was a general outbreak in 1989.

Campylobacter spp. appeared in the food poisoning statistics in the early 1980s for a few years. The Communicable Diseases Surveillance Centre reported over 6000 cases in 1978 (DHSS, 1980) and since that time cases have increased steadily until in 1989 over 32 000 campylobacter enteritis cases were reported. The exclusion of campylobacters from the food poisoning statistics is due primarily to their being regarded as a cause of food-borne infection rather than a true food poisoning (see below).

The anomalies and shortcomings of the statistics collated by different bodies have been stressed by Sheard (1981); the same author has also reviewed the general trends in food poisoning statistics (Sheard, 1986; 1987). He concluded that if the number of cases was to be reduced particular attention should be paid to all areas of the catering industry and to all outlets of raw and cooked meats; the need to monitor imported foods is also stressed as is the importance of food hygiene education.

Before leaving the statistical aspects of food poisoning it should be pointed out that there are about 100 fatal cases per annum in England; this figure is relatively low but it is one which, together with the non-fatal cases, could be substantially reduced if suitable precautions were observed.

2.2.2. Type of Food

The different types of food poisoning generally associate themselves with specific foods (e.g. salmonellas with meats or meat products) although full investigation of these associations may be hampered because in many outbreaks it is impossible to identify the food involved since it has often been disposed of before bacteriological tests can be initiated. However, there is ample evidence to show that the foods most commonly implicated in food poisoning in general are the various meats including poultry. Until recently nearly three-quarters of the cases in which the food has been identified have involved re-heated or cold pre-cooked meats (principally beef, pork, ham and lamb) or poultry (chicken, turkey and duck), stews, minced meats or meat pies.

This situation changed dramatically in the UK in the mid-1980s when *Salmonella enteritidis*, phage type 4 became an ever increasingly significant cause of food poisoning. Evidence suggested that this was due to a spread of infection by this organism in chickens which also contaminated hens' eggs (PHLS Report, 1989); the latter are now apparently a major cause of food poisoning in the UK with nearly 70% of salmonellosis cases being attributed specifically to this phage type.

There are many other foods which are only infrequently implicated in food poisoning incidents; these include unpasteurized milk, other dairy products (e.g. trifles and cream cakes), fried rice and various seafoods.

2.3. BACTERIAL FOOD POISONING

It may be helpful to distinguish between bacterial food poisoning and food-borne bacterial infections which are considered later in this chapter. In the former the causative organism multiplies in the food and by its heavy growth induces illness by one means or another after ingestion of the contaminated food. In food-borne infections the food merely acts as a carrier for the causative organism which does not require to multiply in the food.

It is customary to divide bacterial food poisoning into the infection type and the toxin type; the distinction between them is somewhat ill-defined since presumed infective organisms have, on occasions, subsequently been found to produce exotoxins in the food. An infection food poisoning is characterized by an acute gastroenteritis (i.e. inflammation of the lining of the alimentary canal) following ingestion of food in which multiplication of bacteria has taken place, the ingested viable bacteria continuing to grow within the host's body to produce the typical symptoms. Salmonellas are principally responsible for this type of food poisoning in which endotoxins are released as the bacterial cells disintegrate. The toxin type (often termed an intoxication) is a genuine food poisoning since a poisonous substance, the enterotoxin, is present in the food having been produced by the bacteria which have grown in the food prior to consumption. The toxin also causes an acute gastroenteritis but ingestion of viable bacteria is usually not a prerequisite of the induction of the disease. Bacteria causing toxin food poisoning include *Clostridium perfringens* and *Staphylococcus aureus*.

2.3.1. Salmonellas

2.3.1.1. The Organisms

Salmonellas are short (1–2 μm), motile, Gram negative, non-sporing rods. In fact the genus *Salmonella* contains nearly 2000 different strains (termed serovars or serotypes); the number is continually increasing as new serologically distinct strains are isolated. Many serotypes are named after the place where they were first isolated, e.g. *S. newport, S. derby, S. dublin, S. heidelberg, S. montevideo*, although originally salmonellas were named in a more haphazard manner and often after the disease and affected animal (e.g. *S. typhimurium* causing typhoid in mice). Salmonellas are facultative anaerobes biochemically characterized by their ability to ferment glucose with the production of acid and gas, and their inability to attack lactose and sucrose. Their optimum growth temperature, as with most food poisoning bacteria, is about 38°C; they are relatively heat sensitive being killed at 60°C in 15–20 min and they fail to grow below about 7 or 8°C.

2.3.1.2. The Disease

Salmonellosis is the collective term used for human and animal infections caused by members of the genus *Salmonella*. As mentioned previously salmonellas induce the illness by their death following multiplication in the host's gut and their subsequent lysis with the release of a potent endotoxin. This endotoxin, a lipopolysaccharide, forms part of the membrane of the cell and is primarily responsible for the clinical symptoms. Enterotoxin(s) produced within the human intestine may well have an important role in the disease (D'Aoust, 1991).

In humans the incubation period (i.e. the period between ingestion of the contaminated food and the onset of the symptoms) varies considerably but is usually between 12 and 36 h. The principal symptoms of salmonellosis are nausea, abdominal pain, drowsiness, diarrhoea and a moderate fever; dehydration may occur resulting in great thirst. The stools are watery, greenish in colour, foul smelling and sometimes tinged with blood. The organisms can invade the blood stream and thus cause a septicaemia and in the more extreme cases the patient may go into a coma. The mortality for the population as a whole is low, being well under 1%; susceptibility varies, however, especially with age and infants under 1 year are particularly at risk as are the elderly and infirm. The healthy adult typically needs to ingest at least a half to one million viable salmonellas to produce the symptoms whilst infants and the elderly are susceptible to only a small fraction of this number. There are circumstances, however, where very low doses can initiate the symptoms. For example, it was found that only *ca* 50 viable *S. napoli* cells needed to be ingested in the contaminated chocolate bars which caused the large food poisoning outbreak in the UK in 1982; there were 245 known cases, mainly children, whilst many adults remained unaffected at this level of contamination (Greenwood & Hooper, 1983). It would seem that the chocolate affords some protection to the organism from the stomach acids and rapid passage of the food into the intestines increases this protection.

The illness usually lasts for up to 7 days but some of the symptoms may linger for weeks or even months. During the acute stage large numbers of salmonellas are excreted in the faeces but the number gradually decreases with recovery of the patient until stools are salmonella-free. Nearly 50% of patients still show 'positive' stools after 4 weeks, 10% after 8 weeks and between 0.2 and 5% remain intermittent salmonella excreters for far longer periods. These latter patients are known as 'carriers' but there are other 'carriers' who are found to intermittently excrete salmonellas even though they have no history of any disease symptoms. Because this latter type of carrier is rarely detected before an outbreak occurs, an insidious build-up of carriers may occur within a food concern, particularly if involved with meats and poultry. A typical case was reported by the DHSS (1976).

During the summer of 1974, outbreaks of food poisoning caused by *S. infantis* were reported in South Wales, the Midlands and the West Country. Investigations suggested that the vehicles of infection were cooked meats and pies supplied by one food factory. Sampling of various meat products and extensive swabbing of the premises failed to detect the causative organism. Stool testing of employees revealed 136 excreters of *S. infantis* dispersed throughout the factory, including members of the management and canteen staff. The reason for this high infection rate among the staff may have been the availability of the firm's products for sale both in the canteen and the firm's shop.

The carrier rate amongst the general public is likely to be much higher than expected. Harvey *et al.* (1969) analysed sewage from a modern housing estate in Pontypridd, the sewage containing no waste from food industries nor from butchers' shops. During the 14 month survey period salmonellas were regularly isolated and 35 different serotypes were confirmed, many of them on numerous occasions. In spite of these findings no salmonellosis was reported amongst the population during the survey and therefore it can be concluded that man must frequently be exposed to a wide range of salmonella serotypes albeit at low levels.

2.3.1.3. *Sources*

Mention has already been made of the predominance of meats and poultry as causes of food poisoning and this is primarily a reflection of the association of salmonellas with these foods. Since the intestinal tract of farm animals and birds is the primary habitat of *Salmonella* spp. it is to be expected that carcass meat from these sources may be contaminated with these bacteria. Rates for salmonella contamination of meats are variable but it is probable that contamination is highest in poultry. Roberts (1972) found non-host specific salmonellas in 35% of the raw chickens examined whilst Gibson (1969) quoted rates of 45% for ducklings; more recently, Gilbert (1983) reported that 79% of frozen chickens were contaminated with salmonellas. In addition, turkeys have become a common source of food poisoning which now extends throughout the year rather than being limited to the Christmas period. Such incidents are often associated with larger functions as Vernon (1977) observed:

> Catering facilities for such functions are often overstretched and consequently the turkeys may be cooked 2–3 days before they are due to be eaten. Large frozen banqueting birds, which weigh up to 35 lb and need two days defrosting, are especially likely to be insufficiently thawed and inadequately cooked. They may then be kept at ambient temperature for long periods after cooking and final reheating is often at temperatures too low to destroy bacteria. For example, about 600 persons had turkey for dinner at a club function and 209 were known to have become ill afterwards with *S. typhimurium* infection. The turkeys were large and undercooked and one batch was left overnight in the cooling ovens. They were then refrigerated, subsequently carved and reheated for only half an hour.

In view of the high rates of salmonella contamination on poultry generally it is not surprising to find poultry meat so often the cause of salmonellosis. Thus, Reilly *et al.* (1988) found in Scotland that poultry was responsible for 224 outbreaks affecting 2245 people between 1980 and 1985; this represented 52% of the total number of salmonellosis cases where the food could be identified. Again, Humphrey *et al.* (1988) found that between 1959 and 1985 in England and Wales some 43% of family and general outbreaks of salmonella food poisoning could be attributed to the consumption of poultry.

Beef cattle, pigs and lamb are also important sources of salmonellas. Contamination rates vary enormously but probably fall within the <1–10% range. Much depends on conditions at the abattoirs, many of which may, almost inevitably, encourage the spread of salmonellas from infected to non-infected animals. It is known that the duration of stay in lairages should be short as contamination levels tend to increase with time. Thus Morgan *et al.* (1987) found salmonella contamination of pig carcasses increased from 9% on day 1 to 13% (day 2) and 27% (day 3); overcrowding should also be avoided if cross-contamination is to be minimized. Following slaughter the subsequent dressing and butchery of meats increases the spread of salmonellas on meat surfaces so that by the time the meat is in retail outlets contamination levels have increased to up to 20% (Barrell, 1982; 1987).

Imported meats are often much more heavily contaminated than home-killed meats and contamination rates in excess of 20% can be expected on occasions. Particularly hazardous are the frozen packed boneless meats since the deboning process exposes wider surface areas to cross-contamination via working tops, knives, etc.

Until the 1980s hens' eggs had not been an important cause of salmonellosis in man whereas egg products had long been recognized as serious sources of infection by salmonellas. Outbreaks of salmonellosis were regularly traced to frozen, liquid and dried whole eggs and similar products. However, pasteurization, carried out principally to kill salmonellas, is now mandatory (e.g. the Liquid Egg (Pasteurization) Regulations, 1963, in the UK) and it has proved extremely effective, with very few cases emanating from these sources now being reported. For example, for liquid whole egg, the British require pasteurization at 64·4°C for 2½ min whilst in the United States 60°C for 3 min is deemed necessary; times and temperatures vary for different types of egg product (ICMSF, 1980).

Turning to hens' eggs in more detail, as stated earlier there has been an enormous increase in salmonellosis in many parts of the world

Table 2.2
Salmonella enteritidis, phage type 4 in England and Wales (1981–1988)[a]

Year	Salmonellosis— total number of identifications	S. enteritidis *pt 4* —number of identifications	% S. enteritidis *pt 4* to total
1981	10 251	392	4
1982	12 322	413	3
1983	15 155	823	5
1984	14 727	1 362	9
1985	13 330	1 771	13
1986	16 976	2 979	17
1987	20 532	4 962	24
1988 to end Oct	23 038	10 544	46

[a] From PHLS Report (1989).

specifically associated with *S. enteritidis*, phage type 4. The relevant data for England and Wales are shown in Table 2.2 but similar trends have been reported in many other countries including France, West Germany and Italy (PHLS Report, 1989). In the north-east of the United States there was a sixfold increase in reported *S. enteritidis* infections between 1976 and 1986 and this was attributed primarily to eggs or egg-containing foods (St. Louis *et al.*, 1988). There is, indeed, little doubt that much of this increase is due to salmonella contaminated eggs. A very large outbreak involving fresh eggs was described by Stevens *et al.* (1989). A total of 165/249 guests at a wedding reception together with 8 staff succumbed to *S. enteritidis* phage type 4 infection which was traced to eggs imported from Denmark. The reception menu included the widespread use of these eggs in the preparation of various lightly cooked sauces. The eggs were specially imported to reduce the risk of contamination in foods which, for religious reasons, necessitated the use of fresh eggs in egg-based foods. The authors concluded that 'despite the low risk of an individual egg being contaminated, this episode highlights the magnification of the risk when catering practices involve mixing large numbers of eggs together'.

Contamination of eggs can occur before laying in the ovary (Perales & Audicana, 1989) or after laying due probably to faecal contamination. However, the percentage of eggs that is infected *internally* is extremely low, perhaps 1 in 100 000, in spite of the fact that salmonella isolation rates from the internal organs of chicken can be *ca* 25% (Hopper & Mawer, 1988). Thus substantial contamination of chicken flesh by *S. enteritidis*, phage type 4 is highly likely so that part of the increase in phage type 4 food poisoning could well be due to contaminated chickens *per se*.

An important contributory factor in this contamination generally is salmonella infected poultry feed (see later) exacerbated by the addition of slaughtered chicken remnants, often contaminated, to the feed. Intensive rearing methods and standards of husbandry often leave much to be desired and the culling of infected laying flocks together with tight controls on poultry feeds are essential prerequisites in the control of this problem.

The pasteurization of milk has also resulted in a substantial reduction in the number of salmonellosis cases. However milk is still sometimes implicated, particularly in certain country districts where unpasteurized milk is consumed. That this practice constitutes a considerable risk was illustrated by Barrett (1986) who surveyed milk-related outbreaks occurring in 1983–4. Of 20 such outbreaks, mainly caused by *S. typhimurium*, 19 were associated with the consumption of raw milk whilst the other resulted from post-pasteurization contamination. An earlier outbreak occurred in the Brechin and Montrose districts of Scotland in 1976; over 700 persons were infected after consuming raw milk which had been contaminated with *S. dublin* (Small & Sharp, 1979). Scotland actually introduced legislation in 1983 making pasteurization of milk sold to the public compulsory. The measure has proved remarkably effective; between 1970 and 1982 there were 50 milkborne outbreaks of salmonellosis in Scotland affecting 3518 people, whereas in the 2 years following legislative change there had not been a single incident (Sharp, 1986).

Contaminated powdered milk sometimes causes outbreaks of salmonellosis such as in Britain in 1985. Baby food milk powder was contaminated with low levels of *S. ealing* and caused a big outbreak amongst an extremely susceptible group. Salmonellas are known to survive spray drying but poor factory hygiene was primarily to blame in this instance (DHSS, 1987). Cheese is another food that is occasionally involved in outbreaks of salmonellosis. A major outbreak occurred in Canada in 1984 when an estimated 10 000 persons became ill as a result of eating Cheddar cheese contaminated with low levels of *S. typhimurium* (Ratnam & March, 1986).

The involvement of chocolate as a cause of salmonellosis has already been mentioned but the *S. napoli* outbreak was only one of several where chocolate has been implicated. In the mid-1970s there were more than 200 cases in Canada and the United States; these were mainly in young children and were caused by the relatively rare *S. eastbourne*. This serotype was found to have unusual properties in that it could be isolated in chocolate up to 15 months after manufacture (Tamminga *et al.*, 1977). The origin of the salmonella involved was traced to the cocoa beans used in the preparation of the chocolate. Another outbreak of interest involved gold-foil wrapped chocolate coins imported into Canada and the United States from Belgium (Hockin *et al.*, 1989). Children, 14 years and under, were affected; it was found the chocolate contained low levels (2–24 organisms per 100 g) of *S. nima* which again stressed the extremely low dose (<10 salmonellas) necessary to induce the clinical symptoms.

Imported desiccated coconut, used in the manufacture of biscuits, cakes and other confectionery, has been another unlikely source of these organisms and was implicated in food poisoning outbreaks in the early 1960s. Legislation was successfully introduced which improved processing and hygiene standards in the country of origin and this commodity no longer constitutes a significant health problem.

With two of the chocolate and confectionery products quoted above it was possible to trace the primary source of the salmonellas and in a similar way it has also been possible to trace an earlier source of salmonellas that may, in part, be responsible for the high incidence of these organisms in farm animals and poultry. Animal feedingstuffs, particularly those that are imported, are often contaminated with a wide variety of *Salmonella* serotypes and it has been frequently suggested that these foods are an important source of infection in the live animal (e.g. Wray & Sojka, 1977; Williams, 1981). Examples of salmonella contamination quoted for different feedingstuffs range from 5 to 27% (PHLS Working Group *et al.*, 1972; see Table 2.3) and 19% for meat and bone meal (Stott *et al.*, 1975).

Table 2.3
Growth of *Salmonella* from Different Ingredients of Pig Feed[a]

Raw material	Number of samples	% + ve
Feather meal	99	27
Meat and bone meal	704	23
Fish meal	31	23
Fish pellets	264	20
Herring meal	60	5

[a]From PHLS Working Group *et al.* (1972).

Animal by-products, with salmonella contamination levels of up to 80%, are widely used in animal feeds (ICMSF, 1980); this practice significantly increases the chance of home-reared animals becoming infected. The use of these products in pet foods has been an additional hazard and some cases of salmonellosis have been traced to domestic pets.

2.3.1.4.   The Relationship Between Salmonella Source and Salmonellosis in Man

Many examples could be cited of unusual serotypes found in animal and pet foods being subsequently isolated from the animals feeding thereon. Jones *et al.* (1982) reported an outbreak where *S. mbandaka* was isolated from raw milk obtained at 3 separate farms. It transpired that the dairy herds were all infected by a vegetable supplement feed from the same supplier; samples taken from 13 bags of the unopened feed were all found to contain *S. mbandaka*.

Other links in an contamination/infection chain have also been observed but it is more difficult to demonstrate the total relationship: contaminated animal feed → infected animal → contaminated meat → salmonellosis in man.

The problem is a difficult one with complications arising through: (1) the contamination of feedingstuffs; (2) the movement of animals around the country increasing the chances of cross-contamination and the introduction of newly purchased infected animals to previously salmonella-free herds; and (3) the delay after salmonellosis has occurred before bacteriological investigations can be completed. However, Hobbs & Hugh-Jones (1969) reported on a hospital outbreak of salmonellosis caused by patients consuming infected turkeys which had been fed with *S. senftenberg* contaminated meal; the serotype was isolated at all stages of the chain. Similar evidence involving animal feed, pigs and outbreaks of human salmonellosis in the Cardiff area was found by Harvey (1973). Indirect evidence supporting the relationship comes from the PHLS Working Group *et al.* (1972) who attributed the narrow range of serotypes found in pigs and humans in Denmark to the rigorous standards required in the manufacture of animal feeds in that country. Perhaps the most conclusive evidence for this contamination sequence was that supplied by the remarkable increase in *S. agona* isolations in the early 1970s (DHSS, 1974) which stated:

S. agona is now the second most common salmonella serotype to be isolated from humans in this country. In 1972 there were 570 isolations of this serotype from human cases and excretors and in 1971 there were 765 yet prior to 1970 it was rarely isolated. It is interesting to note the epidemiological background to this remarkable rise. *S. agona* was isolated from a consignment of fishmeal which is widely used in feeds for pigs, chickens and other farm animals imported into this country from South America in 1970. Since then, carriage of *S. agona* in animals and human food, particularly chickens and pork products, has become more common until in 1972 this serotype came second to *S. typhimurium* in isolations from human food. Similar epidemiological features of *S. agona* have been observed over the same period of time in the United States, Israel and the Netherlands. Austria, Finland, Hungary, Italy, Luxembourg and Jugoslavia all reported isolations of *S. agona* for the first time in 1970, though in these cases the source was not known. This epidemiological picture suggests that fishmeal may have been the original vehicle for the wide dissemination of *S. agona* noted since 1969–70, and it demonstrates once again that unusual strains of microorganisms present in animal feedingstuffs can on occasions rapidly become established among farm animals and poultry, with subsequent spread to the human population.

There is clear evidence, therefore, that human and animal salmonellosis can be related to contaminated feeds on occasions. However, there are many other sources of *Salmonella* spp. and many means of spreading them (e.g. rodents and herring gulls) so that the inter-relationships between all these factors are very complicated and the size of any food poisoning outbreak is extremely variable (Oosterom, 1991).

Human salmonellosis is normally seen in the form of small family outbreaks and one of the final stages in the contamination chain is usually the cross-contamination of cooked food by raw food or by dirty working surfaces, the cooked food being subsequently left at room temperature for a number of hours.

2.3.1.5. Control Measures

These are only discussed briefly here but are considered more fully in appropriate chapters later. Some of the recommendations are long-term objectives and may be almost impossible to achieve but if salmonellosis is to be reduced every effort should be made to:

1. Ensure animal feedingstuffs are salmonella-free and imported feeds are suitably heat treated.
2. Eliminate salmonellas in poultry breeding stock.
3. Improve hygiene standards in abattoirs and broiler houses.
4. Avoid cross-contamination risks, particularly of cooked by raw foods, in processing factories and kitchens.
5. Ensure adequate heating of foods, followed by rapid cooling where foods are to be stored.
6. Refrigerate foods at below 5°C where possible and avoid leaving foods at room temperature for lengthy periods.
7. Ensure food handlers are not salmonella carriers.
8. Control rodents, birds and pests in and around factory premises.
9. Increase salmonella surveillance, particularly of cooked foods.

2.3.2. *Clostridium perfringens*

2.3.2.1. *The Organism*
Clostridium perfringens, also known as *Clostridium welchii*, is a large (2–8 μm long and 1 μm wide), non-motile, Gram positive, sporing rod. The spores which are rarely seen are oval and subterminal (Fig. 2.1).

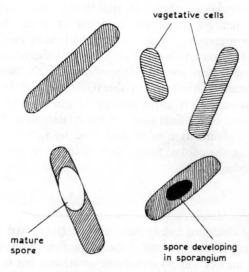

Fig. 2.1. Diagram of *Clostridium perfringens*.

Although classed as an obligate anaerobe, *C. perfringens* will grow in the presence of low levels of oxygen. The organism has a high optimum temperature for growth (45°C) and is capable of growing down to about 15°C. Five types of *C. perfringens*, A, B, C, D and E, are distinguished on the basis of various toxins produced. Of these only type A has been implicated in food poisoning outbreaks in the UK while type C causes a more serious condition called necrotic enteritis. It is the heat-resistant strains of type A, the spores of which can resist boiling for 1 to 5 h, which are responsible for the majority of outbreaks of food poisoning caused by this organism although heat-sensitive strains may sometimes be implicated.

2.3.2.2. The Disease

The symptoms of *C. perfringens* food poisoning first appear after an incubation period of 6 to 22 h. The symptoms are characterized by severe abdominal cramps, nausea and profuse diarrhoea; there is normally no vomiting or fever. Recovery is usually rapid and within 12 to 24 h although the period may be extended in elderly people for a further 1 or 2 days. The mortality rate is again very low and only very few deaths are reported annually in the UK.

For many years it was assumed that this was an infection form of food poisoning although it is now known that an enterotoxin is involved. Relatively large numbers of viable cells, more than one million, need to be ingested and the cells continue to multiply in the small intestine where sporulation occurs. The enterotoxin is produced in the sporulating cells and is released on cell lysis; it is this enterotoxin which is responsible for the typical symptoms (Skjelkvale & Uemura, 1977) although it has been demonstrated that enterotoxin production can be induced by non-sporulating cultures of *C. perfringens* (Goldner *et al.*, 1986).

2.3.2.3. Sources

C. perfringens is very widely distributed and is probably the most common of all pathogenic bacteria on the earth's surface. Type A is the most frequently isolated type and can be readily found in soils, natural waters and the intestinal contents of man and animals. Heat-sensitive strains of *C. perfringens* type A are always found in the intestinal contents of man but heat-resistant strains are much less frequently isolated, less than 10% of normal human faeces samples being positive ; rates are much higher for a short period following a food poisoning outbreak.

C. perfringens type A has been isolated from a wide variety of foods, most frequently from raw meats and poultry. The extensive contamination in meats led Smart *et al.* (1979) to suggest that strains of *C. perfringens* capable of causing food poisoning are present on every commercial carcass of raw meat, even if only in low numbers; similar results have been obtained for poultry (Nakamura & Schulze, 1970).

2.3.2.4. Conditions Leading to Food Poisoning Outbreaks

As might be expected, the majority of *C. perfringens* food poisoning outbreaks result from the ingestion of meat and poultry dishes which have been subjected to incorrect temperature treatment. This type of food poisoning is generally associated with catering malpractices and, as is shown in Table 2.1, the number of individuals affected per outbreak (15–25) is, on average, greater than that for other forms of food poisoning. These outbreaks frequently occur in institutions, hospitals, canteens and similar catering establishments where they are often the principal type of food poisoning (Sharp *et al.*, 1979).

The problem of temperature control needs amplification. Foods can be inadequately heated so that the vegetative cells of the organism are destroyed but the heat-resistant spores remain viable. Even during inadequate cooking oxygen is driven off and conditions in the food become anaerobic thus enabling spore germination to commence once the meat has cooled to 50°C. Slow cooling between 50 and 25°C is particularly hazardous with this organism as growth rates are high at these temperatures and large numbers of vegetative cells may therefore develop within a few hours. The problem is exacerbated if the foods, such as cooked meats, stews and gravies, are then stored at room temperature before being eaten as further growth will occur. The effect of cooking and cooling on *C. perfringens* in meat was studied by Sutton *et al.* (1972) who showed that even vegetative cells may not be killed towards the centre of the larger joints of meat. Cooling at 15°C was ineffective in controlling the growth of the organism and even at 4°C it would take 4½ h to reduce the temperature to 15°C, the minimum growth temperature for *C. perfringens* (Table 2.4). In catering, foods are often prepared on the day prior to consumption and eaten after warming up. This practice can be dangerous as the foregoing shows and, in addition, the final reheating is often only sufficient to stimulate further growth of the organisms rather than destroy them. A typical outbreak was described by Vernon (1977):

Table 2.4

Heat Penetration and Cooling Data for 4.5 kg Joints of Beef Cooked in Oven at 213°C for 3½ h and Cooled at Three Different Temperatures for 7 h[a]

Heat penetration data				
Depth of thermocouple (mm)	76	51	6	3
Time to reach 60°C (min)	220	145	40	37
Temperature at end of cook (°C)	55	64	90	96
Maximum temperature (°C)	65	71	90	96
Cooling data				
Temperature of cooling (°C)	15	4	−30	
Depth of thermocouple (mm)	51	76	76	
Temperature at end of cooling (°C)	22	14	−5	
Time to reach 15°C (min)	>420	395	220	

[a]From Sutton *et al.* (1972).

126 of 230 persons had diarrhoea and abdominal pain 12–14 hours after eating turkey at a staff canteen Christmas dinner. On this occasion frozen turkeys each weighing about 20 lb had been thawed, and after being stuffed were roasted for only 4 hours. They were then left for 24 hours unrefrigerated in a warm kitchen before being carved. The meat was held for a further day at room temperature, reheated to 200°F and then it finally remained on a hotplate for another hour before consumption. Heat resistant *C. welchii* were isolated from all patients tested and from the remains of the turkeys.

Perhaps rather surprisingly, boiled salmon was implicated in 5 large *C. perfringens* food poisoning outbreaks, as reported by Hewitt *et al.* in 1986. In total over a thousand persons were affected and in each instance boiled salmon, served cold as an hors d'oeuvre was incriminated. The two largest outbreaks occurred at a major London hotel over a weekend. In each instance frozen salmon was purchased by the hotel 2 days before consumption. The fish was filleted into slices approximately 1·25 cm thick and placed in a cold water bouillon. The fish was brought to the boil and kept boiling for 15–30 min. It was left to cool for 2–3 h at room temperature, drained, and transferred to trays for overnight refrigerated storage. The following afternoon the salmon fillets were portioned, the dishes made up and again refrigerated until the next night when the fish was served. Approximately 75% of the 1100 guests became ill with typical symptoms after an incubation period of 6–30 h; all guests recovered within 3 days.

2.3.2.5.   Control Measures

Obviously because of its widespread occurrence on foods as well as in the environment generally it would be impractical to eliminate *C. perfringens*. Carcass meat and poultry will inevitably be contaminated with this bacterium and survival of the heat-resistant spores after cooking can also be anticipated. The control measures should therefore be aimed at restricting both spore germination and the subsequent proliferation of vegetative cells whilst the food is cooling and during storage of the food.

Thus the following measures are recommended:

1. Where possible the food should be cooked and eaten immediately.
2. Where food is to be held over after cooking it should be cooled *as rapidly as possible* (to below 20°C within 1 h) and refrigerated; placing hot foods in a refrigerator should be avoided as cooling is likely to be too slow so that blast coolers should be used.
3. Partial or complete cooking of foods on one day with re-heating the next should preferably be avoided but where it is necessary it is important to ensure that the food is thoroughly heated before being consumed.
4. Cross-contamination of cooked foods by raw or by dirty working surfaces, equipment and utensils should be avoided.

2.3.3.   Staphylococcus aureus

2.3.3.1.   The Organism

Staphylococcus aureus is a small (0·5–1 µm in diameter), spherical, Gram positive, non-motile organism typically forming irregular clusters of cells like bunches of grapes. These organisms are facultative anaerobes growing better in the presence of air with an optimum growth temperature around 37°C and capable of growing down to 8°C or slightly below. *S. aureus* is unusual in being able to tolerate low water activity levels (a_w 0·86 minimum) and thus it grows in fairly high salt concentration levels. All *S. aureus* strains are coagulase positive (i.e. possess an enzyme coagulating blood plasma) but only about 30% of strains are able to produce the enterotoxins associated with food poisoning. Six enterotoxins, A, B, C1, C2, D and E, have been identified with types A and D being most commonly involved in food poisoning (Halpin-Dohnalek & Marth, 1989). It has been found that about two-thirds of the enterotoxin-producing strains form one toxin only, usually type A and the majority of the remainder form only two. The enterotoxins are exotoxins, so-called because they are produced within the intact cell and are released into the food or culture medium. The enterotoxins produced by *S. aureus* have the unusual property of being markedly heat-resistant. Most are able to withstand boiling in a food for up to 30 min with type B toxin showing the greatest stability (Bergdoll, 1970). Typically there is a 60–70% loss of activity within a few minutes when enterotoxin B is heated above 80°C but the remaining activity is lost much less rapidly; indeed, reactivation can occur during prolonged storage at room temperature or by heating to a higher temperature (Reichart & Fung, 1976).

According to the statistics, *S. aureus* only accounts for 0·5–1 % of food poisoning cases in the UK. However, it is likely that this is the most common form of poisoning but because of its relatively mild nature few people report the illness. Surprisingly, in the United States it accounted for well over 50% of the reported cases for many years although more recently this figure has dropped to about 10% (Cliver, 1987).

2.3.3.2. The Disease

After ingestion of the contaminated food the symptoms appear quickly, within 1–6 h, with an average of about 3 h. The most predominant and severe symptom is vomiting which is preceded by a feeling of nausea. Vomiting can be very frequent and is followed in the later stages by retching. Other common symptoms are abdominal cramps and diarrhoea. The symptoms usually last for 1–2 days and mortality is extremely low although fatal cases have been recorded.

It is known that large numbers of *S. aureus* must be present in foods for them to be hazardous but the precise number necessary to produce enough enterotoxin to induce the symptoms is not certain; figures in excess of 1 million per gram have been suggested (Halpin-Dohnalek & Marth, 1989), a number that roughly corresponds to an enterotoxin level in the food consumed of 1 μg. This is sufficient to cause illness in man but in school children the dose may only be about 0·2 μg such as was found by Evenson *et al.* (1988) when investigating a large outbreak of staphylococcal food poisoning involving chocolate milk.

2.3.3.3. Sources

The most important source of *S. aureus* is probably the human body, the principal reservoir being the nose. Between 30 and 40% of healthy individuals carry *S. aureus* and many of these nasal carriers inevitably also harbour the organism on their hands and other parts of their body. Many lesions such as boils, carbuncles, septic cuts and abrasions abound with *S. aureus* whilst another fruitful source of the organism is hair where carriage rates may be as high as those for the nose (Bryan, 1976).

Animals can also be important sources of *S. aureus*. Dairy cows commonly carry the organism on the udder and teats, and an infection, a form of bovine mastitis, can be set up by the organism. This close association with the udder inevitably means that milk becomes infected but *S. aureus* can also be spread from the infected region to milking equipment, other utensils and the hands of workers; thus a more general infection of the herd can be initiated. *S. aureus* is also commonly found on poultry and, to a lesser extent, on carcass meat. Its presence is not particularly significant even though fairly high numbers, including enterotoxin producing strains, can be isolated from poultry (Harvey *et al.*, 1982).

2.3.3.4. Foods Involved in Staphylococcal Food Poisoning

Cooked foods that have been handled by a *S. aureus* carrier, especially those with septic lesions, and subsequently stored under warm conditions for lengthy periods are the main cause of this form of food poisoning. Cured cooked meats, especially hams, which are made up into sandwiches or otherwise eaten cold are commonly implicated as are other cold meats and poultry. Other foods less frequently involved in producing outbreaks include egg products such as custards, artificial cream filled confectionery products and trifles, and cooked peeled prawns.

An interesting outbreak associated with dried Italian egg lasagne was reported by Woolaway *et al.* (1986). In January 1984 a food poisoning outbreak involving one family in Florence was traced to lasagne originally produced in Parma. Numerous outbreaks followed in Luxembourg which were traced to the same source. A month later there was a public warning issued by the DHSS in the UK since lasagne of the same brand had been imported recently. Nine food poisoning incidents were subsequently reported, involving 47 cases, following consumption of the lasagne. Examination of 50 unopened packets of lasagne revealed that 41 were heavily contaminated with *S. aureus* (counts of 10^4 to 10^8 per g) and enterotoxin A was also found. The contamination was traced back to inadequately pasteurized liquid egg that had been used in the preparation of the lasagne; the counts of *S. aureus* increased during factory preparation of the product.

Outbreaks of staphylococcal food poisoning from raw or pasteurized milk in the UK are rare but raw milk and raw milk products such as cream and cheese have given rise to outbreaks in many other countries. In the latter cases *S. aureus* can originate from a cow suffering from bovine mastitis but this contamination route is the exception rather than the rule. A more unusual source of staphylococcal food poisoning is canned foods. Canned processed peas, corned beef and other meats have been incriminated in outbreaks after *S. aureus* from infected hands had passed through minute holes in the seams of cans handled whilst still wet with cooling water.

Vernon (1977) described a number of food poisoning outbreaks caused by *S. aureus* as follows:

> In one outbreak, 47 of 95 passengers on a river trip in June developed vomiting and diarrhoea 3–6 hours after a meal which included chicken pieces that had been kept warm after cooking. Similar strains of staphylococci were isolated from a patient, from the chicken and from a dressing on the finger of one of the food-handlers; all the strains produced enterotoxin A. In another outbreak, 30 of 70 guests at a wedding vomited 1–2 hours after a reception at which chicken was served; the cooked chicken had been kept overnight in an unrefrigerated van. A count of 1.4×10^{10} *S. aureus* per g. was obtained from the chicken, which had been prepared by a nasal carrier of staphylococci of a similar type.
>
> There were two large outbreaks in schools. In one 129 of 348 pupils vomited 3 hours after eating custard; similar enterotoxin producing strains of *S. aureus* were isolated from a patient, the custard and two food-handlers. In the second, chicken had been sliced by hand and then left at room temperature: the patients, the chicken and the nose of a food-handler yielded similar strains of *S. aureus* producing enterotoxin A.

The presence of large numbers of *S. aureus* in foods does not necessarily mean that enterotoxin has been produced. Scheusner & Harmon (1973) found that whilst all pre-inoculated samples of various commercial foods which contained enterotoxin also contained high populations of *S. aureus*, enterotoxin was not detected in pre-inoculated samples of meat and fish pies containing similar high final numbers of the organism. There are many factors affecting enterotoxin production including type of food, pH (little if any production below pH 5·0), temperature (optimal production at 37°C but wide range tolerated), presence of oxygen (poor enterotoxin production under anaerobic conditions) and the presence of other organisms whose main effect is to inhibit the growth of *S. aureus* rather than affect enterotoxin production *per se* (Notermans & van Otterdijk, 1985; Halpin-Dohnalek & Marth, 1989). There are, therefore, many inter-related factors determining whether or not *S. aureus* food poisoning occurs and these have been succinctly described by Bryan (1976):

1. A source of an enterotoxigenic strain of *Staph. aureus* must be in a food production, processing or preparation environment.
2. The organism must be transferred from the source to a food.
3. The food must be contaminated with thousands of *S. aureus* per g, or, more usually, the food must be heated before it becomes contaminated, or it must contain high levels of salt or sugar.
4. The organism must survive in the food; it must not be outgrown or inhibited by competing organisms or killed by heat, low pH, or other adverse conditions before it can produce enterotoxin.
5. The food, after it becomes contaminated, must support the growth of *S. aureus*.
6. The contaminated food must stay within the temperature range that is suitable for proliferation of *S. aureus* long enough for this organism to multiply and produce enterotoxin.
7. A sufficient quantity of enterotoxin-bearing food must be ingested to exceed the enterotoxin susceptibility threshold of persons eating the food.

2.3.3.5. Control Measures

As with *C. perfringens*, measures aimed at eliminating *S. aureus* would be totally impracticable due to its widespread occurrence. Therefore control measures should aim to limit contamination and the subsequent growth of the organism in foods. Vital control measures recommended are:

1. Keep handling of cooked foods to a minimum. Particular care should be taken with warm cooked foods which should preferably be cooled to below 20°C where subsequent handling is essential.
2. Personnel with septic lesions should not handle foods; because of the high nasal carriage rate in humans it would be impracticable to prohibit such carriers from handling foods but disposable gloves should be worn by all operatives.

3. Adequate heat treatment of the food is essential followed by prompt cooling to 10°C or below where foods are to be stored.
4. Minimize cross-contamination from raw to cooked foods and from dirty working surfaces, equipment and utensils.

2.3.4. Bacillus cereus

2.3.4.1. The Organism
Bacillus cereus is a large (3–5 μm long and 1 μm wide), Gram positive, spore-forming rod (Fig. 2.2). The spores, which are elliptical and central, are much less heat resistant than those of *C. perfringens* and are destroyed at 100°C within 5 to 30 min. *B. cereus* is a facultative anaerobe typically growing over a temperature range of 10–48°C and strains have an optimum growth temperature of 35–45°C; strains are occasionally isolated growing down to 4°C but failing to grow above 37°C (Van Netten *et al.*, 1990).

Two different toxins can be produced by *B. cereus* inducing two distinct types of food poisoning. The first of these toxins, an enterotoxin, is secreted into the food during logarithmic growth. This enterotoxin is relatively heat sensitive and is destroyed in 30 min at 56°C (Goepfert *et*

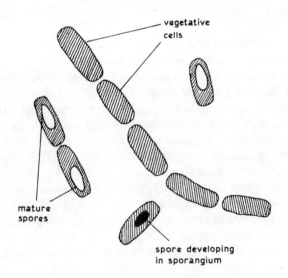

Fig. 2.2. Diagram of *Bacillus cereus*.

al., 1973); the illness induced is a 'diarrhoeal syndrome'. The second toxin is again preformed in the food but is much more heat resistant being unaffected by 90 min heating at 126°C (Melling & Capel, 1978); this toxin induces a 'vomiting syndrome'.

2.3.4.2. The Disease

As stated above, evidence has accumulated that *B. cereus* is responsible for two distinct types of food poisoning. In the first type, which is similar to *C. perfringens* food poisoning, the incubation period is about 8–16 h. The symptoms are characterized by acute abdominal pain and profuse diarrhoea; nausea, vomiting and fever are rare and the symptoms usually last for less than 24 h. This form of the disease has been reported in several European countries, the United States and Canada. Many different foods have been implicated including meat and poultry products and sauces.

The second type of food poisoning, which is similar in character to that produced by *S. aureus*, has a much shorter incubation period of only 1–5 h. The symptoms are acute nausea and vomiting; diarrhoea is comparatively rare. The duration of the illness is from 6 to 24 h. This form of the disease was first reported in the UK in 1971 and since then there have been many cases always associated with the consumption of cooked rice; similar incidents have been reported in other parts of the world (Gilbert, 1979).

2.3.4.3. Foods Involved in B. cereus Food Poisoning

B. cereus is found commonly in soil and water and can be readily isolated from a wide variety of plant foods including cereal dishes, mashed potatoes, vegetables and vegetable soups. These foods together with various cooked meat dishes have been responsible for the diarrhoeal form of food poisoning common in several European countries. It has been reported as a common form of food poisoning in the Netherlands (Beckers, 1988) and has been associated with meat dishes that are frequently well seasoned with spices; it should be noted that spices often contain large numbers of aerobic spore-bearing bacilli including *B. cereus*. The cooking of these foods was clearly inadequate allowing the spores to survive so that the warm storage conditions after cooking resulted in spore germination and subsequent heavy growth of vegetative cells (Ormay & Novotny, 1969). In most outbreaks large numbers of *B. cereus*, between 10^5 and 10^8 per gram, were isolated from the incriminated foods; examples of outbreaks have been cited by Gilbert & Taylor (1976).

As already mentioned, the second type of *B. cereus* food poisoning appears to be associated almost exclusively with the consumption of fried rice prepared in Chinese restaurants or 'take-away' shops. In these establishments portions of boiled rice have been allowed to 'dry-off' at room temperature for periods of up to 24 h or even longer. In this way boiled rice may take many hours to cool thus creating ideal conditions for the germination of any surviving spores and for the subsequent multiplication of vegetative cells. When required the rice is fried with beaten egg for about 1 min and kept warm until served. The frying process reduces bacterial numbers substantially but the final holding period, preserving, can again be extensive and further heavy growth of bacteria may take place (Gilbert *et al.*, 1974).

Beckers (1988) reported a typical food poisoning episode in which large numbers of *B. cereus* were isolated from fried rice:

> Three people, a man and his two children aged 2 and 6, became ill about 3–5 h after eating fried rice. The symptoms were nausea, vomiting, abdominal cramps, palpitations and severe tightness of the chest. The mother had not eaten any of the fried rice and did not fall ill. Tests carried out on the remains of the fried rice showed it to contain about 10^7 *B. cereus*/g and 0.96% monosodium glutamate. No fecal samples were tested. The case history and the test results, and the fact that the remains of the fried rice had been stored in the refrigerator, pointed to *B. cereus* as the cause of the food poisoning. Its effect may have been reinforced by the presence of a considerable quantity of monosodium glutamate, which probably caused the palpitations and tightness of the chest.

2.3.4.4. Control Measures

Gilbert *et al.* (1974) have suggested the following preventative measures:

1. Rice should be boiled in smaller quantities on several occasions during the day, thereby reducing the storage time before frying.
2. After boiling the rice should either be kept hot, at not less than 63°C, or cooled quickly and transferred to a refrigerator within 2 h of cooking. The cooling of rice, especially large bulks of boiled rice, will be hastened by dividing the product into separate portions or by spreading the bulk in clean shallow containers.
3. Boiled or fried rice must not be stored under warm conditions and never at a temperature between 15° and 50°C. Under no

circumstances, therefore, should cooked rice be stored at kitchen temperature for more than 2 h.

4. The beaten egg used in the preparation of fried rice should be freshly prepared.

2.3.5. *Vibrio parahaemolyticus*

2.3.5.1. The Organism

Vibrio parahaemolyticus is a short, slender (1·5–2·5 μm by 0.5 μm), curved, Gram negative rod which is motile by means of a single polar flagellum (Fig. 2.3). The organism is facultatively anaerobic. It is slightly halophilic growing best in the presence of 2–4 % NaCl but tolerating salt concentrations of up to 8%. The temperature range of growth is 10–44°C and optimum growth occurs at 37°C at which temperature cell division can occur extremely rapidly (every 10–12 min). A further unusual property of this organism, and related vibrios, is its preference for alkaline conditions and some strains exhibit optimal growth even as high as pH 9·0. *V. parahaemolyticus* is very heat-sensitive and is readily

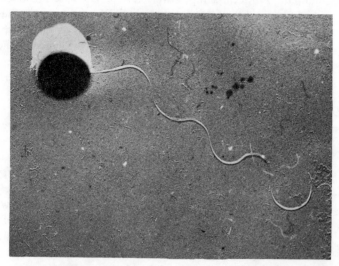

Fig. 2.3. Electron micrograph of a cell of *Vibrio parahaemolyticus* showing single polar flagellum (× 15 000).

killed at temperatures above 50°C. Not all strains are pathogenic for man but those that are almost invariably give a positive Kanagawa reaction (i.e. lyse red blood cells). These pathogenic strains appear to possess a 'toxic factor' which is released on lysis of the bacterial cells (Brown *et al.*, 1977); however, in spite of extensive studies into the properties of the Kanagawa haemolysin and pathogenic mechanisms in *V. parahaemolyticus*, no explanation as to how the illness is induced has been forthcoming (Sakazaki & Shimada, 1986).

2.3.5.2. The Disease

The symptoms of *V. parahaemolyticus* food poisoning usually appear 10–18 h after ingestion of the contaminated food although incubation periods varying from 2 to 48 h have been reported. The main symptoms are nausea, vomiting, abdominal pain and diarrhoea. Mild fever occurs on occasion but the duration of the illness is usually short, from 1 to 4 days; the mortality rate is very low, most deaths being associated with the elderly and infirm. Carriers of the organism have not been reported.

2.3.5.3. Foods Associated with V. parahaemolyticus Food Poisoning

This organism is almost exclusively found in the marine environment. It is most abundant during the summer months in inshore areas such as shallow coastal and estuarine waters when the temperature of these waters is at its warmest. It has also been isolated from coastal marine sediments and from various marine animals, especially bivalve shellfish but also from crustaceans and fish. *V. parahaemolyticus* is particularly prevalent in Far Eastern waters and hence natural levels of contamination can be high; Chan *et al.* (1989) found mussels and oysters to contain, on average, 4·6 and 3·4 × 10^4 per g respectively.

In Japan, where raw fish is commonly eaten, *V. parahaemolyticus* accounts for approximately 60% of food poisoning cases. For many years it was thought that the organism was restricted to the Far East but more recently it has been isolated from many other regions of the world. A survey (Ayres & Barrow, 1978) undertaken to determine the distribution of *V. parahaemolyticus* around Britain showed isolation rates for the organism of 11, 18 and 14%, respectively, from coastal waters, sediments and shellfish, i.e. oysters, cockles and mussels; however, all isolates gave a Kanagawa negative reaction and the authors concluded that these organisms were unlikely to pose any serious health hazard. Similar isolation rates for European waters have been reported elsewhere although van den Broek *et al.* (1979) found that as many as

26% of their samples of mussels from estuarine waters in The Netherlands were contaminated with *V. parahaemolyticus*. Higher isolation rates have been reported in other parts of the world, e.g. Thompson & Vanderzant (1976) found 60% of oysters from the northern Gulf of Mexico to be contaminated.

V. parahaemolyticus is thus principally associated with seafoods and it is these foods which have been responsible for most of the outbreaks of this type of food poisoning. In European countries and the United States, unlike Japan, most seafoods are cooked before consumption and therefore the organism is invariably killed. However, oysters are an exception but they are rarely implicated in food poisoning outbreaks, probably because the numbers of *V. parahaemolyticus* present in this mollusc are too low to initiate the illness. Foods that have been involved in food poisoning outbreaks in the UK and United States include dressed crab, lobster, prawns and shrimps, usually imports from the Far East. Typically there used to be about 100 cases a year reported in the UK but this number was reduced to almost zero by the latter half of the 1980s.

Outbreaks appear to have been caused in two ways, either by inadequate cooking or by recontamination after cooking, both of which left viable organisms in the food which were capable of developing into vast numbers during subsequent storage at room temperature. A typical outbreak was reported by Hooper *et al.* (1974):

In July 1973, the nurse at a holiday camp on the south coast of England told the general practitioner in charge of the medical services that five people had reported to her that morning with acute gastroenteritis and that there seemed to be a common source of infection ... [as] ... all had eaten dressed crab the previous evening. Each of the four families concerned had purchased crab meat on the same day from the same fishmonger in the town.

Symptoms developed 16–18 hours after the crab meat had been eaten. Severe cramp-like abdominal pains were followed soon afterwards by violent diarrhoea accompanied by weakness and fever. Two patients also reported vomiting before diarrhoea started. All recovered in 24–48 hours. ...

The outbreak was described in the local, and later in the national, press, and as a result six people in a further three families reported that they had also had a similar illness on the same date after meals of dressed crab ... purchased from the same fishmonger.

The crab had been caught locally and none remained available for laboratory examination but faecal samples obtained from six patients yielded *V. parahaemolyticus*. Further investigations led to the conclusion that cross-contamination of cooked by live crabs or from dirty working surfaces was the primary cause of this outbreak.

2.3.5.4. Control Measures
Once again this is primarily a storage and cross-contamination problem so control measures should aim to:

1. Ensure adequate heat treatment of cooked seafoods.
2. Ensure rapid cooling and refrigeration of cooked seafoods if they are to be stored and ensure raw seafoods are refrigerated at all times.
3. Avoid cross-contamination of cooked by raw seafoods or by surfaces which have come into contact with the raw food.
4. Where possible, attempt to import seafoods in the raw state so that final processing can be monitored.

2.3.6. Botulism

2.3.6.1. The Organism
Clostridium botulinum, the causative organism of botulism, is an obligately anaerobic, motile, Gram positive, spore-forming rod; the rods are large (4–8 μm by 0·9–1·2 μm), although type E (see below) is more slender (width 0·3–0·7 μm). Seven types of *C. botulinum* are now recognized, designated A to G on the basis of the toxins they produce; however strains producing two toxins (e.g. A and B or A and F) have been isolated so that this strict classification no longer holds true. Types A, B and E produce exotoxins that are almost entirely responsible for botulism in man. Types C and D are important as causes of botulism in animals, including birds, although type C has been implicated in a few outbreaks of human botulism. Types F and G have also caused human botulism on a few occasions.

The optimum temperature for growth of the organism varies from 25° to 37°C depending upon the strain; the minimum growth temperature for the majority of the types varies between 10°C and 20°C but psychrotrophic strains of types B,E and F have minimum growth temperatures of 3–5°C. The spores of *C. botulinum* are markedly heat-resistant and those of types A and B can resist boiling for up to 6 h. The spores of types C and D are rather less heat-resistant whilst type E spores are inactivated at 80°C in 15 min.

The exotoxins produced by *C. botulinum* are extremely poisonous to man and a fatal dose for an adult of type A toxin, the most potent, can be as little as 1×10^8 g; in other words, 1 g could kill 100 million people. The toxins which are preformed in the food before ingestion are normally absorbed in the small intestine, transferred via the bloodstream and finally absorbed into the nervous system. The toxins are classed as neurotoxins since their action is specifically directed to nerves, especially the peripheral nerves of the involuntary muscles of the body. The toxins are not very resistant to heat, types A and B being inactivated by exposure to 80°C for 10 min whilst type E is destroyed within 5 min at 60°C (Riemann, 1969; Sakaguchi, 1969).

2.3.6.2. The Disease

The first symptoms usually appear some 18–36 h after ingestion of food containing the toxin but they may be delayed a further 36 h or even longer, considerable variation occurring in individual cases. In rare cases the first symptoms may occur within a few hours, the earlier the onset the less favourable the prognosis. The first symptoms are usually nausea, vomiting and possibly diarrhoea accompanied by fatigue, headache and dizziness. Persistent constipation is a common feature following the early symptoms and this is usually accompanied by blurred vision and difficulty in swallowing and speaking. Extreme muscular weakness is another common feature. There is no fever and the temperature is frequently subnormal. In the more severe cases the involuntary muscles become paralysed and the paralysis spreads to the respiratory system and heart, death finally resulting from respiratory failure or cardiac arrest. The severest symptoms last for up to 10 days with death most common between the third and sixth day after ingestion of the food. With survivors recovery is extremely slow but is complete within 6 months. The mortality rate is high being quoted at 65% in the United States but the rate is generally much lower in European countries.

Treatment of patients includes the administration of antitoxin specific for the type of toxin involved, together with the use of modern life-support techniques which have helped to reduce the previously high mortality rates.

2.3.6.3. Sources

C. botulinum is widely distributed in soils although the types causing human botulism are more prevalent in the nothern hemisphere than in the southern. Concentrations are low, probably less than 1 per gram of soil. *C. botulinum* can be isolated readily from many fresh and marine waters, their sediments and from fish, but such isolates are typically the psychrotrophic strains of types B, E and F; their original source may well be terrestrial. In the UK about 5 % of soil samples contain *C. botulinum*, mainly type B, but isolation rates from aquatic environments are substantially higher. Smith *et al.* (1978) found 35% of samples to be positive; type B was again the most widespread (30% positive) whilst types C, D and E were isolated occasionally (1–3% positive).

The predominance of type B is common in many European countries but in Scandinavia and the Baltic type E is widespread. Type E is also relatively common in marine waters, sediments and fish in other areas of the northern hemisphere such as Alaska and the Pacific coast of the United States, the Soviet Union and Japan (Hobbs,1976). In the United States mesophilic rather than psychrotrophic strains tend to predominate but geographical location affects the distribution pattern. Thus from the Rocky Mountains to the Pacific coast type A is more common than B whereas the reverse is true to the east of the Mississippi River (Sugiyama & Sofos, 1988).

2.3.6.4. Foods Implicated in Botulism

A wide variety of foods has been involved in human botulism reflecting the world distribution of the predominant types and the eating habits of the populations affected.

In the UK no substantiated outbreak of botulism was recorded until 1922 when the Loch Maree, Gairloch, tragedy occurred. This has been described in detail by Dewberry (1959) but the following is a summary:

> A party of eight visitors staying at a local hotel went out fishing on the 14th August and during the day ate duck paste sandwiches. About 3 am on the 15th the first of the party was taken ill and late that evening three died; in the following week the remaining members of the party also died. Samples of the duck paste, as well as one of the sandwiches, were examined and found to contain *C. botulinum*, type A, together with lethal quantities of type A toxin.

After that outbreak there were no further cases in the UK until 1935 when four deaths occurred in London. Three cases occurred in 1944 and five in 1947. In 1955 two Mauritian students became victims of type A poisoning after eating pickled fish brought from Mauritius; both patients recovered after treatment with type A antitoxin. Four old-age

pensioners in Birmingham contracted type E botulism in the 1978 outbreak described by Ball *et al.* (1979) of which the following is a summary:

> After eating a tin of canned salmon, originally processed in Alaska,
> for tea on Sunday, 30th July the first couple began to show symp-
> toms of vomiting and severe diarrhoea at 2 am on the Monday.
> They were moved to hospital in acute distress a few hours later
> where they were examined and found to have blurred vision and
> had difficulty in speaking. Police were alerted to the second couple
> who shared the meal and called at their house to find them desper-
> ately ill and too weak to seek help. They were rushed to hospital
> where they deteriorated rapidly and by 10 am they were unable to
> breathe as their muscles became paralysed. All four victims were
> placed in the intensive care unit and treated with antitoxin but in
> spite of the latest life-support techniques the first couple admitted
> died 17 and 23 days after eating the contaminated food. Death was
> due to cardiac arrest and respiratory failure. *C. botulinum* type E
> was isolated from the remains of the salmon in the tin.

It was later ascertained that the can had a construction fault which
enabled it to become contaminated during the air cooling of cans at the
Alaskan factory. It was alleged that operatives handling raw fish dried
their overalls on the cooling cans during breaks thus allowing *C. botulinum* spores to gain entry into any faulty cans; if true, this is a perfect
example of the results of cross-contamination of cooked food by raw.

The latest outbreak of botulism in the UK occurred in the summer of
1989 associated with hazlenut yoghurt (O'Mahony *et al.*, 1990). Twenty-
seven cases were recorded all showing the typical symptoms; although
many required intensive care treatment there was only one fatality. *C.
botulinum* type B toxin was detected in cans of hazlenut puree, used in
the preparation of the yoghurt, but toxin levels must have been minimal
for only one death to have been recorded.

Outbreaks in Europe have been principally associated with various
meats and meat products, often home-preserved, in which type B toxin
has been formed. Historically botulism is associated with blood and
liver sausages and the name botulism is derived from the Latin word for
sausage *botulus*; smoked ham, pork brawn and various pâtés have also
been implicated. Fish, and to a lesser extent other marine foods, in
which type E toxin has been produced have been a common cause of

botulism in northern European countries, the Soviet Union and Japan. In the latter country during the period 1951–83 there were 189 outbreaks, almost all type E botulism associated with fish (Sugiyama & Sofos, 1988). Such outbreaks can usually be explained on the basis of poor home-processing or the consumption of uncooked foods.

In the United States the picture is rather different and most cases of botulism have been caused by canned or bottled fruits or vegetables which have been inadequately processed in the home. In the period from 1930 to 1949 when home-canning was at its height there were 255 outbreaks attributed to that cause. Green beans and sweetcorn have been the foods most frequently implicated and where toxin identification has been possible type A has predominated. Botulism continues to be relatively common in the United States with 203 cases occurring between 1978 and 1983; home prepared foods were again mainly responsible (Sugiyama & Sofos, 1988). Commercially canned foods have a much better, although not blameless, record with 60 outbreaks and just over 100 fatalities occurring from the turn of the century to 1970. Occasional outbreaks still occur and commercial canned products such as vichyssoise, beef stew and canned mushrooms were responsible for small outbreaks with fatalities in the 1970s (Lynt *et al.*, 1975).

As already indicated, *C. botulinum* type E is primarily isolated from the aquatic environment and hence fish and other seafoods predominate in outbreaks of this type. This is also true for the United States where the majority of type E outbreaks have been associated with these foods (Sugiyama & Sofos, 1988). Particular interest was aroused in the early 1960s by two outbreaks of type E botulism in which 12 people died following the consumption of vacuum-packed smoked fish. Until these episodes it had been assumed that the combination of vacuum-packaging and refrigerated storage was a safe method of preserving fish products. It was found that toxin production in fish contaminated with type E spores occurred not only under vacuum-packaged conditions but may also be produced in certain species of fish (e.g. herring) regardless of the method of packaging (Abrahamsson *et al.*, 1965).

Instances of an alternative form of botulism, 'infant botulism' have been frequently reported in more recent years (Sugiyama & Sofos, 1988). This infection is caused by *C. botulinum*, typically either type A or type B, growing in the intestinal tract of infants aged under 9 months; whilst growing the toxin is produced and this form of illness is therefore similar to *C. perfringens* food poisoning in that respect. The organism is transmitted in honey, a food that has been found to contain low numbers of *C. botulinum* spores on occasions (Hauschild *et al.*, 1988); such spores are probably initially picked up by bees. It has been recommended that honey should not be given to infants under 1 year old since below that age the infant has not yet developed a competitive gut microflora which would prevent the germination of the *C. botulinum* spores (Sugiyama & Sofos, 1988).

2.3.6.5. Factors Affecting Toxin Formation in Foods

Toxin formation in foods is obviously dependent on the spores withstanding the processing treatment, or on a post-process contamination. Subsequently conditions must allow for spore germination, growth of the vegetative cells and toxin production. Conditions affecting this chain of events include the composition of the food, the temperature and period of storage, as well as the moisture content, pH, oxygen availability and OR potential, and salt content of the food.

Although many types of food have been implicated in outbreaks, the foods do have common features. Most have been subjected to some form of preservation, they have been stored under conditions permitting toxin to be formed and they have been either eaten without further cooking or the reheating has been insufficient to destroy the toxin.

Temperature is an important factor but toxin formation is possible over the whole growth temperature range. However other factors such as the degree of acidity and the salt content of the food become more effective growth inhibitors as the temperature is reduced so that lower storage temperatures (15°C as a maximum) are to be preferred. Again, toxin production is possible at all moisture and pH values at which growth occurs although toxin is unstable at pH values above 7 (Baird-Parker, 1971). No growth or toxin production can occur at pH 4·5 or below under normal conditions.

In some foods, such as canned vegetables, toxin production is unlikely unless air is excluded. Foods like meats and fish have a strong reducing tendency so that growth and toxin formation are quite possible even in the presence of air; the example of herring has been mentioned previously.

Strains of *C. botulinum* differ considerably in their sensitivity to salt in foods. Some, mainly strains of types A and B, will tolerate levels of up to 10% sodium chloride whilst type E strains are inhibited at 4·5–5 % (Sugiyama & Sofos, 1988). The situation is complex because sensitivities to sodium chloride alter at different pH values and at different sodium nitrite levels if the latter is included as a preservative in the food.

2.3.6.6. Control Measures

There are many facets to this problem but the following points are crucially important:

1. Ensure that the heat treatment for canned and bottled foods (pH >4·5) is sufficient to destroy the most heat-resistant *C. botulinum* spores.

2. Use the cleanest possible chlorinated water for cooling cans after processing. Where air cooling is used ensure that cross-contamination from raw materials or dirty equipment, or by handling is avoided during cooling.

3. Where milder heat processes are applied to foods ensure that suitable inhibitory compounds are introduced or that the pH is low enough to prevent growth of *C. botulinum*.

4. Ensure adequate heat treatment with semi-processed or processed vacuum-packed foods and store at a maximum temperature of 3°C.

5. Never taste a food that is suspect unless it is pre-heated to 100°C.

2.3.7. Listeriosis

2.3.7.1. The Organism

Listeria monocytogenes, the causative organism of listeriosis, is a facultatively anaerobic, motile (at 20–25°C), short Gram positive rod of regular appearance; cells are 0·5–2·0 μm in length and 0·4–0·5 μm in diameter. The optimum temperature of growth of the organism is typically about 35–37°C but strains can exhibit wide growth temperature ranges of 1–45°C; in fact the majority of strains may well grow at *ca* 1°C (Junttila *et al.*, 1988). However *L. monocytogenes* also exhibits a surprising resistance to heat and it has been suggested that it survives the minimum pasteurization heat treatment (72°C for 15 s) required by many countries for raw milk (Fernandez Garayzabel *et al.*, 1987). Apparent differences in heat sensitivity have been highlighted depending on the methodologies employed. Thus Bradshaw *et al.* (1987) claimed that normal pasteurization treatments were sufficient to destroy *L. monocytogenes*. This has been questioned by Doyle *et al.* (1987) who stressed the need to use suitable enrichment procedures in order to facilitate the recovery of metabolically injured cells (see Chapter 3). This point is illustrated in Table 2.5 which shows that *L. monocytogenes* was only isolated after 48 h refrigeration of the milk following pasteurization.

Table 2.5
Counts of *Listeria monocytogenes* in Milk (Initial Count 10^8 per ml) Stored at 4°C for 5 d Following Pasteurization (72°C/15 s)[a]

	Days	Count per ml
Immediately after pasteurization	0	0
Storage at 4°C	1	0
	2	40
	3	150
	4	800
	5	2 500

[a]From Fernandez Garayzabel *et al.* (1987).

The virulence mechanism has to be elucidated but *L. monocytogenes* strains lyse red blood cells (i.e. haemolytic strains) and this appears to be associated with pathogenicity; likewise *L. monocytogenes* kills chick embryos when cells are injected, which non-pathogenic *Listeria* spp. fail to do (Schönberg, 1989).

2.3.7.2. The Disease
Listeria monocytogenes is a low grade pathogen, the ingestion of low numbers of viable cells causing no clinical manifestations in healthy adults. Certain groups of humans are more susceptible to the organism and these include infants, the elderly, persons immunocompromised by drugs or disease and pregnant women. With such groups *L. monocytogenes* causes, in particular, septicaemia and meningitis but such cases are also reported occasionally in apparently healthy adults, possibly where large numbers of organisms have been ingested. Listeriosis is of especial concern in pregnant women; infections produce 'flu-like symptoms and may cause premature labour, stillbirth or early death of the newly-born infant. Mortality is very high, 50% in infants, and at least 25% with the other groups. The incubation period is lengthy, usually about 14 days, although it can vary from 4 to 21 days.

2.3.7.3. Sources of Listeria and Foods Involved in Listeriosis
Listeria spp. are widely distributed in the environment and the same is true, to a lesser extent, of *L. monocytogenes*. Thus the latter can be isolated from soil, silage, water, vegetation, a wide range of foods and the faeces of many animals including humans. Cox *et al.* (1989), in an examination of listerias in food processing and other environments, found 3·5% of samples taken to contain *L. monocytogenes* whilst Charlton *et*

al. (1990) found *Listeria* spp. in 75/597 samples (12·6%) taken from milk processing plants; a half of these isolates were identified as *L. monocytogenes*. These typical findings stress the comparative ubiquity of this organism.

Raw milk is recognized as an important source of *L. monocytogenes*. Hayes *et al.* (1986) found 12% of over 100 such samples examined in the United States to harbour the organism; contamination rates of up to 55% for listerias in general have also been reported in certain parts of that country (Dirksen & Flagg, 1988). Thus Gitter *et al.* (1980) had stressed earlier that the excretion of *L. monocytogenes* in milk was an important factor in the transmission and epidemiology of listeriosis, whilst Slade *et al.* (1989) concluded that 'the ubiquity of *Listeria* spp. in the rural environment makes their control and eradication from the agricultural niche, and hence from raw milk supplies, a daunting, somewhat improbable, task'.

It has already been mentioned that there is conflicting evidence concerning the ability of *L. monocytogenes* to withstand pasteurization treatments used for milk. This problem was highlighted earlier, in 1983, in a big outbreak of listeriosis in the United States which was directly attributed to the consumption of pasteurized milk. However, recently it has been shown unequivocally that the *L. monocytogenes* strain involved in that outbreak could not survive correct pasteurization treatment (Lovett *et al.*, 1990). The presence of *L. monocytogenes* in 1% of over one thousand pasteurized milk samples examined in the UK indicates that post-pasteurization contamination or inadequate processing remain a problem (Greenwood *et al.*, 1991).

Soft Mexican-style cheese was incriminated in a further outbreak in the United States in which there were 103 reported cases of whom, 18 adults and 29 unborn infants died. Some doubt exists about the exact cause but it was most likely due to some form of post-pasteurization contamination (Dirksen & Flagg, 1988). This incident illustrates the inevitable association of dairy products in general being causes of listeriosis; thus cheeses, particularly soft cheeses, made either from pasteurized or unpasteurized milk have been implicated. An isolated case in the UK was reported by Bannister (1987):

> A 36-year-old woman was admitted to hospital on 9th January with an 18 hours' history of fever, back pain, aching legs and increasing headache with neck stiffness...... She was not pregnant. On examination she was distressed and drowsy with a temperature of 38·7°C..... *Listeria monocytogenes* appeared in CSF (i.e. cerebral spinal fluid) culture..... (It was found that) she had purchased two

Food Microbiology and Hygiene

soft French cheeses for Christmas, and had kept them refrigerated for about 2 weeks. She had eaten both types of cheese during the week before her illness but her husband and son had not done so. Remaining portions of the cheese were examined .... One of them, a Camembert, did not yield any growth of significance, but a soft country cheese produced a heavy growth of *L. monocytogenes*. Both the strain isolated from the CSF and the one isolated from the cheese .....were shown to be of (the same serotype).

The frequency of listeriosis has increased in England, Wales and Northern Ireland (see Table 2.6) and concern has been expressed about this trend. More recent data suggest the peak may have been reached. However, it is reasonable to assume that, since listeriosis is not notifiable in the UK, the official figures are a considerable underestimate of the true position (Lacey & Kerr, 1989). Foods are the most likely means of acquiring the infection (WHO Working Group, 1988) so that much attention is necessarily being paid to the possible role of food in the spread of listeriosis. Most studies have been published on cheeses: these have included surveys on the natural incidence of *L. monocytogenes* in cheeses and the behaviour of deliberately inoculated strains during the manufacture and ripening of cheeses. Beckers *et al.* (1987) found 14·5% of soft cheeses imported into the Netherlands to contain *L. monocytogenes* with recoveries of the organism of up to 10^6 per g cheese. Pini & Gilbert (1988) in a survey of home produced and imported soft

Table 2.6
Listeriosis in England, Wales and Northern Ireland (1973–1991)[a]

Year	No. Cases[b]	Year	No. Cases[b]
1973	25	1983	119 (115[c])
1974	28	1984	91 (115)
1975	39	1985	124 (149)
1976	31	1986	113 (137)
1977	47	1987	204 (259)
1978	87	1988	239 (291)
1979	70	1989	(250)
1980	75	1990	(117)
1981	86	1991	(130)
1982	77		

[a]Based on data from PHLS Communicable Disease Reports.
[b]Unedited annual totals.
[c]Numbers in parentheses show data enhanced by additional information from the PHLS Laboratory of Microbiological Reagents.

cheeses in the UK found similar isolation rates (10%) and counts of up to 10^5 per g; one interesting conclusion drawn from this study was that it made little difference whether the cheeses had been prepared from pasteurized or non-pasteurized milk.

When *L. monocytogenes* is deliberately inoculated into pasteurized milk used in the manufacture of different cheeses the fate of the organism is determined by the type of cheese and, in particular, its pH. Thus Ryser & Marth (1987) have shown that, when *L. monocytogenes* is inoculated at 5×10^2 per ml, with Camembert cheese there is an initial slight increase in numbers followed by a decrease to < 10–100 per g during the first 18 days' ripening; counts of the organism then increase to 1 $\times 10^6 - 5 \times 10^7$ per g, these increases reflecting the higher pH (6–7) of the cheese. No such increases are seen with Cheddar cheese whilst with cottage cheese decreases and possible elimination of the organism are possibly due to the low pH (*ca* 5) of the latter (Ryser *et al.*, 1985). One can safely conclude from these studies that the presence of significant numbers of *L. monocytogenes* in milk used in the manufacture of soft cheeses is a potential health hazard.

Concern has been expressed about raw chickens and meats as an important source of *L. monocytogenes*. Pini & Gilbert (1988) found 60% of fresh and frozen birds in the UK to be contaminated with the organism; Bailey *et al.* (1989) reported that 23% of broiler carcasses harboured *L. monocytogenes* in the United States. Isolation rates for *L. monocytogenes* from raw meats tend to be somewhat lower (0–45%) although for such a ubiquitous organism it is not surprising that it can normally be isolated from raw beef, pork and lamb (Johnson *et al.*, 1990).

Studies involving the deliberate inoculation of the organism onto chicken carcasses have included that of Harrison & Carpenter (1989). They used high inoculum levels (10^6–10^7 per g) onto chicken breasts and found that heating them to an internal temperature of 82°C was insufficient to eliminate totally *L. monocytogenes*, which recovered to 10^7 per g following post-heating storage at 10°C for 10 days; storage at 4°C for 4 weeks following heating to an internal temperature of 77°C allowed the organism to recover to 10^5 per g. These results indicate that a recommended final temperature of 71°C will not eliminate *L. monocytogenes* although the authors point out that the inoculum level was far higher than would be expected. Glass and Doyle (1989) studied the survival and growth of the organism in a range of foods during storage at 4°C for up to 12 weeks. Growth of *L. monocytogenes* was best on sliced chicken and turkey, counts increasing from 10^3 to 10^5 per g in 4 weeks;

there were smaller increases on other foods (e.g. ham, bratworst) with levels of 10^4 per g being established after 6 weeks whilst on cold roast beef only slight growth occurred over the full 12 weeks. The authors again concluded that pH was crucial in determining growth patterns; above pH 6 growth was most rapid whereas *L. monocytogenes* grew poorly or not at all on products with a <5 pH. However, George *et al.* (1988), using laboratory media rather than foodstuffs, found *L. monocytogenes* was capable of growing down to pH 4·4 (20°C incubation), pH 4·6 (10°C) and only down to pH 5·2 at 4°C suggesting that storage temperature is an important determinant of growth control when pH is considered.

Coleslaw was responsible for an outbreak of listeriosis in Canada in 1981 and celery, tomatoes and lettuce have been implicated in hospital outbreaks; in fact, *L. monocytogenes* can be isolated readily from fresh vegetables (Heisick *et al.*, 1989). Growth of food poisoning bacteria on these and similar foods would be unexpected and little work has been published. However, Steinbruegge *et al.* (1988) found that *L. monocytogenes* inoculated onto lettuce increased 10-fold in numbers over 14 days' storage at 5°C and from 10^3 to 10^6 per g over the same period at 12°C. These results again stress the remarkable facility of this pathogen to grow at relatively low temperatures on unexpected foodstuffs. The safety of cook-chill foods has therefore been questioned. Lacey & Kerr (1989) found 25% of such foods at retail outlets to be contaminated with *L. monocytogenes* and, in consequence, proposed that the maximum storage time for cook-chill foods should be limited to 3 days at a temperature below 3°C.

2.3.7.4. Control Measures

Control measures should surmount the problems associated with the ability of *L. monocytogenes* to withstand relatively high temperatures, its ability to grow down to just over 0°C and its ability to so effectively repair from the effects of metabolic injury at low temperatures. It is therefore essential to:

1. Ensure that milk is adequately pasteurized and there is no post-pasteurization contamination.
2. Use only *L. monocytogenes*-free milk in the manufacturing of soft cheeses.
3. Ensure foods, particularly poultry, are heated to a minimum internal temperature of 72°C.
4. Where chill storage of pre-heated foods like poultry is required;

(a) ensure higher internal temperatures (minimum 75°C) are reached,
(b) limit the shelf life of all products, heated or unheated, where growth of *L. monocytogenes* is feasible to a maximum of 5 days,
(c) ensure storage temperatures are maintained at a maximum of 4°C for such products.

2.3.8. Enteritis due to *Campylobacter* spp.

2.3.8.1. The Organism

Two *Campylobacter* species have been implicated as enteric pathogens in humans. These are *C. jejuni* and *C. coli* although the former is responsible for over 90% of cases. Campylobacters used to be classified as vibrios but valid differences between these two groups are now recognized. Campylobacters are small, spiral or curved Gram negative rods which are motile with a single polar flagellum at one or both ends of the cell. They differ from vibrios in being strictly microaerophilic and will thus only grow in reduced oxygen levels. *C. jejuni* and *C. coli* are able to grow at 42°C but not at 25°C and they are often loosely described therefore as 'thermophilic campylobacters'. Their pathogenic mechanism has yet to be fully elucidated but it is known that campylobacters elaborate a lipopolysaccharide endotoxin rather like that found with salmonellas; in addition, it has been claimed that an enterotoxin, similar to cholera toxin, may be formed (Hoffman & Blankenship, 1986) as well as a cytotoxin which assists the invasiveness of the organism (Guerrant *et al.*, 1987). Feeding experiments by Robinson (1981) have demonstrated that the infective dose in healthy adults is below 500 cells; a fraction of this number may suffice since numbers in foods are low.

2.3.8.2. The Disease

Campylobacters are a relatively recent addition to the list of organisms causing gastroenteritis; in fact, Skirrow (1977) described it as a 'new' disease although infections in man caused by these organisms had been recognized some years earlier. The incubation period varies from 2 to 11 days. The most common symptoms are diarrhoea and abdominal cramps but these are often preceded by a fever which lasts for up to 24 h. Diarrhoea is profuse for up to 3 days and may be accompanied by bloody stools; vomiting is unusual. The duration of the illness varies from 3 to 14 days.

Foods Involved

)w recognized that campylobacters are responsible for more cases
oı ~ute infective diarrhoea in developed countries than any other bacteria. In the UK there are now over 30 000 reported cases per year but,
because these bacteria do not multiply to any extent in the food, campylobacter infections are not regarded as a true bacterial food poisoning.

There is little doubt that both *C. jejuni* in particular, and *C. coli* are
widely distributed both in foods and the environment generally but contaminated poultry and raw milk have been the main causes of infections
in humans. Campylobacters can be readily isolated from the gut contents of poultry and poultry carcasses are, in consequence, often contaminated with these organisms. Thus Lammerding *et al.* (1988), in a
survey of different raw foods in Canada, found 74% of turkeys and 38%
of chickens harboured campylobacters; lower isolation rates for *C. jejuni* and *C. coli* were found for beef (23%) and pork (17%) whilst veal
(43%) was more heavily contaminated. These results were largely substantiated by Fricker & Park (1989) who, in a detailed study covering a
two-year period, again found the highest recoveries of these campylobacters from poultry; these and recoveries from other sources are
shown in Table 2.7. Both Lammerding *et al.* (1988) and Fricker & Park
(1989) also reported that whereas *C. jejuni* was predominant (> 90%) in
most foods and the environmental samples *C. coli* predominated (95%)
on pork.

Genigeorgis *et al.* (1986) detected that the level of contamination
in poultry increased steadily during growth of the birds. Only 2% of

Table 2.7
Recoveries of *Campylobacter* spp. from Environmental and Food Sources[a]

Type of Sample	No. examined	No. positive	Percentage positive
Sewage	436	424	96·6
River water	345	105	30·4
Poultry	758	421	55·5
Beef	127	30	23·6
Pork	158	29	18·4
Lamb	103	16	15·5
Seafood	89	13	14·6
Cooked meats	86	2	2·3
Salads	106	0	—

[a]From Fricker & Park (1989).

10-day old chicks carried *C. jejuni* but this figure had risen to 82% by day 50. These authors suggested that transmission from one generation to another was feasible through contaminated litter or equipment but they argued that the development of *C. jejuni*-free flocks was now possible.

Food poisoning outbreaks involving poultry are usually due to the campylobacters surviving cooking. An unusual outbreak involving cadets at a military camp in the Netherlands was described by Brouwer *et al.* (1979):

> ... live chickens were obtained for the evening meal of 123 first-year cadets ... (on a survival exercise). The chickens were killed and skinned by each cadet for himself and prepared over a wood fire.... A large number (89) of the cadets became ill in the following week with symptoms of enteritis.... Faeces samples from 104 cadets ... (yielded 34 positive cultures) ... for *Campylobacter fetus* ssp *jejuni*. Seven of these were isolated from cadets who never showed symptoms whereas 27 originated from the 89 men who became ill.
>
> We believe that this extensive outbreak of enteritis was a consequence of eating chicken contaminated with *Campylobacter fetus* ssp *jejuni*. The method of preparation left the chickens partially unheated so that the campylobacters could survive the preparation of the food. Moreover, by handling, cooking and eating the chickens with bare hands, reinfection of the adequately cooked parts of the chickens could not be excluded.

Another important source of infection is unpasteurized milk and in this case the bacteria originate from the bovine faeces or possibly infected udders (Hoffman & Blankenship, 1986). As well as poultry many other types of animal faeces contain campylobacters. Humphrey & Beckett (1987) found 72% of bovine faecal samples to be positive whilst Beumer *et al.* (1988), in the Netherlands, reported that 22% of such samples yielded *C. jejuni*; however only 4·5% of raw milks were positive and these authors suggested that an enzyme, lactoperoxidase, naturally present in milk tends to kill-off the campylobacters. However, because of the low infective dose, enteritis outbreaks emanating from unpasteurized milk are often reported. Robinson *et al.* (1979) described an outbreak involving 63 people in Cumbria and 14 in Bradford, and Porter & Reid (1980) reported on an outbreak involving 148 patients in Kincardineshire and Angus. A large outbreak involving 2500 primary school children in the Luton area was described by Jones *et al.* (1981);

the children became seriously ill after drinking free school milk which was almost certainly contaminated. Large numbers of campylobacters were isolated from the faeces of patients in these outbreaks although direct isolations from the milk were not effected. In the last example the milk should have been pasteurized but faulty operation of a by-pass valve probably allowed the milk to pass through unheated; campylobacters are readily killed by pasteurization (Waterman, 1982).

Although other animal foods often harbour campylobacters they are not regarded as a significant source of infection. Dogs probably account for nearly 5% of human sporadic cases and puppies, in particular, have been implicated (Miller *et al.*, 1987).

2.3.8.4. Control Measures

C. jejuni and *C. coli* tend to die-off in foods held at ambient temperatures and growth, if at all possible, is always slow. But campylobacters are important enteric pathogens and control methods are very necessary. These include:

1. Milk pasteurization.
2. Introduction of campylobacter-free poultry flocks.
3. Thorough cooking of poultry.
4. Separation of raw and cooked meats and poultry.

2.3.9. Miscellaneous Bacterial Food Poisoning

2.3.9.1. Escherichia coli

E. coli is a short, typically motile, Gram negative rod with many characteristics similar to those of salmonellas. One of the principal features differentiating *E. coli* from salmonellas is its ability to attack lactose and sucrose with the production of acid and gas. Non-pathogenic *E. coli* strains can be isolated in massive numbers from the intestines of warm-blooded animals. However, many strains of *E. coli* are pathogenic to humans to varying degrees and these strains can be divided into four groups (Olsvik *et al.*, 1991).

The first group, enteropathogenic *E. coli* (EPEC) cause severe diarrhoea in infants but the pathogenic mechanism remains unknown although certain EPEC strains produce one or more cytotoxins. The second group, enterotoxigenic *E. coli* (ETEC), also cause diarrhoea in humans, both infants and adults, the latter group usually succumbing to the world-wide illness known as traveller's diarrhoea. ETEC strains produce enterotoxins of two distinct types: these are a heat-labile toxin, in-activated at 60°C in 30 min, and a heat-stable toxin, resistant to 100°C for 15 min (Scotland, 1988). ETEC strains have been implicated in sev-eral large outbreaks of infection with water as well as a wide range of foods (e.g. various meats and poultry, mashed potatoes, milk and cheese) as sources. MacDonald *et al.* (1985) described a multistate out-break in the United States caused by ETEC contaminated imported Brie and Camembert cheeses produced at the same factory. One hun-dred and sixty adults were affected by nausea, vomiting, diarrhoea and, typically, abdominal cramps following an incubation period of roughly 44 h after consuming the cheese; similar outbreaks involving the same factory were also reported in many European countries. Riordan *et al.* (1985) reported an outbreak in England where cold curried turkey may-onnaise was implicated; 27 staff and visitors at a school function were affected by severe diarrhoea some 36 h after consuming the food which had been left at ambient temperature for 8 h following preparation. Faeces samples taken from patients contained strains producing both heat-labile and heat-stable toxins.

The third group of pathogenic *E. coli*, enteroinvasive *E. coli* (EIEC) produce a cytotoxin and often induce rather more severe illnesses like colitis and a form of dysentery, accompanied by fever and bloody stools. Many outbreaks have been attributed to EIEC strains, with cheeses, milk and meats the foods most frequently incriminated. Imported French cheese was again responsible for a large outbreak in the United States (Marrier *et al.*, 1973); on this occasion at least 387 persons contracted gastroenteritis.The EIEC strain was isolated from the stools of patients, from various samples of the cheese and from the French factory where the cheese had been prepared. The final group, enterohaemorrhagic *E. coli* (EHEC) also produce cytotoxins to give more severe symptoms. EHEC strains have been implicated in a number of outbreaks in nursing homes in the United States, with hamburgers the food most often involved. Ryan *et al.* (1986) described such an out-break affecting 34 patients, 19 of whom had severe haemorrhagic colitis. A more serious disease caused by EHEC strains is haemorrhagic-uraemic syndrome, an often fatal condition in young and elderly pa-tients characterized by anaemia and renal failure.

2.3.9.2. Yersinia enterocolitica
Y. enterocolitica is a small Gram negative rod which has the unusual property of being non-motile at 37°C but motile, with peritrichous

flagella, below this temperature. Another unusual feature is its ability to grow at 4°C with most strains growing down to 1°C or even below although its optimum growth temperature is probably about 33°C.

Y. enterocolitica has been associated with a variety of clinical conditions, particularly a form of gastroenteritis in which abdominal pain and diarrhoea are the main symptoms; less frequently observed are vomiting and a mild fever. In another enteric condition the symptoms mimic acute appendicitis and in this case fever and vomiting are more common (Nilehn, 1969).

Y. enterocolitica is widely distributed and can be isolated from many foodstuffs (Swaminathan *et al.*, 1982); however, only a small number of strains (i.e. serotypes) are associated with human disease. The main source of *Y. enterocolitica* is undoubtedly pigs which are known to carry the specific pathogenic strains on occasions. De Boer *et al.* (1986), in a detailed survey of a variety of foods, found the highest isolation rates on raw pork (73%), unpasteurized egg (43%), raw vegetables (43%) and raw beef (42%), with a surprisingly low recovery (10%) of *Y. enterocolitica* from raw milks, often implicated in outbreaks of yersiniosis. These authors stressed that only 4% of the isolates from foods were recognized pathogenic strains, these all being obtained from pig samples; on the other hand, Anderson (1988) reported that 25% of pig carcasses were contaminated with human pathogenic *Y. enterocolitica*.

There has been a marked increase in the number of reported cases of yersiniosis world-wide in the last decade, but this may be at least partly attributable to our greater awareness of the pathogenic nature of the organism. A very large outbreak involving 16000 people occurred in the United States in 1982 (Dirksen & Flagg, 1988); contamination was traced to a dairy in Tennessee from which outdated milk was transported, in crates, to a pig farm where the milk was unloaded. On return the crate washing proved inadequate and the crates contaminated milk carton exteriors on which the yersinias were shown to survive for up to 21 days.

Many yersinias produce a heat-stable toxin similar to that produced by ETEC strains. Toxin production by *Y. enterocolitica* has been demonstrated in milk at 25°C but it is not a common property amongst yersinias; furthermore, such production is suppressed at chill temperatures (Walker & Gilmour, 1990). It has therefore been suggested that enterotoxin production may not be important in pathogenesis (Schiemann, 1988) but our knowledge of the epidemiology of this organism is less well understood than with most food poisoning bacteria.

2.3.9.3. Enterococci

The enterococci, also known as the faecal streptococci, comprise two species *Streptococcus faecalis* and *S. faecium* although it is the former which has been mainly implicated in occasional food poisoning outbreaks. Both species are small (0·5–1 μm in diameter), spherical, Gram positive organisms which occur mostly in pairs or short chains. Both species contain a few strains which are motile, both exhibit a fairly wide temperature range of growth, from 10 to 45°C and both are isolated from human and other warm-blooded animal faeces.

The incubation period of the illness ranges from 2 to about 20 h and the main symptoms are vomiting, abdominal pain and diarrhoea. Foods involved in outbreaks include cheeses, various beef dishes, pasteurized canned hams, Vienna sausage, turkey products and evaporated milk. Whilst high numbers of enterococci were found in these foods, much of the evidence implicating these organisms as food poisoning agents is circumstantial. No toxins have ever been identified and attempts to reproduce the symptoms in human volunteers fed massive doses of enterococci have produced inconsistent and unconvincing results (Bryan, 1979); doubt therefore remains as to whether or not these organisms are capable of causing food poisoning.

S. zooepidemicus, although not an enterococcus, has been implicated in a small number of isolated incidents and in a severe food-borne outbreak in Yorkshire due to unpasteurized milk (Edwards *et al.*, 1988); of 11 persons taken to hospital 7 died and it is almost certain that others were taken ill in the area. All patients were affected with septicaemia and a proportion with meningitis. The authors concluded that this was a clear example of the perils of drinking unpasteurized milk!

2.3.9.4. Aeromonas spp.

The genus *Aeromonas* is composed of species that are Gram negative motile rods with, typically, a single polar flagellum. Two aeromonads, *A. hydrophila* and *A. sobria* are now recognized as potential pathogens which can cause diarrhoea in humans. *A. hydrophila* is widely distributed and is very common in water so that, in the past, it has often been assumed that infections are water-borne. However, many foods carry this bacterium and *A. sobria* to a smaller extent; Fricker & Tompsett (1989) reported carriage rates for the two aeromonads of 69% (for poultry), 21% (for both pork and raw salads), 17% (for beef) and 15% (for fish) and they concluded that both raw and cooked foods were potential sources of human infection. Both the aeromonads produce enterotoxins and cytotoxins which are similar to those of *E. coli* but there is no clear evidence that they have a role in pathogenicity (Todd *et al.*, 1989).

It has not been demonstrated that growth of either *S. zooepidemicus* or the aeromonads in food is a necessary prerequisite for induction of the illness. Thus these organisms cannot be regarded as true food poisoning bacteria. However in the remaining illnesses described the food does act primarily as a carrier of the causative organism, growth being unnecessary. Examples of these food-borne illnesses are briefly described.

2.3.9.5. Bacillary dysentery

The disease known as bacillary dysentery or shigellosis is caused by *Shigella* species of which *S. sonnei* and *S. flexnuri* are the most important. Shigellas are similar to salmonellas in many respects but are non-motile and fail to produce gas from carbohydrates. They are often transmitted by means other than the ingestion of food but where food-borne shigellosis is concerned the incubation period is relatively short, some 7–36 h. The principal symptoms are diarrhoea (often heavily blood-stained), fever, nausea and abdominal cramps. Bacillary dysentery is usually caused by contaminated drinking water but, where foods have been implicated, milk and various salads are the most common vehicles. It is likely that poor personal hygiene is a contributory cause in the spread of this illness and most outbreaks have incriminated infected food handlers at some point in the chain of infection.

2.3.9.6. Cholera

Cholera is a disease originating in the East and its symptoms were first described many centuries ago. The causative organism, *Vibrio cholerae*, also known as *V. comma*, is a curved Gram negative rod similar to *V. parahaemolyticus*. A recent form of cholera, El Tor cholera, has become widespread in the last 30 years or so. It originated in Indonesia and spread through the Indian sub-continent into the Middle East and Africa and reached Western Europe in the early 1970s. The incubation period varies from 6–8 h up to 2 or 3 days and the illness is characterized by a severe inflammation of the intestinal tract resulting in the almost continuous discharge of liquid ('rice-water') stools containing blood and mucus. As a consequence of this fluid loss the patient becomes seriously dehydrated and death may follow if the fluid is not replaced; an enterotoxin is primarily responsible for the symptoms although non-toxin producing strains can cause a relatively mild form of gastroenteritis. Generally cholera is a spasmodic infection transmitted by person-to-person contact but where water, in particular, is contaminated with sewage epidemics may occur. Where foods have been incriminated it has usually followed contact with contaminated water. Fish, oysters and mussels, infected by the water in which they live and which have been eaten raw or after inadequate cooking, have been responsible for outbreaks; salad foods, fruits and vegetables washed in contaminated water have also been implicated.

2.3.9.7. Typhoid

Typhoid fever is caused by *Salmonella typhi*, a member of the genus *Salmonella* described earlier but distinguished by its inability to produce gas from carbohydrates; in many respects, therefore, it is similar to the shigellas. Typhoid is the most severe of the diseases caused by salmonellas. After a lengthy incubation period of 7 to 21 days the disease sets in with a general feeling of malaise and during the first week the body temperature steadily rises. About the seventh to tenth day a rash appears and during the second week the fever is at its highest. Death may occur from the severity of the disease at this stage but in the less severe cases there is a gradual improvement in the third or fourth week. *S. typhi* is excreted in the faeces in large numbers during the illness and, as with salmonellosis, a carrier problem exists with patients who have recovered and with symptomless excreters. Sewage-contaminated water is the most common source of infection but chlorination of water supplies has largely eliminated this source in developed countries. The last big water-borne outbreak in the UK was in Croydon in 1937 when 431 people were infected by water that had been contaminated by a symptomless excreter repairing the mains system.

Two foods that used to be frequently incriminated in typhoid fever outbreaks were raw milk and ice-cream but the heat treatment regulations introduced in the UK in the 1940s overcame this source of infection. Raw milk was incriminated in an extensive outbreak in Bournemouth in 1938 when over 700 people were infected and 70 died. Ice-cream was the vehicle of infection in Aberystwyth in the same year when 210 cases were recorded. In both these examples the infection again stemmed from symptomless carriers and this stresses the problems that such people often unknowingly pose. Other foods associated with the disease have been shellfish including oysters, infected by the contaminated water in which they were living. Various canned meats have been responsible for a number of outbreaks in more recent years in the UK, the last major one being in Aberdeen in 1964 when 515 cases were reported. The original source of the organism was a South American tin of corned beef which had been cooled in contaminated water after cooking; *S. typhi* was undoubtedly spread to other meats in the shop by handling and by equipment — this outbreak was an excellent example of the problems created by cross-contamination!

2.3.9.8. Brucellosis

Members of the genus *Brucella* cause brucellosis which is also known as undulant fever and Malta fever. The causative organisms, *B. abortus, B. suis* and *B. melitensis*, are all small Gram negative ovoid rods although coccal forms are common. These three species are predominant in cattle, pigs, and goats or sheep, respectively, although other animals such as horses, rabbits, chickens, dogs and cats may become infected. These animals therefore serve as a source of infection and human brucellosis almost always results from contact with infected animals or from ingestion of their products.

Clinically the incubation period lasts for 1–3 weeks and the onset of the illness is gradual with increasing fever, general weakness and pains, and chills. The fever tends to vary in intensity, often over a number of weeks or even months; hence the description of the illness as 'undulant fever'. The normal length of the illness is between 2 and 3 months but in many cases weakness lasting from the infection can persist for a year or more. The disease is widely distributed and infection usually occurs following the consumption of raw milk from infected cows or goats, or cheese prepared from similarly infected milk. At one time about 20% of dairy cows in Europe and the United States were infected with *B. abortus* but incidents of brucellosis have been substantially reduced by the introduction of techniques which have enabled the infected animal to be detected and removed from the herd. Another important factor contributing to the decrease is the general consumption of pasteurized rather than raw milk, the brucellas being readily killed by this heat treatment.

2.4. MYCOTOXICOSES

Mycotoxicoses are caused by the ingestion of poisonous metabolites (mycotoxins) which are produced by fungi growing in food. The list of fungi and the mycotoxins produced by them which may be detrimental to human health is an extensive one; much of the evidence linking mycotoxins with illness is circumstantial and only those where a clear relationship has been established will be discussed. Of these, the aflatoxins are the most important and are considered first.

2.4.1. Aflatoxins

In 1960 in the UK a mouldy peanut meal fed to turkeys produced a loss of appetite and general weakness followed by death within a week; because the cause was unknown it was called 'turkey X disease'. The feed was found to be infected with the mould *Aspergillus flavus* which had formed a poisonous toxin subsequently given the name 'aflatoxin' in view of its origin. Almost simultaneously there were similar diseases involving ducklings in Kenya and Uganda, and infections of hatchery-reared rainbow trout in Europe and the United States. *A. flavus* contaminated feeds were found to be responsible for the incidents and toxic factors were successfully isolated and identified. There are four main aflatoxins designated B_1, B_2, G_1 and G_2 by the blue (B) or green (G) fluorescence given when viewed under a UV lamp. Chemically the four compounds are very similar and they are classed as heterocyclic compounds, or, more specifically, substituted coumarins. All are heat-resistant and prolonged heating at 100°C is necessary to destroy their potency; they are unaffected by long-term storage, remaining stable in peanut butter stored at 23°C for over two years. The structure of aflatoxin B_1 is given in Fig. 2.4.

It is now known that about 30% of *A. flavus* strains produce aflatoxins (mainly B_1 and B_2) and that a second species, *A. parasiticus*,

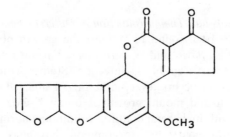

Fig. 2.4. Chemical structure of aflatoxin B_1.

produces both B and G aflatoxins. Both these species are commonly isolated from the soil and are found on living and dead plants and animals
throughout the world.

2.4.1.1. Biological Effects

The main target organ for aflatoxins is the liver where they can cause either tissue damage or tumours. In many animal species aflatoxin is a
highly active liver carcinogen and tumours may be produced by the administration of minimal amounts of aflatoxin. On the other hand, other
animal species are relatively resistant to the cancer-inducing effects but
may be sensitive to the acute effects. Thus the rainbow trout affected in
the early outbreaks were found to have hepatomas (i.e. cancer of the
liver) whereas only tissue damage was found in the livers of the turkeys
(Barnes, 1970).

The main interest in aflatoxins at the present time is in determining
whether or not they are a cause of hepatomas in humans and evidence
is accumulating in support of this relationship (Bullerman, 1979). It has
been known for many years that there is a high incidence of hepatomas
in African Bantu tribes who rely to a considerable extent on fermented
cereal-based foods, often mould-contaminated, as a protein source.
Extensive liver cancer also occurs in many Asiatic communities. The incidence of hepatomas in Indonesia is amongst the highest in the world
and a fermented peanut 'cake', a very popular food in this area, usually
contains measurable amounts of aflatoxin. Many deaths occur annually
amongst Thai children and the tissues have been found to contain high
levels of aflatoxin. In 1974 in India there were several hundred deaths
linked to the consumption of corn contaminated with aflatoxin and it is
becoming increasingly apparent that the global distribution of liver car-
cinoma is related to the consumption of foodstuffs which support the
growth of aflatoxin-producing moulds.

2.4.1.2. Aflatoxins and Their Production in Different Foods

Many foods have been found to support the growth of aflatoxin-producing strains of *A. flavus* and *A. parasiticus*. Various nuts, and in particular peanuts, are susceptible to infection. Studies with peanuts have
shown that infection occurs after harvesting. Damaged nuts are most
prone to infection and mould growth is optimal under conditions of
high humidity and high temperature, conditions normally found in
those areas where hepatomas are common. Nowadays great care is
taken to prevent infected nuts from being included in consignments to

manufacturers. Electronic screening to eliminate discoloured and potentially infected nuts has been introduced, and many other testing and selection procedures are routinely performed. Control measures in the field include the possible use of insecticides to minimize insect damage to nuts. Storage of nuts below 5°C or in modified atmospheres (e.g. 20% O_2: 60% CO_2: 20% N_2; see p.124) totally suppresses aflatoxin production by the aspergilli (Paster & Bullerman, 1988). Antifungal agents such as sorbic acid, potassium sorbate and propionic acid have been shown to exert control of growth and toxin production by *A. flavus* and *A. parasiticus* (Ray & Bullerman, 1982; Bullerman *et al.*, 1984); however much of the experimental work has been performed under laboratory conditions rather than in the field where reduced efficiency of these mould inhibitors may be experienced.

There are many foods on which mould growth is common, such as cheeses, and concern has been expressed about the possibility of toxin formation in such foods. Aspergilli have been readily isolated from commercial Cheddar and Swiss cheeses, but aflatoxin-producing strains were rarely noted (Bullerman & Olivigni, 1974; Bullerman, 1976). Toxin formation was only demonstrated under laboratory conditions and no aflatoxin was demonstrated in any purchased cheese. Whilst it is possible to induce aflatoxin formation by artificially inoculating aflatoxin-producing strains into cheeses, the risk of aflatoxin being present in commercial products appears remote; nevertheless, checks must be continued in view of the occasional isolation of aflatoxin-producing strains. With regard to cheeses produced by 'starter' moulds such as Roquefort and Camembert the evidence suggests that mycotoxins, if formed, are not a health hazard in such cheeses (Scott, 1981); furthermore, aflatoxins have never been detected. Mould growth is also common on cured and fermented meats and the situation with these foods is similar to that of cheese. Aflatoxin-producing moulds have been isolated from aged cured meats (country cured hams and 'salami type' sausages) but aflatoxins and other mycotoxins have not been demonstrated in these foods (Wu *et al.*, 1974). Soya beans, various ground spices and dried fruits, rice, maize and spaghetti are other foods in which aflatoxins have been found.

In recent years concern has been expressed about the presence of hydroxylated aflatoxins in milk and milk products. These toxins are produced by lactating mothers and animals which have previously ingested foods contaminated with aflatoxin B_1, this compound being converted to the hydroxylated form (aflatoxin M_1). In underdeveloped countries there have been reports of very young children being exposed to the effects of these toxins through their mothers' milk before weaning (Moss, 1989). In developed countries it is cows' milk and products derived from it which have been under the closest scrutiny and evidence has accumulated that milk often contains extremely low levels of aflatoxin M_1 (e.g. Blanco *et al.*, 1988). The efficiency of aflatoxin conversion in cows is poor; Frobish *et al.* (1986) reported that under 2% of aflatoxin B_1, deliberately added to grain fed to lactating animals, was converted to the hydroxylated form. Again, although many different types of cheeses can contain aflatoxin M_1, it is almost invariably at such low levels that no danger to human health exists. Detailed studies have been performed to determine the fate of aflatoxin M_1 in different cheeses but similar results have been reported. For example, Brackett & Marth (1982) used aflatoxin M_1 contaminated milk to prepare cheddar cheese and processed cheese slices; they found that although there were variations in toxin levels during a 1-year storage period little overall change occurred in both types of cheese. The cheese slices were prepared from the semi-matured cheddar cheese and processing included heating the cheese to 87°C, a temperature that had no effect on the toxin levels; this shows that the M forms of aflatoxin are again heat stable. This stability is again in evidence with yoghurts. When prepared from skimmed milk naturally contaminated with aflatoxin M_1, Wiseman & Marth (1983) demonstrated that the toxin remained stable in yoghurt over a 6-week storage period at 7°C.

Since the source of aflatoxin M_1 is foods contaminated with aflatoxin B_1, stringent standards have either been introduced or at least recommended in developed countries for animal feedingstuffs. For example, within the European Community regulations for different categories of animal food vary from 10 to 50 μg per kg maximum, specifically for aflatoxin B_1; there are also regulations concerning the movement of contaminated animal feeds between Member States (see Chaper 13).

In conclusion, there is little doubt that the risk of exposure to aflatoxin in the western world is remote. The most likely source is peanuts or peanut butter but control measures are extremely rigorous and apparently effective. However, peanut butter obtained from health food shops does contain low levels of aflatoxin on occasion, the absence of suitable preservatives in such foods permitting mould growth. Legislation has been recommended in the UK allowing an upper limit of 10 μg per kg for *total* aflatoxin in nuts and nut products; limits of 5–20 μg per kg have been enforced in many other countries. Unfortunately the situation in developing countries is very different and there is little doubt that much aflatoxin is being unwittingly consumed to produce its serious long-term effects. It was even suggested (Marth & Calanog, 1976) that 'toxic metabolites of moulds constitute the single greatest food-borne hazard to human health worldwide'; this remains true today.

2.4.2. Miscellaneous Mycotoxins

2.4.2.1. *Ergot Alkaloids*

Probably the first mycotoxicosis described was ergotism, an illness resulting from the ingestion of rye and less commonly other grains infected with the mould *Claviceps purpurea*. The fungus tissue grows into a hard mass, the ergot, which contains a number of alkaloids, all derivatives of lysergic acid, which are used medicinally when purified. One of the derivatives is lysergic acid diethylamide (LSD), a hallucinatory drug. Although ergots have a medicinal value, the unpurified alkaloids are poisonous and may produce two kinds of ergotism, the gangrenous and the convulsive. Convulsive ergotism is the more common form and is characterized by a variety of symptoms including vomiting, diarrhoea, disorderly speech, convulsive fits, hallucinations and delirium with death ensuing in severe cases. Nowadays outbreaks are rare and the last recorded in the western world, which resulted in five deaths, was in France in 1951; in the UK there was a mild although extensive outbreak in Manchester in 1927.

2.4.2.2. *Fusarial Toxins*

Outbreaks of a mycotoxicosis called alimentary toxic aleukia (ATA) have been recorded in the USSR since the last century but there are no reports of its occurrence elsewhere. ATA is caused principally by toxin-producing strains of *Fusarium sporotrichioides* which grow in stored grains (wheat, millet and barley) held in poor conditions during winter months. Many of these strains grow down to −10°C and toxin formation is most active just above 0°C. The toxin produced, sporofusarin, is remarkably heat-stable withstanding 125°C for 30 min. There was a series of outbreaks in the USSR between 1942 and 1947, probably the result of war and its aftermath, when tens of thousands of people are thought to have died from ATA. Initially the symptoms are relatively mild and consist of vomiting and diarrhoea but after an asymptomatic period of 1–12 months the patient's condition becomes serious. The later symptoms include a severe rash, mouth and throat lesions, angina and leukaemia, and mortality can be in excess of 50%.

2.4.2.3.   Ochratoxins

Aspergillus ochraceus and *Penicillium viridicatum* are able to produce ochratoxins which, chemically, are substituted coumarins like the aflatoxins. Both species have been frequently isolated from mouldy grain, particularly barley, and nuts amongst other foods and concern has been expressed about possible health hazards. It has been found that ochratoxin A, the most potent of these toxins, causes severe tissue damage in human and animal kidneys. There is evidence that a fatal chronic kidney disease, Balkan endemic nephropathy, common in rural areas of Bulgaria, Romania and Yugoslavia is caused by the consumption of cereal products which are frequently contaminated (10%) with ochratoxin A. This level of contamination is higher than that reported in other areas of the world and strongly suggests that the toxin is responsible for this disease (Pavlovic *et al.*, 1979).

2.4.2.4.   The 'Yellow Rice' Toxins

Members of the genus *Penicillium* produce a number of mycotoxins in food, particularly rice. This commodity is liable to rapid mould spoilage in the humid, warm conditions prevailing in the East and rice that is heavily contaminated with mould appears yellow. Three species of *Penicillium* are known to produce mycotoxins in rice which are hazardous to health; these are *P. islandicum* (mycotoxins cause liver damage), *P. citrinum* (kidney damage) and *P. citreoviride* (paralysis of central nervous system). Although no large-scale outbreaks have been reported and the evidence relating to such illness remains circumstantial, the vast quantity of rice consumed together with its frequent contamination suggest that it is probably the cause of much human disease; in fact, acute cardiac beriberi, common throughout Asia, may well be linked to the consumption of 'yellow rice' toxins.

2.5.   VIRUS FOOD POISONING

Viruses are unable to multiply in food and therefore food can only act as a vehicle for the transmission of these organisms, much in the same way that it does with typhoid fever and bacillary dysentery. Many different viruses have been isolated from foods of different kinds but their significance in the food is often poorly understood.

Viruses which infect by being ingested are termed intestinal viruses. They multiply in the intestines of the infected person and large numbers of them (up to 10^{12}/g of faeces) may be excreted. In view of their high degree of host specificity (Chapter 1) it is obvious that the primary source of infection in food-borne outbreaks must be of human origin. Two viruses which are important agents of human infection and which are transmitted via food are infective hepatitis and poliomyelitis.

2.5.1. Infective Hepatitis

Infective hepatitis, caused by hepatitis A virus, is the most common of the food-associated virus diseases; some 100 outbreaks involving many thousands of cases have been reported in the last 25 years. The most common vehicle of transmission has been shellfish (oysters and clams), previously contaminated by polluted water, which have been eaten either raw or after very light cooking. In many cases it has proved impossible to ascertain the food implicated since the disease has a lengthy incubation period, varying from 10 to 50 days, and therefore by the time an outbreak is detected the food is no longer available for examination. Many foods other than shellfish have been implicated occasionally and these include raw milk, dairy products, various cold meats and watercress, probably all contaminated initially by food handlers or polluted water (Cliver, 1979).

2.5.2. Poliomyelitis

It has long been recognized that poliovirus, the agent responsible for poliomyelitis, can be spread by food and in nearly every case the food implicated has been raw milk. The original source of the virus, where it was determined, proved to be man, again underlining the high degree of host specificity of these agents. One large outbreak occurred in Broadstairs, Kent, in 1927 when 62 cases were reported over a 16-day period. Children in local boarding schools contracted the illness after consuming raw milk obtained from a single dealer although the original source of the virus was never traced. Because of the introduction of pasteurization and the widespread use of vaccines, outbreaks of poliomyelitis attributable to milk are rarely reported in developed countries although pasteurized milk has been implicated in two outbreaks; this was presumably caused by improper pasteurization or post-pasteurization contamination.

2.5.3. Gastroenteritis

There are certain viruses, including enteroviruses and parvoviruses as well as some which are as yet uncharacterized, which cause gastroenteritis in man. In recent years there have been frequent outbreaks of gastroenteritis of this type following the consumption of raw oysters but other shellfish have also been implicated on occasions. One such outbreak was recorded as follows (DHSS, 1977):

> Shortly before Christmas the Department was informed of a number of outbreaks of food poisoning after the eating of cockles.... In an incident in one restaurant, of a large number of people who ate cockles, over 90% became ill while other people in the same restaurant who had eaten no cockles, but otherwise had the same food, remained well. Twenty to thirty hours after eating the cockles the victims developed diarrhoea, vomiting, abdominal pains and giddiness; there were no deaths but a number of cases were admitted to hospital.... The evidence suggested that all the cockles concerned came from Leigh-on-Sea in Essex and that the infection was due to a virus which had been picked up by the cockles from sewage-contaminated water and had not been destroyed in processing, due to inadequate heat treatment. Over 800 people were eventually known to have been affected.

The shellfish often grow in polluted waters but it is possible to purify them by transplanting them to clean water; in the clean water they cleanse themselves of harmful bacteria and most viruses by their normal feeding and respiratory activities.

2.5.4. Bovine Spongiform Encephalopathy

Bovine spongiform encephalopathy (BSE) was first described in the UK in 1987 and it has been mainly restricted to this country although there have been isolated incidents reported in Eire and France. There were 46 cases in the UK in 1987 but since that time numbers have risen steeply so that in 1991 a total of 18 000 BSE cases was reported (House of Commons answer, 20 January, 1992).

BSE is an infection of cattle manifested by total loss of coordination of movement, and extreme apprehension. BSE is probably a form of scrapie, a virus-like (perhaps a prion) infection of sheep, which has been transmitted to cattle orally, that is, by the inclusion of infected sheep brains in cattle feedingstuffs or possibly through infected grazing land. The host specificity so typical of viruses has thus been lost undermining the so-called 'species-barrier'. There is mounting evidence that the 'prion' can be transmitted from certain animals to others but vulnerability to infection appears to be extremely variable (Dealler & Lacey, 1990). Furthermore the 'prion' may change its protein structure when passing from one animal species to another and in so doing the range of animal species it can subsequently infect may alter. Thus the scrapie particle cannot infect rhesus monkeys directly from sheep but if an intermediate host (e.g. mink) is infected with the particle, the 'prion' from mink is then able to infect rhesus monkeys. It should be pointed out that many of the 'species-barrier' experiments are artificial in that the various infective particles are often inoculated by abnormal routes (e.g. massive doses directly into the brains of test animals) rather than orally.

In cattle there is a lengthy incubation period of 2–5 years during which time the scrapie-like particles spread to the brain and viscera, a prelude to the appearance of the symptoms. Thus the disease only manifests itself in adult animals, i.e. mainly dairy cows. However, it is likely that symptomless younger livestock carries the infective particle. This carriage could be due to transmission from one generation to another, but infection via pasture land is perhaps more feasible.

The fundamental question to be asked is whether a further species barrier can be breached allowing humans to become infected. A number of epidemiological studies have failed to produce evidence of any link between exposure to scrapie and Creutzfeld-Jakob disease, the form of infection in humans (Taylor, 1989); this is a progressive and fatal encephalitis with an extremely lengthy incubation period of 5–15 years. But the spectrum of animals that can be affected apparently depends on the animal the particle last infected so that conclusions drawn on scrapie to human infection may not hold true for BSE to human infection. Furthermore, infection of humans by the oral route has been clearly suggested by earlier studies on another human spongiform encephalopathy, kuru, which was common in remote parts of New Guinea earlier this century (Critchley *et al.*, 1972). This encephalopathy was apparently caused by the traditional consumption of human brains, mainly by young women, who predominantly suffered from the illness in later years. Thus the suggested preventative measures that all animal brains should be excluded from human food together with prohibition of milk and certain tissues from BSE infected cattle in such food are eminently sensible.

2.5.5. General Control Measures

The source of infection in food-borne virus infections is normally human faeces. Faecal contamination can occur directly by means of an infected food handler and in this case control could prove difficult. Alternatively, contamination of food can be indirect principally by means of polluted water. This source of infection can be controlled by cleansing shellfish in chlorinated or pure water, the shellfish purging themselves of the contaminating virus particles. Furthermore, most viruses are relatively heat-sensitive, being inactivated at 80°C in a few minutes, so adequate cooking can be relied upon as an additional safeguard.

2.6. ANIMAL TOXINS AND PARASITIC INFECTIONS

Certain freshwater and marine animals are toxic to man even when eaten fresh (natural toxicity). In addition, many marine foods may become toxic either after consuming other forms of marine life which are themselves toxic or as a result of microbial action after death (secondary toxicity). Finally, a wider range of foods, including pork and beef as well as marine foods, may be infested with animal parasites which are harmful to man and which can be readily transmitted by ingestion of the food (parasitic infections).

2.6.1. Animals that are Naturally Toxic to Man

2.6.1.1. Puffer Fish Poisoning

Poisonous fish are widely distributed in warm oceans but are particularly numerous in the Central and South Pacific Oceans and in the Caribbean. Probably the most poisonous are the puffer fish of which there are many different species. They are commonly found along the coasts of Japan, China and Indonesia and they have been responsible for many deaths in these countries. The toxin responsible is termed tetrodotoxin and it is found in the highest concentration in the viscera of the fish. It is fairly heat-stable and will resist boiling for 10 min. Most deaths occur following the consumption of fish roe rather than the flesh itself. The onset of symptoms is rapid occurring within a few minutes of consuming the food; the symptoms include nausea, vomiting, and numbness or a tingling sensation in the throat and limbs. Within a further short period muscular weakness and progressive paralysis become apparent and, in over 60% of cases, this is followed by death, all within a few hours of eating the food.

2.6.1.2. Moray Eel Poisoning

Moray eels are large animals, about 3 m long, which inhabit tropical reefs. They are frequently caught and eaten by natives of the Far East where they have caused widespread poisoning. In Moray eel poisoning the toxins are present in the flesh of the fish and about 10% of the cases prove fatal. The symptoms are similar to puffer fish poisoning although the paralysis is normally less severe.

2.6.1.3. Miscellaneous Poisoning

Certain indigenous species of whelks, squids, crabs and turtles, which are eaten in certain parts of the world, are known to be inherently toxic and have caused food poisoning outbreaks.

2.6.2. Secondary Toxicity

2.6.2.1. Paralytic Shellfish Poisoning (PSP)

Certain marine dinoflagellates, a type of plankton, synthesize a toxin, saxitoxin, that is harmful to man. These toxin-containing dinoflagellates are normally present in low numbers in the sea and the toxin level is consequently low. However in certain climatic conditions planktonic blooms can occur and shellfish (oysters, mussels and clams) feeding in affected water then accumulate the toxin to such an extent that humans who consume the shellfish contract PSP. Outbreaks of PSP have been reported in many parts of the world but most commonly in North America where numerous deaths have been attributed to this form of poisoning. The symptoms of the illness generally develop within 5–30 min after ingestion of the affected food and they are again similar to, although less severe than, those associated with puffer fish poisoning. The duration of the illness is from 1 to 3 days and the mortality rate has been variously reported as between 1 and 20%. The first recorded outbreak in the UK was described (DHSS, 1969) as follows:

> On 30th May, 1968 a general practitioner with a woman patient who complained of tingling in her hands and fingers, a feeling in her mouth like a dental anaesthetic and weakness and inco-ordination of her arms and legs, sent her to hospital with a diagnosis of mussel poisoning. She had indeed cooked and eaten, 1½ hours

before her symptoms began, some fresh mussels which were part of a large consignment gathered between 19th and 29th May. In all 78 people were affected. ... happily nobody was seriously ill, possibly because the retailer discarded the liquid in which the mussels were boiled. Samples of the mussels were examined ... to determine the concentration of toxin in the flesh of the shellfish. The toxin was shown ... to be derived from a minute dinoflagellate organism ... found in high concentration in sea water over more than 300 miles off the East Coast. Its toxin had been taken up in the muscle tissue of mussels, cockles, scallops and clams, whose method of feeding is by filtering out the plankton from sea water. The mussels concentrated the most toxin and soft clams the least.

Control measures should ensure that shellfish are not harvested in waters containing heavy growth of dinoflagellates.

2.6.2.2. Ciguatera Poisoning

This form of poisoning was first recognized in the West Indies. It often results from a sequence of events commencing with toxic algae commonly found in warm waters near coral reefs or sea shores. Herbivorous fish feed on the algae and they in turn are eaten by carnivorous fish such as sea bass and barracuda. It is these latter fish which typically cause ciguatera poisoning in man although over 300 species of fish have been incriminated. The symptoms, which develop between 30 min and 4 h after ingestion of the food, are, once again, similar to puffer fish poisoning except that paralysis is seen only in severe cases. The mortality rate is from 2 to 7% and in these cases death occurs within 1–24 h. It is of little importance in the UK although one case has been reported following the consumption of imported fish.

2.6.2.3. Scombrotoxin Poisoning

Scombrotoxin poisoning is generally regarded as a chemical intoxication resulting from eating scombroid fish, particularly mackerel and tuna; the exact causative agent has yet to be determined. Scombroid fish contain unusually large amounts of histidine, an amino acid, which is converted to histamine by certain spoilage bacteria including, of greatest significance, *Proteus morganii*. Histamine production only occurs by this means at fairly high temperatures (above 10°C) so that it is in fish stored at higher temperatures (15–25°C) that histamine accumulates rapidly (Hobbs, 1987). Symptoms, which usually appear a few minutes after eating the food, are characteristic of histamine poisoning, and include dizziness, vomiting, diarrhoea, difficulty in swallowing and large itchy red blotches on the skin; the combination of symptoms manifested can vary in each case but they usually disappear within 8–12 h and fatalities are rare. The relationship between histamine levels and clinical symptoms is rather tenuous and there have been many cases recorded where non-scombroid fish (e.g. herring and sardines) have been implicated; thus it has been suggested that more than one causative agent may be involved (Hobbs, 1987).

It has been claimed that commercially canned scombroids are safe to eat although it is known that the toxin is heat-resistant: a major outbreak in the United States in 1973, involving 232 people who had eaten canned tuna, appears to refute the claim. Between 1979 and 1986 there were over 250 suspected incidents of scombrotoxic fish poisoning in the UK but the number has decreased substantially in the last few years suggesting that greater care is being taken over storage temperatures for fish.

2.6.3. Parasitic Infections

Many foods may act as carriers of parasites which can be ingested by man to initiate infections. Many of these infections have a worldwide distribution whilst others are limited to certain geographical regions. The parasites responsible for these infections are varied in character and many have complex life cycles passing through different stages in different animal hosts. They commonly exhibit a free-living stage in soil or water and an internal parasitic stage in the gut or other tissues of man and/or other animals. It is not within the scope of this book to discuss the life cycles of these parasites; more detailed information on these and other aspects can be obtained from appropriate references listed in the Bibliography. The principal parasitic infections of man are outlined in Table 2.8 and some additional information on these infections is given in the following paragraphs.

2.6.3.1. *Protozoa*
Amoebic dysentery, caused by *Entamoeba histolytica*, is a widespread disease but primarily associated with tropical and sub-tropical areas. It has been claimed that 10% of the world's population is affected but outbreaks in temperate zones, usually resulting from sewage-contaminated

Table 2.8
Principal Parasitic Infections of Man

Phylum	Species	Illness produced in man	Principal symptoms	Principal source of infection	Intermediate animal hosts	Number of cases in England (1980)[a]
Protozoa (single-celled organisms)	*Entamoeba histolytica*	Amoebic dysentery	Variable. May be limited to diarrhoea but in severe cases ulceration of organs of the body and possible death.	Vegetables and fruits contaminated by handlers, flies etc. Contaminated water.	None	543
	Toxoplasma gondii	Toxoplasmosis	Hydrocephalus and blindness in children. Less severe in adults. often chronic with possible involvement of the eyes.	Undercooked or raw meats (pork, lamb, beef, poultry). Domestic cats.	Domestic cats	790
Trematodes (non-segmented flatworms, the flukes)	*Fasciola hepatica*	Fascioliasis	Inflammation, abscesses and haemorrhages in the intestines.	Watercress.	Snails and sheep or cattle	4
	Clonorchis sinensis	Clonorchiasis	Infections of bile duct, gall bladder and liver.	Undercooked or raw freshwater fish.	Snails and freshwater fish	28
Cestodes (segmented flat - worms, the tapeworms)	*Taenia solium* and *T. saginata*	Taeniasis	Loss of weight or more severe symptoms of vomiting, abdominal pain and, in extreme cases, death.	Undercooked pork (*T. solium*) or beef (*T. saginata*)	Pigs and cattle, respectively	2 74
	Diphyllobothrium latum	Diphyllobothriosis	Asymptomatic in 50% of cases. Diarrhoea, vomiting, dizziness and general weakness in remainder.	Undercooked or raw fish.	Freshwater fish and water-fleas	3
Nematodes (long, cylindrical unsegmented worms, the roundworms)	*Trichinella spiralis*	Trichinosis	Majority of cases asymptomatic. In others vomiting and diarrhoea in first few days. Followed by muscular pains, oedema and fever.	Undercooked pork or pork sausages.	Pigs	—
	Anisakis sp.	Anisakiasis or 'herring-worm' disease	Variable. May be limited to stomach pains and vomiting. Stomach ulcers in severe cases.	Undercooked or raw fish, often lightly salted.	Marine fish including herring	—

[a] Communicable Disease Reports, Quartely Editions (PHLS).

drinking water, are uncommon (Table 2.8). The transmissible stage of the life cycle is termed the 'cyst' but it is not very resistant and remains viable only if kept moist. It is destroyed at 55°C and will not resist freezing for more than 24 h.

2.6.3.2. *Toxoplasma gondii*
This is the causative organism of toxoplasmosis and can be found in a wide range of animals including wild cats, rodents, dogs, foxes, hedgehogs and birds, as well as those listed in Table 2.8. The worldwide incidence of this disease is not known but estimates put the number of cases well in excess of 1 million. In the United States infection rates in the newborn are 1 in 10000 and nowadays it is the most commonly reported parasitic infection in the UK. The transmissible stage is a cyst which is destroyed at 60°C and also by freezing. Transmission of the disease to humans is normally either by ingestion of contaminated raw or undercooked meats or through the domestic cat. The former mode can be effectively blocked by adequate cooking and the more likely method of transmission is by contamination with the infected faeces of domestic cats through direct handling of cats or cat litter.

2.6.3.3. *Trematodes*
The fluke *Fasciola hepatica* is a common parasite of sheep and cattle and causes human infections, principally in Mediterranean and South American countries. Occasional outbreaks in Western European countries including Britain have been reported. The life cycle of this fluke is complex and includes snails as an intermediate host alternating with the above-named herbivores or man. Human cases almost always result from the contamination of watercress beds by snails infected with the fluke. The snail releases a free-living stage of the fluke which becomes encysted and in this form is ingested, with the raw watercress, by man.

Clonorchiasis, caused by *Clonorchis sinensis*, is the most important human disease caused by flukes. More than 40 species of freshwater fish may become infected by the fluke and humans are, in turn, infected by consuming inadequately cooked, salted, dried or raw fish. This practice is common in the Far East where this disease is prevalent, and it has been estimated that, in this region, some 20 million people are infected. The cysts are readily destroyed at temperatures over 50°C but they are resistant to salting and drying.

2.6.3.4.   Cestodes

Two cestode worms are responsible for taeniasis, *Taenia solium*, the pork tapeworm, and *T. saginata*, the beef tapeworm; both have a world-wide distribution. Their life cycles are similar with man acting as the primary host. Infection results from the ingestion of undercooked or raw meats containing the cysts. Infected pork or beef is often described as 'measly'. The cyst develops by hooking onto the intestinal wall when growth of the tapeworm begins. The adult tapeworm, which may reach several metres in length, is composed of hundreds of segments which in-termittently break off and are voided in the faeces. The eggs released from the segments are comparatively resistant to environmental ex-tremes and the life cycle is completed by the infection of animals graz-ing in pastures which have been treated with contaminated manure. In a small proportion of cases with *T. solium* the life cycle may follow a different pathway. Humans can accidently ingest the voided eggs which later hatch in the intestines to produce oncospheres which, in turn, are carried to all parts of the body by circulating blood. The oncospheres develop into mature worms causing human cysticercosis; many cases of this form of infection associated with the eye, heart, liver, lungs and brain have been reported although cerebral cysticercosis is probably the most common. Prevention of taeniasis, and hence cysticercosis, is easily achieved by heating infected meat to 60°C or by freezing.

The third tapeworm infecting man is *Diphyllobothrium latum* which is found in a variety of freshwater fish including trout, perch and pike. Two intermediate hosts are required in the life cycle, the fish becoming infected by eating copepods (water-fleas) which are themselves infected from human sewage containing the eggs. Other fish-eating animals in-cluding dogs, cats and pigs can also be infected. Like the other tape-worms it is readily killed by heating or freezing and like them is widely distributed.

2.6.3.5.   Nematodes

Trichinosis, caused by *Trichinella spiralis*, is probably the most impor-tant of all the food-borne parasitic infections although cases in the UK are now rare; its distribution is worldwide and many animals can act as a host and cause an infection in man although pigs are recognized as the major source of infection. The life cycle is relatively simple and, un-like many of the other parasites, the whole life cycle can be completed in one host. After ingestion of the cyst-contaminated meat, the round-worms develop in the intestine finally producing larvae which pass by way of the blood vessels into muscles where fresh cysts are formed. Adequate cooking of pork to a temperature of 58°C will destroy the parasite; so also will freezing (-15°C for 20 days) or curing.

Anisakiasis, caused by *Anisakis* spp. and related genera, has only been recognized as a parasitic infection for the past 25 years or so. It was first described in The Netherlands when several hundred people were infected after eating raw, slightly salted herring and similar outbreaks have been reported subsequently in Japan and South America; mackerel and cod have also been implicated. Heating to 50°C, freezing, curing or marinating to a pH below 4 will kill the organisms.

2.7. POISONOUS PLANTS

Many plants are known to be inherently toxic to man but the majority are not normally eaten and therefore need not be considered here. Only a small number of examples will be mentioned and readers requiring more information should consult the appropriate references listed in the Bibliography.

Cultivated mushrooms have become an important part of the human diet but in Europe, in particular, many exotic edible fungi are collected and eaten. Unfortunately, poisonous mushrooms are sometimes mistaken for edible forms and food poisonings of this kind are the most frequent attributed to plants.

There are about a dozen poisonous fungi found in the UK of which *Amanita phalloides*, the death cap or death angel mushroom, is by far the most important; it is responsible for about 90% of deaths caused by poisonous fungi. *A. phalloides* contains a number of toxins (phallotoxins and amatoxins) which act on the liver and kidney causing severe tissue damage. After ingestion of the mushroom the first symptoms, vomiting, diarrhoea and severe abdominal pain, appear after 6–24 h and are followed by a rapid loss of strength and convulsions; death occurs in over 50% of cases, usually within 2–8 days. The toxins are extremely resistant to boiling and frying so that normal cooking processes afford little protection.

Amanita muscaria, the fly agaric or scarlet fly cap mushroom, is another of the amanitas which is poisonous to man. The toxic principle is muscarin, an alkaloid producing hallucinatory effects which are evident in the most severe cases. However, this fungus is less toxic to humans than *A. phalloides*.

Probably the next most important form of plant poisoning is that resulting from the ingestion of berries, usually by children. The dark purple berries of *Atropa belladonna*, the deadly nightshade, contain a mixture of poisonous alkaloids including atropine and the consumption of three or four berries may be sufficient to cause death, usually within 24 h. Symptoms include vomiting, a rapid heartbeat, double vision and hallucinations. *Solanum dulcamara*, the woody nightshade, forms red berries containing solanine, another alkaloid. Vomiting, diarrhoea and convulsions in the more serious cases are the principal symptoms and fatalities have been occasionally reported in children. The berries produced by bryony *(Bryonia dioica)*, black nightshade (*Solanum nigrum*), wood laurel (*Daphne laureola*) and spurge olive (*Daphne mezereum*) and the seeds of the laburnum (*Laburnum anagyroides*) are amongst others that have caused fatalities in children.

The consumption of raw red kidney beans (*Phaseolus vulgaris*) was responsible for a number of food poisoning outbreaks reported in the UK in the late 1970s. Kidney beans and certain others (e.g. lima beans) contain a number of potential toxic factors of which lectins are the most important; they sensitize tissues in the small intestine leading to gastrointestinal disturbances. In the above outbreaks, following a short incubation period of 1–3 h, those affected suffered severe nausea and vomiting followed by diarrhoea at a later stage; recovery was rapid. It was reported that by eating 4–5 raw beans the symptoms can be induced (Noah *et al.*, 1980).

Finally, mention should be made of three plants that are commonly eaten by man which have poisonous parts to them. Thus: (1) the seeds of apples can be fatal if consumed in large numbers; (2) the tuber sprouts and peelings of potatoes contain solanine (see above) and can also prove fatal; and (3) the leaves and stalks of rhubarb, which contain oxalic acid, may again result in death.

2.8.   CHEMICAL POISONING

Chemical poisoning through the ingestion of food is rare in the UK. In spite of this, foods are regularly consumed that contain chemicals that would prove harmful if ingested in substantially greater amounts. Thus there is a threshold level in man for these chemicals and provided that this level is not exceeded no harm is done as the chemicals are excreted or rendered innocuous in some other way.

The most common form of chemical poisoning is metallic poisoning; it is characterized by an extremely rapid onset of symptoms, often within a minute or two of consumption of the food, the principal features being vomiting and acute abdominal pain. In the past, metallic poisoning was relatively common but the introduction of legislation controlling the use of poisonous metals has virtually eliminated this hazard. The metals mainly implicated have been arsenic, lead, mercury, antimony, cadmium and zinc, and these will be discussed briefly. Other information including details of outbreaks of historical interest can be found in Dewberry (1959).

Arsenic has probably caused most metallic poisoning in humans and yet it occurs in trace amounts in marine foods. A large outbreak of arsenical food poisoning occurred in Lancashire and surrounding counties at the turn of the century when over 600 people were poisoned, 70 fatally, after drinking beer contaminated with arsenic. Until the introduction of organically based pesticides, poisonings were occasionally reported which resulted from the ingestion of inadequately washed fruits treated with arsenical sprays.

Lead is another metal that occurs naturally in many foods in minute quantities but it causes toxic symptoms if the threshold level is exceeded. At one time lead poisoning was not uncommon due to the use of lead in cooking utensils, food processing machinery, pipework, pottery glazes, artificial food-colouring materials and fruit sprays. However, this form of poisoning has been virtually eliminated by restrictive legislation.

Poisoning caused by *mercury* is rare but there have been one or two serious outbreaks in comparatively recent times. Mercury, discharged as effluent into the sea around Japan, was converted by bacteria into methyl mercury and deposited in the flesh of fish. Methyl mercury attacks the nervous system and in the Minamata Bay outbreak, which covered a period from 1953 to 1961, there were 48 fatalities resulting from the ingestion of contaminated fish; this form of poisoning is now known as 'Minamata disease'.

The remaining three metals are all associated with a similar poisoning sequence, i.e. the release of the toxic agent through the reaction of an acid food with a plated food utensil. Thus *antimony* caused outbreaks of food poisoning, particularly in the 1920s and 1930s, when cheap enamel-coated utensils were attacked by the acids contained in such foods as lemon juice and vinegar; in this way the antimony was dissolved out of the enamel and consumed with the food. Similarly, *cadmium,* another metal that has been used to plate utensils in which acid foods have been held, has caused some poisoning outbreaks. Finally, *zinc,* an element that is widely distributed in foods in trace amounts, has been responsible for isolated outbreaks of metallic food poisoning which have been attributed to the use of galvanized iron vessels for the storage of fruits, e.g. apples, plums and rhubarb.

BIBLIOGRAPHY

Fenlon, D. R. (1981). Seagulls (*Larus* spp.) as vectors of salmonellae: an investigation into the range of serotypes and numbers of salmonellae in gull faeces. *Journal of Hygiene, Cambridge,* **86**, 195–202

Franco, D. A. (1988). *Campylobacter* species: considerations for controlling a foodborne pathogen. *Journal of Food Protection,* **51**, 145–53.

Genigeorgis, C. A. (1989). Present state of knowledge on staphylococcal intoxication. *International Journal of Food Microbiology,* **9**, 327–60.

Gross, R. J. & Rowe, B. (1985). *Escherichia coli* diarrhoea. *Journal of Hygiene, Cambridge,* **95**, 531–50.

Healy, G. R. & Juranek, D. (1979). Parasitic infections. In: *Food-Borne Infections and Intoxications,* 2nd edn. Eds H. Riemann and F. L. Bryan, New York, Academic Press.

Mackey, B. M. & Bratchell, N. (1989). A review. The heat resistance of *Listeria monocytogenes. Letters in Applied Microbiology,* **9**, 89–94.

McLauchlin, J. (1987). A review. *Listeria monocytogenes,* recent advances in the taxonomy and epidemiology of listeriosis in humans. *Journal of Applied Bacteriology,* **63**, 1–11.

Moreau, C. (1979). *Moulds, Toxins and Food.* Chichester, John Wiley & Sons.

Moss, M. O. (1987). Microbial food poisoning. In: *Essays in Agricultural and Food Microbiology.* Eds J. R. Norris and G. L. Pettipher, Chichester, John Wiley and Sons.

Park, R. W. A., Griffiths, P. L. & Moreno, G. S. (1991). Sources and survival of campylobacters: relevance to enteritis and the food industry. In: *Pathogens in the Environment.* Ed. B. Austin. Journal of Applied Bacteriology Symposium Supplement No. 20, 97S–106S.

Pusztai, A. (1986). The role in food poisoning of toxins and allergens from higher plants. In: *Developments in Food Microbiology—2.* Ed. R. K. Robinson, London, Elsevier Applied Science Publishers.

Roberts, D. (1982). Factors contributing to outbreaks of food poisoning in England and Wales 1970–1979. *Journal of Hygiene, Cambridge,* **89**, 491–8.

Samarajeewa, U., Sen, A. C., Cohen, M. D. & Wei, C. I. (1990). Detoxification of aflatoxins in foods and feeds by physical and chemical methods. *Journal of Food Protection,* **53**, 489–501.

Scott, H. G. (1969). Poisonous plants and animals. In: *Food-Borne Infections and Intoxications.* Ed. H. Riemann, New York, Academic Press.

Smith, J. L. (1987). *Shigella* as a foodborne pathogen. *Journal of Food Protection*, **50**, 788–801.

REFERENCES

Abrahamsson, K., De Silva, N. N. & Molin, N. (1965). Toxin production by *Clostridium botulinum* type E, in vacuum-packed, irradiated fresh fish in relation to changes of the associated microflora. *Canadian Journal of Microbiology*, **11**, 523–9.

Anderson, J. K. (1988). Contamination of freshly slaughtered pig carcasses with human pathogenic *Yersinia enterocolitica*. *International Journal of Food Microbiology*, **7**, 193–202.

Ayres, P. A. & Barrow, G. I. (1978). The distribution of *Vibrio parahaemolyticus* in British coastal waters: report of a collaborative study 1975–6. *Journal of Hygiene, Cambridge*, **80**, 281–94.

Bailey, J. S., Fletcher, D. L. & Cox, N. A. (1989). Recovery and serotype distribution of *Listeria monocytogenes* from broiler chickens in the southeastern United States. *Journal of Food Protection*, **52**, 148–50.

Baird-Parker, A. C. (1971). Factors affecting the production of bacterial food poisoning toxins. *Journal of Applied Bacteriology*, **34**, 181–97.

Ball, A. P. *et al.* (1979). Human botulism caused by *Clostridium botulinum* type E: the Birmingham outbreak. *Quarterly Journal of Medicine*, **48** (New Series), 473–91.

Bannister, B. A. (1987). *Listeria monocytogenes* meningitis associated with eating soft cheese. *Journal of Infection*, **15**, 165–8.

Barnes, J. M. (1970). Aflatoxin as a health hazard. *Journal of Applied Bacteriology*, **33**, 285–98.

Barrell, R. A. E. (1982). Isolations of salmonellas from human, food and environmental sources in the Manchester area: 1976–1980. *Journal of Hygiene, Cambridge*, **88**, 403–11.

Barrell, R. A. E. (1987). Isolations of salmonellas from humans and foods in the Manchester area: 1981–1985. *Epidemiology and Infection*, **98**, 277–84.

Barrett, N. J. (1986). Communicable disease associated with milk and dairy products in England and Wales: 1983–1984. *Journal of Infection*, **12**, 265–72.

Beckers, H. J.(1988). Incidence of foodborne diseases in the Netherlands: annual summary 1982 and an overview from 1979 to 1982. *Journal of Food Protection*, **51**, 327–34.

Beckers, H. J., Soentoro, P. S. S. & Delfgou-van Asch, E. H. M. (1987). The occurrence of *Listeria monocytogenes* in soft cheeses and raw milk and its resistance to heat. *International Journal of Food Microbiology*, **4**, 249–56.

Bergdoll, M. S. (1970). Enterotoxins. In: *Microbial Toxins III. Bacterial Protein Toxins*. Eds T. C. Montie, S. Kadis and S. J. Ajl, New York, Academic Press.

Beumer, R. R., Cruysen, J. J. M. & Birtantie, I. R. K. (1988). The occurrence of

Campylobacter jejuni in raw cows' milk. *Journal of Applied Bacteriology*, **65**, 93–6.

Blanco, J. L. *et al.* (1988). Presence of aflatoxin M_1 in commercial ultra-high-temperature treated milk. *Applied and Environmental Microbiology*, **54**, 1622–3.

Brackett, R. E. & Marth, E. H. (1982). Fate of aflatoxin M_1 in cheddar cheese and in process cheese spread. *Journal of Food Protection*, **45**, 549–52.

Bradshaw, J. G. *et al.* (1987). Thermal resistance of *Listeria monocytogenes* in dairy products. *Journal of Food Protection*, **50**, 543–4.

Brouwer, R., Mertens, M. J. A., Siem, T. H. & Katchaki, J. (1979). An explosive outbreak of *Campylobacter* enteritis in soldiers. *Antonie van Leeuwenhoek*, **45**, 517–9.

Brown, D. F., Spaulding, P. L. & Twedt, R. M. (1977). Enteropathogenicity of *Vibrio parahaemolyticus* in the ligated rabbit ileum. *Applied and Environmental Microbiology*, **33**, 10–14.

Bryan, F. L. (1976). *Staphylococcus aureus*. In: *Food Microbiology: Public Health and Spoilage Aspects.* Eds M. P. Defigueiredo and D. F. Splittstoesser, Westport, Avi Publishing Co.

Bryan, F. L. (1979). Infections and intoxications caused by other bacteria. In: *Food-Borne Infections and Intoxications,* 2nd edn. Eds H. Riemann and F. L. Bryan, New York, Academic Press.

Bullerman, L. B. (1976). Examination of Swiss cheese for incidence of mycotoxin producing moulds. *Journal of Food Science*, **41**, 26–8.

Bullerman, L. B. (1979). Significance of mycotoxins to food safety and human health. *Journal of Food Protection*, **42**, 65–86.

Bullerman, L. B. & Olivigni, F. J. (1974). Mycotoxin producing potential of moulds isolated from cheddar cheese. *Journal of Food Science*, **39**, 1166–8.

Bullerman, L. B., Schroeder, L. L. & Park, K-Y. (1984). Formation and control of mycotoxins in food. *Journal of Food Protection*, **47**, 637–46.

Chan, K.-Y., Woo, M. L., Lam, L. Y. & French, G. L. (1989). *Vibrio parahaemolyticus* and other halophilic vibrios associated with seafood in Hong Kong. *Journal of Applied Bacteriology*, **66**, 57–64.

Charlton, B. R., Kinde, H. & Jensen, L. H. (1990). Environmental survey for *Listeria* species in California milk processing plants. *Journal of Food Protection*, **53**, 198–201.

Cliver, D. O. (1979). Viral infections. In: *Food-Borne Infections and Intoxications*, 2nd edn. Eds. H. Riemann and F. L. Bryan, New York, Academic Press.

Cliver, D.O. (1987). Foodborne disease in the United States, 1946–1986. *International Journal of Food Microbiology*, **4**, 269–77.

Cox, L. J. *et al.* (1989). *Listeria* spp. in food processing, non-food and domestic environments. *Food Microbiology*, **6**, 49–61.

Critchley, M., O'Leary, J. L. & Jennett, B. (1972). *Scientific Foundations of Neurology.* London, Heinemann.

D'Aoust, J-Y. (1991). Pathogenicity of foodborne *Salmonella. International Journal of Food Microbiology*, **12**, 17–40.

Dealler, S. F. & Lacey, R. W. (1990). Transmissable spongiform encephalopathies : the threat of BSE to man. *Food Microbiology*, **7**, 253–79.

De Boer, E., Seldam, W. M. & Oosterom, J. (1986). Characterization of *Yersinia enterocolitica* and related species isolated from foods and porcine tonsils in the Netherlands. *International Journal of Food Microbiology*, **3**, 217–224.

Dewberry, E. B. (1959). *Food Poisoning*, 4th edn, London, Hill.

DHSS (1969). *On the State of Public Health for the Year 1968*. London, HMSO.

DHSS (1974). *On the State of Public Health for the Year 1972*. London, HMSO.

DHSS (1976). *On the State of Public Health for the Year 1974*. London, HMSO.

DHSS (1977). *On the State of Public Health for the Year 1976*. London, HMSO.

DHSS (1980). *On the State of Public Health for the Year 1978*. London, HMSO.

DHSS (1987). *On the State of Public Health for the Year 1986*. London, HMSO.

Dirksen, J. & Flagg, P. (1988). Pathogenic organisms in dairy products; cause, effects and control. *Food Science and Technology Today*, **2**, 41–3.

Doyle, M. P. *et al.* (1987). Survival of *Listeria monocytogenes* in milk during high-temperature, short-time pasteurization. *Applied and Environmental Microbiology*, **53**, 1433–8.

Edwards, A. T., Roulson, M. & Ironside, M. J. (1988). A milk-borne outbreak of serious infection due to *Streptococcus zooepidemicus* (Lancefield Group C). *Epidemiology and Infection*, **101**, 43–51.

Evenson, M. L., Hinds, M. W., Bernstein, R. S. & Bergdoll, M. S. (1988). Estimation of human dose of staphylococcal enterotoxin A from a large outbreak of staphylococcal food poisoning involving chocolate milk. *International Journal of Food Microbiology*, **7**, 311–6.

Fernandez Garayzabel, J. F. *et al.* (1987). Survival of *Listeria monocytogenes* in raw milk treated in a pilot plant pasteurizer. *Journal of Applied Bacteriology*, **63**, 533–7.

Fricker, C. R. & Park, R. W. A. (1989). A two-year study of the distribution of 'thermophilic' campylobacters in human, environmental and food samples from the Reading area with particular reference to toxin production and heat-stable serotype. *Journal of Applied Bacteriology*, **66**, 477–90.

Fricker, C. R. & Tompsett, S. (1989). *Aeromonas* spp. in foods: a significant cause of food poisoning? *International Journal of Food Microbiology*, **9**, 17–23.

Frobish, R. A. *et al.* (1986). Aflatoxin residues in milk of dairy cows after ingestion of naturally contaminated grain. *Journal of Food Protection*, **49**, 781–5.

Genigeorgis, C., Hassuneh, M. & Collins P. (1986). *Campylobacter jejuni* infection on poultry farms and its effect on poultry meat contamination during slaughtering. *Journal of Food Protection*, **49**, 895–903.

George, S. M., Lund, B. M. & Brocklehurst, T. F. (1988). The effect of pH and temperature on initiation of growth of *Listeria monocytogenes*. *Letters in Applied Microbiology*, **6**, 153–6.

Gibson, E. A. (1969). Salmonella infection in farm livestock in Britain. In: *Bacterial Food Poisoning*. Ed. J. Taylor, London, Cox and Wyman.

Gilbert, R. J. (1979). *Bacillus cereus* gastroenteritis. In: *Food-Borne Infections and Intoxications*, 2nd edn. Eds H. Riemann and F. L. Bryan, New York, Academic Press.

Gilbert, R. J. (1983). Food-borne infections and intoxications—recent trends and prospects for the future. In: *Food Microbiology: Advances and Prospects*. Eds T. A. Roberts and F. A. Skinner, London, Academic Press.

Gilbert, R. J., Stringer, M. F. & Peace, T. C. (1974). The survival and growth of *Bacillus cereus* in boiled and fried rice in relation to outbreaks of food poisoning. *Journal of Hygiene, Cambridge*, **73**, 433–44.

Gilbert, R. J. & Taylor, A. J. (1976). *Bacillus cereus* food poisoning. In: *Microbiology in Agriculture, Fisheries and Food*. Eds F. A. Skinner and J. G. Carr, London, Academic Press.

Gitter, M., Bradley, R. & Blampied, P. H. (1980). *Listeria monocytogenes* infection in bovine mastitis. *The Veterinary Record*, **107**, 390–3.

Glass, K. A. & Doyle, M. P. (1989). Fate of *Listeria monocytogenes* in processed meat products during refrigerated storage. *Applied and Environmental Microbiology*, **55**, 1565–9.

Goepfert, J. M., Spira, W. M., Glatz, B. A. & Kim, H. U. (1973). Pathogenicity of *Bacillus cereus*. In: *The Microbiological Safety of Food*. Eds B. C. Hobbs and J. H. B. Christian, London, Academic Press.

Goldner, S. B., Solberg, M., Jones, S. & Post, L. S. (1986). Enterotoxin synthesis by nonsporulating cultures of *Clostridium perfringens*. *Applied and Environmental Microbiology*, **52**, 407–12.

Greenwood, M. H. & Hooper, W. L. (1983). Chocolate bars contaminated with *Salmonella napoli*: an infective study. *British Medical Journal*, **286**, 1394.

Greenwood, M. H., Roberts, D. & Burden, P. (1991). The occurrence of *Listeria* species in milk and dairy products: a national survey in England and Wales. *International Journal of Food Microbiology*, **12**, 197–206.

Guerrant, R. L. *et al.* (1987). Production of a unique cytotoxin by *Campylobacter jejuni*. *Infection and Immunity*, **55**, 2526–30.

Halpin-Dohnalek, M.I. & Marth, E. H. (1989). *Staphylococcus aureus*: production of extracellular compounds and behavior in foods—a review. *Journal of Food Protection*, **52**, 267–82.

Harrison, M. A. & Carpenter, S. L. (1989). Survival of large populations of *Listeria monocytogenes* on chicken breasts processed using moist heat. *Journal of Food Protection*, **52**, 376–8.

Harvey, J., Patterson, J. T. & Gibbs, P. A. (1982). Enterotoxigenicity of *Staphylococcus aureus* strains isolated from poultry: raw poultry carcases as a potential food-poisoning hazard. *Journal of Applied Bacteriology*, **52**, 251–8.

Harvey, R. W. S. (1973). Salmonella contaminated animal feed in relation to infection in animals and man. In: *The Microbiological Safety of Food*. Eds B. C. Hobbs and J. H. B. Christian, London, Academic Press.

Harvey, R. W. S., Price, T. H., Foster, D. W. & Griffiths, W. C. (1969). Salmonellas in sewage. A study in latent human infection. *Journal of Hygiene, Cambridge*, **67**, 517–23.

Hauschild, A. W. H., Hilsheimer, R., Weiss, K. F. & Burke, R. B. (1988). *Clostridium botulinum* in honey, syrups and dry infant cereals. *Journal of Food Protection*, **51**, 892–4.

Hayes, P. S. *et al.* (1986). Isolation of *Listeria monocytogenes* from raw milk. *Applied and Environmental Microbiology*, **51**, 438–40.

Heisick, J. E., Wagner, D. E., Nierman, M. L. & Peeler, J. T. (1989). *Listeria* spp. found on fresh market produce. *Applied and Environmental Microbiology*, **55**, 1925–7.

Hewitt, J. H. *et al.* (1986). Large outbreaks of *Clostridium perfringens* food poisoning associated with the consumption of boiled salmon. *Journal of Hygiene, Cambridge,* **97**, 71–80.

Hobbs, B. C. & Hugh-Jones, M. E. (1969). Epidemiological studies on *Salmonella senftenberg.* 1. Relations between animal foodstuff, animal and human isolations. *Journal of Hygiene, Cambridge,* **67**, 81–8.

Hobbs, G. (1976). *Clostridium botulinum* and its importance in fishery products. *Advances in Food Research,* **22**, 135–85.

Hobbs, G. (1987). Microbiology of fish. In: *Essays in Agricultural and Food Microbiology.* Eds J. R. Norris and G. L. Pettipher, Chichester, John Wiley & Sons.

Hockin, J. C. *et al.* (1989). An international outbreak of *Salmonella nima* from imported chocolate. *Journal of Food Protection,* **52**, 51–4.

Hoffman, P. S. & Blankenship, L. C. (1986). Significance of *Campylobacter* in foods. In: *Developments in Food Microbiology—2.* Ed. R. K. Robinson, London, Elsevier Applied Science Publishers.

Hooper, W. L., Barrow, G. I. & McNab, D. J. N. (1974). *Vibrio parahaemolyticus* food poisoning in Britain. *Lancet,* **1**, 1100–2.

Hopper, S. A. & Mawer, S. (1988). *Salmonella enteritidis* in a commercial laying flock. *Veterinary Record,* **123**, 351

Humphrey, T. J. & Beckett, P. (1987). *Campylobacter jejuni* in dairy cows and raw milk. *Epidemiology and Infection,* **98**, 263–9.

Humphrey, T. J., Mead , G. C. & Rowe, B. (1988). Poultry meat as a source of human salmonellosis in England and Wales. *Epidemiology and Infection,* **100**, 175–84.

International Commission on Microbiological Specifications for Foods (ICMSF) (1980). *Microbial Ecology of Foods,* Vol 2 (Food Commodities). New York, Academic Press.

Johnson, J. L., Doyle, M. P. & Cassens, R. G. (1990). *Listeria monocytogenes* and other *Listeria* spp. in meat and meat products. A review. *Journal of Food Protection,* **53**, 81–91.

Jones, P. H. *et al.* (1981). *Campylobacter* enteritis associated with the consumption of free school milk. *Journal of Hygiene, Cambridge,* **87**, 155–62.

Jones, P. W., Collins, P., Brown, G. T. H. & Aitken, M. (1982). Transmission of *Salmonella mbandaka* to cattle from contaminated feed. *Journal of Hygiene, Cambridge,* **88**, 255–63.

Junttila, J. R., Niemela, S. I. & Hirn, J. (1988). Minimum growth temperatures of *Listeria monocytogenes* and non-haemolytic listeria. *Journal of Applied Bacteriology,* **65**, 321–7.

Lacey, R. W. & Kerr, K. G. (1989). Opinion. Listeriosis—the need for legislation. *Letters in Applied Microbiology,* **8**, 121–2.

Lammerding, A. A. *et al.* (1988). Prevalence of *Salmonella* and thermophilic *Campylobacter* in fresh pork, beef, veal and poultry in Canada. *Journal of Food Protection,* **51**, 47–52.

Lovett, J. *et al.* (1990). High-temperature short-time pasteurization inactivates *Listeria monocytogenes. Journal of Food Protection,* **53**, 734–8.

Lynt, R. K., Kautter, D. A. & Read, R. B. (1975). Botulism in commercially canned foods. *Journal of Milk and Food Technology,* **38**, 546–50.

MacDonald, K. L. *et al.*(1985). A multistate outbreak of gastrointestinal illness caused by enterotoxigenic *Escherichia coli* in imported semisoft cheese. *Journal of Infectious Diseases*, **151**, 716–20.

Marrier, R. *et al.* (1973). An outbreak of enteropathogenic *Escherichia coli* foodborne disease traced to imported French cheese. *Lancet*, **2**, 1376–8.

Marth, E. H. & Calanog, B. G. (1976). Toxigenic fungi. In: *Food Microbiology: Public Health and Spoilage Aspects*. Eds M. P. Defigueiredo and D. F. Splittstoesser, Westport, Avi Publishing Co.

Melling, J. & Capel, B. J. (1978). Characteristics of *Bacillus cereus* emetic toxin. *FEMS Microbiology Letters*, **4**, 133–5.

Miller, I. S., Bolton, F. J. & Dawkins, H. C. (1987). An outbreak of *Campylobacter* enteritis transmitted by puppies. *Environmental Health*, **95** (July), 11–4.

Morgan, I. R., Krautil, F. L. & Craven, J. A. (1987). Effect of time in lairage on caecal and carcass contamination of slaughter pigs. *Epidemiology and Infection*, **98**, 323–30.

Moss, M. O. (1989). Mycotoxins of *Aspergillus* and other filamentous fungi. In: *Filamentous Fungi in Foods and Feeds*. Eds M. O. Moss, B. Jarvis & F. A. Skinner. Journal of Applied Bacteriology Symposium Supplement No. 18, 69S–81S.

Nakamura, M. & Schulze, J. A. (1970). *Clostridium perfringens* food poisoning. *Annual Review of Microbiology*, **24**, 359–72.

Nilehn, B. (1969). Studies on *Yersinia enterocolitica* with special reference to bacterial diagnosis and occurrence in human acute enteric disease. *Acta Pathologica et Microbiologica Scandinavica*, Supplement, **206**, 1–48.

Noah, N. D., Bender, A. E., Reaidi, G. B. & Gilbert, R. J. (1980). Food poisoning from raw red kidney beans. *British Medical Journal*, **281**, 236–7.

Notermans, S. & van Otterdijk, R. L. M. (1985). Production of enterotoxin A by *Staphylococcus aureus* in food. *International Journal of Food Microbiology*, **2**, 145–9.

Olsvik, Ø., Wasteson, Y., Lund, A. & Hornes, E. (1991). Pathogenic *Escherichia coli* found in food. *International Journal of Food Microbiology*, **12**, 103–14.

O'Mahony, M. *et al.* (1990). An outbreak of foodborne botulism associated with contaminated hazelnut yoghurt. *Epidemiology and Infection*, **104**, 389–95.

Oosterom, J. (1991). Epidemiological studies and proposed preventive measures in the fight against human salmonellosis. *International Journal of Food Microbiology*, **12**, 41–52.

Ormay, L. & Novotny, T. (1969). The significance of *Bacillus cereus* food poisoning in Hungary. In: *The Microbiology of Dried Foods*. Eds E. H. Kampelmacher, M. Ingram and D. A. A. Mossel, Bilthoven, International Association of Microbiological Societies.

Paster, N. & Bullerman, L. B. (1988). Mould spoilage and mycotoxin formation in grains as controlled by physical means. *International Journal of Food Microbiology*, **7**, 257–65.

Pavlovic, M., Plestina, R. & Krogh, P. (1979). Ochratoxin A contamination of foodstuffs in an area with Balkan (endemic) nephropathy. *Acta Pathologica et Microbiologica Scandinavica,* **87B**, 243–6.

Perales, I. & Audicana, A. (1989). The role of hens' eggs in outbreaks of salmonellosis in north Spain. *International Journal of Food Microbiology*, **8**, 175–80.

PHLS Report. (1989). Memorandum of evidence to the Agriculture Committee inquiry on salmonella in eggs. *PHLS Microbiology Digest*, **6**, 1–9.

PHLS Working Group, Scovgaard, N. & Nielsen, B. B. (1972). Salmonellas in pigs and animal feeding stuffs in England and Wales and in Denmark. *Journal of Hygiene, Cambridge*, **70**, 127–40.

Pini, P. N. & Gilbert, R. J. (1988). The occurrence in the UK of *Listeria* species in raw chickens and soft cheeses. *International Journal of Food Microbiology*, **6**, 317–26.

Porter, I. A. & Reid, T. M. S. (1980). A milk-borne outbreak of *Campylobacter* infection. *Journal of Hygiene, Cambridge*, **84**, 415–19.

Ratnam, S. & March, S. B. (1986). Laboratory studies on salmonella—contaminated cheese involved in a major outbreak of gastroenteritis. *Journal of Applied Bacteriology*, **61**, 51–6.

Ray, L. L. & Bullerman, L. B. (1982). Preventing growth of potentially toxic molds using antifungal agents. *Journal of Food Protection*, **45**, 953–63.

Reichart, C. A. & Fung, D. Y. C. (1976). Thermal inactivation and subsequent reactivation of staphylococcal enterotoxin B in selected liquid foods. *Journal of Milk and Food Technology*, **39**, 516–20.

Reilly, W. J. *et al.* (1988). Poultry-borne salmonellosis in Scotland. *Epidemiology and Infection*, **101**, 115–22.

Riemann, H. (1969). Botulism-types A, B and F. In: *Food-Borne Infections and Intoxications*. Ed. H. Riemann, New York, Academic Press.

Riordan, T. *et al.* (1985). An outbreak of food-borne enterotoxigenic *Escherichia coli* diarrhoea in England. *Journal of Infection*, **11**, 167–71.

Roberts, D. (1972). Observations on procedures for thawing and spit-roasting frozen dressed chickens, and post-cooking care and storage: with particular reference to food-poisoning bacteria. *Journal of Hygiene, Cambridge*, **70**, 565–88.

Robinson, D. A. (1981). Infective dose of *Campylobacter jejuni* in milk. *British Medical Journal*, **282**, 1584.

Robinson, D. A. *et al.* (1979). *Campylobacter* enteritis associated with consumption of unpasteurized milk. *British Medical Journal*, **1**, 1171–3.

Ryan, C.A. *et al.* (1986). *Escherichia coli* 0157:H7 diarrhoea in a nursing home: clinical, epidemiological, and pathological findings. *Journal of Infectious Diseases*, **154**, 631–8.

Ryser, E. T. & Marth, E. H. (1987). Fate of *Listeria monocytogenes* during the manufacture and ripening of Camembert cheese. *Journal of Food Protection*, **50**, 372–8.

Ryser, E. T., Marth, E. H. & Doyle, M. P. (1985). Survival of *Listeria monocytogenes* during manufacture and storage of cottage cheese. *Journal of Food Protection*, **48**, 746–50.

Sakaguchi, G. (1969). Botulism-type E. In: *Food-Borne Infections and Intoxications*. Ed. H. Riemann, New York, Academic Press.

Sakazaki, R. & Shimada, T. (1986). *Vibrio* species as causative agents of food-borne infection. In: *Developments in Food Microbiology—2*. Ed. R. K. Robinson, London, Elsevier Applied Science Publishers.

Scheusner, D. L. & Harmon, L. G. (1973). Growth and enterotoxin production by various strains of *Staphylococcus aureus* in selected foods. *Journal of Food Science*, **38**, 474–6

Schiemann, D. H. (1988). Examination of enterotoxin production at low temperature by *Yersinia* spp. in culture media and foods. *Journal of Food Protection*, **51**, 571–3.

Schönberg, A. (1989). Method to determine virulence of *Listeria* strains. *International Journal of Food Microbiology*, **8**, 281–4.

Scotland, S. M. (1988). Toxins. In: *Enterobacteriaceae in the Environment and as Pathogens*. Eds B. M. Lund, M. Sussman, D. Jones and M. F. Stringer. Journal of Applied Bacteriology Symposium Supplement No. 17, 109S–129S.

Scott, P. M. (1981). Toxins of *Penicillium* species used in cheese manufacture. *Journal of Food Protection*, **44**, 702–10.

Sharp, J. C. M. (1986). Milk and dairy products. *PHLS Microbiology Digest*, **3**, 28–9.

Sharp, J. C. M., Collier, P. W. & Gilbert, R. J. (1979). Food poisoning in hospitals in Scotland. *Journal of Hygiene, Cambridge*, **83**, 231–6.

Sheard, J. B. (1981). Food poisoning in England during 1979. Another year of high incidence. *Environmental Health*, **79**, 204–9.

Sheard, J. B. (1986). Food poisoning in England and Wales during 1983. A new title but still the same problems. *Environmental Health*, **94**, 57–62

Sheard, J. B. (1987). Food poisoning cases nearly double the level of 1970. *Environmental Health*, **95**, 10–15.

Skirrow, M. B. (1977). *Campylobacter* enteritis: a 'new' disease. *British Medical Journal*, **2**, 9–11.

Skjelkvale, R. & Uemura, T. (1977). Experimental diarrhoea in human volunteers following oral administration of *Clostridium perfringens* enterotoxin. *Journal of Applied Bacteriology*, **43**, 281–6.

Slade, P. J., Fistrovici, E.C. & Collins-Thompson, D. L. (1989). Persistence at source of *Listeria* spp. in raw milk. *International Journal of Food Microbiology*, **9**, 197–203.

Small, R. G. & Sharp, J. C. M. (1979). A milk-borne outbreak due to *Salmonella dublin*. *Journal of Hygiene, Cambridge*, **82**, 95–100.

Smart, J. L., Roberts, T. A., Stringer, M. F. & Shah, N. (1979). The incidence and serotypes of *Clostridium perfringens* on beef, pork and lamb carcasses. *Journal of Applied Bacteriology*, **46**, 377–83.

Smith, G. R., Milligan, R. A. & Moryson, J. (1978). *Clostridium botulinum* in aquatic environments in Great Britain and Ireland. *Journal of Hygiene, Cambridge*, **80**,431–8.

St.Louis, M.E. *et al.* (1988). The emergence of Grade A eggs as a major source of *Salmonella enteritidis* infections. *Journal of the American Medical Association*, **259**, 2103–7.

Steinbruegge, E. G., Maxcy, R. B. & Liewen, M. B. (1988). Fate of *Listeria monocytogenes* on ready to serve lettuce. *Journal of Food Protection*, **51**, 596–9.

Stevens, A. *et al.* (1989). A large outbreak of *Salmonella enteritidis* phage type 4 associated with eggs from overseas. *Epidemiology and Infection*, **103**, 425–33.

Stott, J. A., Hodgson, J. E. & Chaney, J. C. (1975). Incidence of salmonellae in animal feed and the effect of pelleting on content of Enterobacteriaceae. *Journal of Applied Bacteriology*, **39**,41–6.

Sugiyama, H. & Sofos, J. N. (1988). Botulism. In: *Developments in Food Microbiology—4*. Ed. R. K. Robinson, London, Elsevier Applied Science.

Sutton, R. G. A., Kendall, M. & Hobbs, B. C. (1972). The effect of two methods of cooking and cooling on *Clostridium welchii* and other bacteria in meat. *Journal of Hygiene, Cambridge*, **70**,415–24.

Swaminathan, B., Harmon, M. C. & Mehlman, I. J. (1982). A review: *Yersinia enterocolitica. Journal of Applied Bacteriology*, **52**, 151–83.

Tamminga, S. K., Beumer, R. R., Kampelmacher, E. H. & Van Leusden, F. M. (1977). Survival of *Salmonella eastbourne* in milk chocolate prepared with artificially contaminated milk powder. *Journal of Hygiene, Cambridge*, **79**, 333–7.

Taylor, D. M. (1989). Bovine spongiform encephalopathy and human health. *Veterinary Record*, **125**, 413–15.

Thompson, C. A. & Vanderzant, C. (1976). Relationship of *Vibrio parahaemolyticus* in oysters, water and sediment, and bacteriological and environmental indices. *Journal of Food Science*, **41**, 117–22.

Todd, L. S., Hardy, J. C., Stringer, M. F. & Bartholomew, B. A. (1989). Toxin production by strains of *Aeromonas hydrophila* grown in laboratory media and prawn puree. *International Journal of Food Microbiology*, **9**, 145–56.

Van Den Broek, M. J. M., Mossel, D. A. A. & Eggenkamp, A. E. (1979). Occurrence of *Vibrio parahaemolyticus* in Dutch mussels. *Applied and Environmental Microbiology*, **37**,438–42.

Van Netten, P. *et al.* (1990). Psychrotrophic strains of *Bacillus cereus* producing enterotoxin. *Journal of Applied Bacteriology*, **69**, 73–9.

Vernon, E. (1977). Food poisoning and salmonella infections in England and Wales, 1973–75. *Public Health, London*, **91**, 225–35.

Walker, S. J. & Gilmour, A. (1990). Production of enterotoxin by *Yersinia* species isolated from milk. *Journal of Food Protection*, **53**, 751–4.

Waterman, S. C. (1982). The heat sensitivity of *Campylobacter jejuni* in milk. *Journal of Hygiene, Cambridge*, **88**, 529–33.

WHO Working Group. (1988). Food-borne listeriosis. *Bulletin of the World Health Organisation*, **66**, 421–8.

Williams, J. E. (1981). Salmonellas in poultry feeds—a worldwide review. Part I. *World's Poultry Science Journal*, **37**, 6–19.

Wiseman, D. W. & Marth, E. H. (1983). Behavior of aflatoxin M_1 in yogurt, buttermilk and kefir. *Journal of Food Protection*, **46**, 115–18.

Woolaway, M. C. *et al.* (1986). International outbreak of staphylococcal food poisoning caused by contaminated lasagne. *Journal of Hygiene, Cambridge*, **96**, 67–73.

Wray, C. & Sojka, W. J. (1977). Reviews of the progress of Dairy Science: bovine salmonellosis. *Journal of Dairy Research*, **44**, 383–425.

Wu, M. T., Ayres, J. C. & Koehler, P. E. (1974). Toxigenic aspergilli and penicillia isolated from aged, cured meats. *Applied Microbiology*, **28**, 1094–6.

Chapter 3

FOOD SPOILAGE

3.1. INTRODUCTION

Spoilage of food involves any change which renders food unacceptable for human consumption and may result from a variety of causes. It is often difficult to decide when a food is actually spoiled since views differ on what is and is not acceptable and fit or unfit to eat. These differences of opinion are particularly evident when viewed on a worldwide basis as can be illustrated by the following well-known example. The British prefer game meat to be 'hung' for several days to allow organoleptic changes to take place which encourage the development of a 'strong' flavour. Whilst the British consider such flavoured meat to be a delicacy other nationalities, including Americans, regard it as spoiled and unacceptable.

Food spoilage may be caused by one or more of the following:

1. Insect damage.
2. Physical injury due to bruising, pressure, freezing, drying and radiation.
3. The activity of indigenous enzymes in animal and plant tissues. If these enzymes are not destroyed they continue to function during processing and storage. Thus peroxidases naturally present in green vegetables can cause off-flavours during storage.
4. Chemical changes not induced by microbial or naturally occurring enzymes. These changes usually involve oxygen and, other than microbiological spoilage, are the most common cause of

spoilage. Examples of chemical spoilage are the oxidative rancidity of fats and oils, and the discolouration of cured meats.

5. The activity of microorganisms, particularly bacteria, yeasts and moulds.

Spoilage caused by microorganisms is undoubtedly the most important of the above and this chapter is primarily concerned with this aspect.

On the basis of susceptibility to spoilage, foods may be classed as stable or non-perishable (e.g. flour), semi-perishable (e.g. apples), and perishable (e.g. raw meats). Which of these classes a particular food is placed in depends on the many often inter-related factors which have been discussed in Chapter 1. Thus flour is intrinsically a stable food because of its low a_w, but poor storage conditions, which facilitate the absorption of moisture, could convert it into a perishable commodity.

In considering the spoilage of raw foods it can be safely assumed that a wide variety of microorganisms is initially present on the food and that when microbial growth commences certain species will find conditions more favourable than others and in consequence the former will outgrow the latter. In fact, competitive growth amongst the favoured strains usually results in one or two types becoming predominant and often being responsible for any spoilage produced. Thus even though spoilage microorganisms may represent only a very small part of the initial flora they will consistently become predominant in a given type of raw food under a given set of storage conditions; as a result a specific type of microbial spoilage can normally be predicted for a given raw food under specified storage conditions. At this juncture it should, however, be pointed out that the predominant microbial type at the time of spoilage may not always be responsible for that spoilage; for example, it has been shown that in the case of fish the true 'spoilers' only represent about 30% of the total flora at the time of spoilage. Although the predominant types and 'spoilers' may be closely related taxonomically (i.e. members of the same genus or even of the same species), only the latter have the capacity to induce the chemical changes associated with spoilage.

It was pointed out by Ingram (1971) that 'the amount of chemical change caused by a single microbial cell is very small, so that alterations detectable by orthodox chemical means can only be produced by cell populations approaching the maximum density possible'. Ingram calculated that some 10^8 bacterial cells per gram may be necessary to induce measurable spoilage in foods over a number of days and he suggested that spoilage produced in association with bacterial numbers substantially less than this figure is non-bacterial in origin. Microbial and, in particular, bacterial populations in foods rarely exceed *ca* 10^{10} cells per gram so it can be seen that the organisms responsible for spoilage, although not always predominant, are likely to represent an important part of the flora at the time of advanced spoilage.

The foregoing has been concerned mainly with the spoilage of raw foods. Processed foods which have been heated are subject to special types of spoilage due to the selective action of heat on the microorganisms present in the food. Obviously the extent of the microbial selection is dependent upon the time and temperature at which the foods have been heated, the harsher the treatment the fewer the number and variety of organisms surviving. Thus spoilage in canned foods can be caused by a pure culture of the one microorganism resistant to the process conditions; this is typically a bacterium producing a markedly heat-resistant spore.

3.2. SPOILAGE OF FRESH MEATS

3.2.1. Contamination of Tissues by Microorganisms

The housing conditions in which animals are reared determine, to a degree, the bacterial load on their exterior surfaces, the cleaner the conditions the lower the load. During transport from the farm to the slaughterhouse livestock are herded together and frightened so that cross-contamination occurs more readily. Whatever the cleanliness of the livestock large numbers of microorganisms and a great variety of types are inevitably found on the exterior surface and in the intestinal tract of cattle, sheep and pigs before death. Counts in excess of 10^5 per cm^2 are common on the hides of cattle and substantially higher recoveries (*ca* 10^8 per cm^2) have been observed for pigs and from the unwashed wool of sheep. However, the underlying muscle tissue is generally assumed to be sterile except in infected animals.

The slaughter of cattle with a captive bolt pistol and the subsequent procedures such as sticking, skinning, evisceration and butchery common to all animals have the effect of contaminating the previously sterile underlying muscle tissues (sheep and pigs conventionally are stunned electrically and this does not involve microbial contamination). Obviously it is the exposed newly cut surface flesh which is going to carry the majority of the contaminating organisms but deep tissue may in due time become contaminated via the visceral blood supply. Total bacterial counts for freshly cut meat surfaces are likely to vary between 10^3 and 10^5 organisms per cm^2. These organisms are derived mainly from the exterior and the gut of the animal but also from knives, other utensils, butchery tables, etc. so that variations in count often reflect the hygienic conditions in the abattoir (Nottingham, 1982).

3.2.2. Control of Microbial Growth

Meat is an ideal environment for the growth of microorganisms, particularly bacteria, and rapid growth can be expected unless control is effected. The numbers of microorganisms on meats can be controlled in a variety of ways which are briefly considered below. It should be pointed out that most studies on the spoilage of meats have involved beef but the spoilage characteristics of lamb, pork and other meats are essentially similar.

3.2.2.1. Initial Contamination
As will be seen later, the onset of off-odours and other spoilage characteristics in meats are associated with a particular level of bacteria. As the rate of bacterial growth on meats at a given temperature follows a pre-scribed pattern (see growth curve, Fig. 1.8, p. 18), the lower the initial contamination of the meat the longer it will take for the bacterial flora to achieve spoilage levels. Thus, with beef stored at 5°C, if the initial count exceeds 10^5 per cm^2 spoilage can be detected within 6 days whereas with a count of 10^3 per cm^2 spoilage would not occur until the 10th or 11th day (Ayres, 1960). Since the hide is probably the source of most microorganisms on dressed carcasses, animals brought to the abattoir should be freed of adhering dirt by washing before slaughter. The reduction in bacterial numbers achieved by spray or other forms of washing is typically between 50 and 90% (Crouse *et al.*, 1988). Further reductions in the microbial load of the freshly killed animal can be achieved by spraying the dressed carcasses with hot water or by the inclusion of chlorine, lactic acid or other chemicals in the water (Kelly *et al.*, 1981; Smulders *et al.*, 1986) but meat discolouration and the production of unpleasant odours have limited the scope of this work. Good hygienic practices and the implementation of rigorous sanitation standards in the abattoir will help; the latter should include thorough cleaning of items such as walls, floors, cutting tables, knives and other utensils, operatives' clothes, etc. Unfortunately it remains true that there is still no recognized procedure that significantly reduces the microbial load on the carcass surface.

3.2.2.2. Glycogen Reserve

When animals are slaughtered, glycogen stored in the muscles is converted to lactic acid. Under normal conditions this causes a fall in the pH of the muscle from about 7 to 5·6 and this drop is important since it is responsible for a reduction in the growth rate of the bacterial contaminants. However, if the animal is stressed before slaughter (e.g. by excitement, fatigue or starvation) the glycogen reserves become depleted so that a reduced amount of lactic acid is produced and the ultimate pH of the meat is nearer to neutrality; such meat spoils more rapidly and therefore it is imperative that animals are in a sound physiological condition immediately before slaughter.

3.2.2.3. Oxidation-Reduction (OR) Potential

After slaughter oxygen stored in the muscle is depleted thus causing a fall in OR potential to extremely low levels. The strong reducing capacity of the medium (see Section 1.7.4) together with the high initial temperature (38°C) create an ideal environment for the growth of anaerobic bacteria. The predominant spoilage bacteria are *Clostridium* spp, particularly *C. putrefaciens*, which grow within rather than on the meats breaking down the tissues with the production of offensive decomposition products such as hydrogen sulphide and ammonia. This process is known as putrefaction and must be avoided by cooling the meat rapidly before the OR potential falls sufficiently to allow the growth of these organisms (Dainty, 1971). Furthermore, it is now recognized that the presence of certain putrefactive anaerobes in large numbers can be a cause of food poisoning. One of the organisms predominant in the early stages of putrefaction is *C. perfringens*, and *C. botulinum* can be isolated occasionally from putrefying meats (Ingram & Dainty, 1971). The rapid cooling of meat post-slaughter is also advantageous in that it reduces the growth of other food poisoning bacteria such as salmonellas which are also frequent contaminants.

3.2.2.4. The Rate of Cooling

It is obvious from the foregoing that rapid cooling of carcasses is essential to reduce the early bacterial growth and thus increase potential storage times. However if the rate of cooling pre-rigor is too rapid 'cold-shortening' of the meat will occur; this is caused by the contraction and hardening of muscle and results in tough meat. 'Conditioning' of the meat at *ca* 15°C for a few hours effectively counteracts this phenomenon and the meat can then be further cooled to chill temperatures for tenderization over a number of days' storage.

The storage temperatures used have a profound effect upon the microbial spoilage pattern and this aspect is now discussed in some detail.

3.2.3.   Effect of Storage Temperature

3.2.3.1.   Spoilage Under Warm Conditions
Carcass meat and meat joints held at temperatures of 20°C or over will inevitably undergo putrefactive spoilage as indicated above. However, if the surface area to volume ratio is increased by mincing or slicing the raw meat the OR potential also increases thus creating conditions that are less favourable for the growth of putrefactive anaerobes. Growth at or near the surface of the meat is now most rapid but the raised OR potential enables a mixed flora to develop. The bacterial flora at the time of spoilage still contains clostridia but it is now dominated by mesophilic, facultatively anaerobic, Gram negative rods. Most of these organisms are enteric in origin and will include genera such as *Escherichia, Aeromonas, Proteus* and *Enterobacter*. Other genera also represented should include *Staphylococcus* and *Micrococcus* (Gram positive cocci) and *Bacillus* (aerobic and facultatively anaerobic spore forming bacteria).

Spoilage of sliced or minced fresh meats at 20°C is rapid and maximum numbers are reached within 3 to 4 days (Fig. 3.1). The first signs of spoilage (i.e. off-odours) are detected within 2 days and surface slime

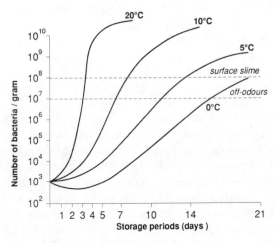

Fig. 3.1. Numbers of bacteria on fresh minced beef stored at different temperatures.

is evident at 3 days. It is interesting to note that, whatever the storage temperature, off-odours and slime production are always first evident when the total counts have reached *ca* 10^7 per cm^2, and 10^8 per g, respectively; in fact, this relationship holds true for meats in general.

The higher OR potential of minced or sliced meat favours proteolytic rather than putrefactive spoilage. The off-odours that appear are often referred to as 'sour' and are due to the formation of volatile acids such as formic and acetic acids; the surface slime is caused by heavy bacterial growth and the softening of meat structural proteins. The nature of the biochemical changes occurring at these higher temperatures has been little studied and far more work has been done on these changes at the chill storage temperatures used commercially.

3.2.3.2. Spoilage Under Cool Conditions

With storage temperatures below 20°C there is a tendency for the mesophilic bacteria to be overgrown by psychrotrophs although a small proportion of the former may still be capable of growth at 5°C (see Section 1.7.2). Sliced and minced meats held at 15 or 10°C develop off-odours after 4 to 5 days' storage and surface slime is evident at about 7 days (Fig. 3.1); the bacterial flora becomes progressively dominated by *Pseudomonas* spp. which represent over 95% of the flora at the time of spoilage (Gardner, 1965). A bacterium of interest which can grow to high numbers on the surfaces of meats is *Brochothrix thermosphacta*. It is unusual in being Gram positive in a largely Gram negative flora and being able to grow well on fatty tissues as well as lean down to 5°C or below. Some of the off-odours which develop in the early stages of spoilage are often attributed to this organism (Skovgaard, 1985).

3.2.3.3. Spoilage Under Refrigeration Conditions

At temperatures of 5°C and below a definite lag phase is apparent. The length of this phase depends on the storage temperature and extends for *ca* 24 h at 5°C and for 2 to 3 days at 0°C. In addition, at temperatures close to 0°C there is an initial fall in the numbers of viable bacteria which is probably due to the death or injury of many types at these low temperatures. As temperatures approach 0°C bacterial growth, once initiated, is much slower and progressively fewer types are capable of growing. Thus the period before the onset of the first signs of spoilage is extended and off-odours and slime production take *ca* 8 and 12 days, respectively, to develop at 5°C and *ca* 16 and 22 days at 0°C (Fig. 3.1).

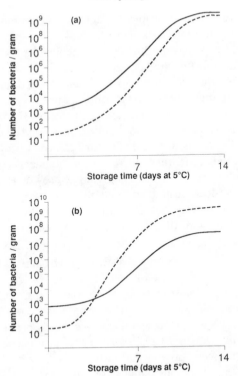

Fig. 3.2. Growth of bacteria on fresh meat stored at 5°C showing: (a) increasing proportion of pseudomonads (———, total count; – – – –, *Pseudomonas*); (b) changes in total psychrotrophs (– – – –) and total mesophiles (———).

Qualitatively, the spoilage flora is dominated to an even greater extent by pseudomonads in the later stages, due to their ability to grow faster than all other competing species at these low temperatures (Gill & Newton, 1977). The pseudomonads present at this stage are mainly (*ca* 75%) non-fluorescent types with *Ps. fragi* being by far the most common species isolated (Shaw & Latty, 1984); the main fluorescent species isolated are *Ps. fluorescens* and *Ps. lundensis*, the latter a recently recognized species apparently only isolated from meat (Molin & Ternström, 1986). At these low temperatures true mesophiles now only represent a small fraction of the total flora but the fact that the numbers of bacteria appearing on media incubated at 37°C increase during storage indicates that some mesophilic types must be growing on meats held at 5°C (Fig. 3.2).

Because of the strongly aerobic character of pseudomonads, growth is limited to the surface and to a depth of 3–4 mm in the underlying tissues. The spoilage pattern is thus largely independent of the size of cut or joint of meat and the spoilage of carcasses is likewise limited to the superficial layers; the growth of clostridia is inhibited at these low temperatures so that anaerobic putrefaction does not occur.

Under normal conditions of carcass meat storage the humidity is high and the surface layers remain moist. Over prolonged storage periods or at lower humidity levels drying of the surface layers becomes more pronounced and the consequent drop in a_w renders conditions more favourable for fungal growth. When fungal growth is induced in this way it is largely localized and only involves the most superficial layers; it can be trimmed away without any harm to the rest of the meat. Spoilage associated with the growth of moulds includes:

1. 'Whiskers': members of the genera *Mucor, Rhizopus* and *Thamnidium* produce mycelia giving a whiskery, white to grey appearance to the surface of beef carcasses.
2. 'Black spot': *Cladosporium herbarum* and *C. cladosporoides* can grow on a variety of meats even at temperatures as low as −5°C. It produces black spots of growth due to the darkly pigmented mycelium.
3. *Penicillium* spp. and *Cladosporium* spp. produce large numbers of yellow to green spores when growing on meats; these cause similarly coloured spoilage patches on the meat.
4. 'White spot': normally caused by the growth of *Sporotrichum carnis*.

3.2.4. Chemical Changes Produced by Bacteria in Chilled Meats

In considering the chemical changes which occur during spoilage at low temperatures it is necessary to differentiate between changes induced by enzymes naturally present in animal tissues and those induced by bacterial enzymes. In fact, such a differentiation is difficult to make and this has meant that until the 1970s the changes induced by bacteria *per se* were poorly understood. Changes occur in free amino acid levels in stored meats and these are brought about by either or both of these sets of enzymes. Bacteria initially attack glucose, amino acids and other low molecular weight compounds such as nucleotides rather than the primary meat proteins (Gill, 1976). These changes are accompanied by a marked rise in pH from *ca* 5·6 to as high as pH 8·5 due, primarily, to the formation of ammonia and certain amines by bacterial degradation of amino acids; in consequence pH values have found a use in the assessment of the keeping qualities of meats.

Proteolysis, the breakdown of primary meat proteins, only occurs at a relatively late stage of storage and only becomes evident after the onset of other signs of spoilage. This protein breakdown results from the activities of bacterial proteases and is first noted near the surface of the meat; in time, however, these enzymes penetrate deeper into the tissues (Tarrant *et al.*, 1971). The pseudomonads are primarily responsible for this proteolysis which occurs when their numbers are already in excess of 10^9 per cm^2 (Dainty *et al.*, 1975). Large numbers of volatile compounds can be produced as a result of this growth. Compounds detected most consistently with stored meats include methanol, acetone, methyl ethyl ketone, dimethyl sulphide and dimethyl disulphide; ammonia and certain amines (*viz.* methylamine, dimethylamine and trimethylamine) are also produced whilst hydrogen sulphide is also a likely common end-product (Dainty *et al.*, 1983). It is now recognized that the diamines putrescine and cadaverine are also formed during the breakdown of certain amino acids, the production of the former being specifically attributed to pseudomonads (Edwards *et al.*, 1983). *Ps. fragi* and certain other pseudomonads also produce 'fruity' odours by the formation of methyl and ethyl esters of short chain fatty acids (Edwards *et al.*, 1987a). Many pseudomonads are also active lipase producers at chill temperatures and they have been implicated in the hydrolysis of fats, a process which results in the production of undesirable flavours consequent upon the formation of fatty acids. Spoilage bacteria may also produce lipoxidases which accelerate the oxidation of unsaturated fatty acids to produce aldehydes and in this way they contribute to the problem known as 'oxidative rancidity'. Oxidative rancidity is caused normally by the slow uptake of oxygen and is not primarily of microbiological origin; it has also yet to be established conclusively that pseudomonads have a significant role in the hydrolysis of fats. It is known that the bacterial spoilage of surface fat tissues on fresh meat follows a similar pattern to that of protein breakdown with bacterial attack on lipids again being delayed until an advanced stage of spoilage. However, because of the low glucose concentration in fats, amino acids are dissimilated earlier so that off-odours are produced with lower numbers of bacteria (*ca* 10^6 per cm^2) than with lean meat (Gill & Newton, 1980). At the same time as amino acids are being utilized glycerol, a major component of fats, is also being metabolized with the production of acetoin; *Brochothrix thermosphacta* may well have an important role in this reaction (Macaskie *et al.*, 1984). However fatty tissues have a lower a_w than lean and this reduces the growth of bacteria so that, on balance, the fat portions are no more perishable than lean.

3.3. SPOILAGE OF CURED MEATS

3.3.1. Curing Agents

The principal ingredients of any traditional cure are salt, sodium nitrate and sodium nitrite although nitrite has largely replaced nitrate in order to effect greater control over the processing; other possible additions to the cure include sucrose and ascorbic acid. Curing can be combined with other processing techniques including smoking, heating and fermentation so that so-called cured meats include a variety of diverse products prepared under different conditions (Egan & Roberts, 1987).

Salt is present as a preservative agent which acts by reducing the a_w of the meat. *Pseudomonas* spp., important in the spoilage of refrigerated meats, are particularly sensitive to lowered a_w levels (see Table 1.5) and this partly accounts for the relative stability of cured meats. The role of nitrate in spoilage control is not entirely clear although it proved useful in the development of the red colour of meats and it is reduced by bacteria to nitrite. The nitrite also helps to maintain the colour and flavour of meat but its principal role is to prevent growth of germinating spores. On its own nitrite is not very active but its effectiveness is enhanced by factors such as the salt concentration, pH and storage temperature which are all important in determining the stability of cured meats.

3.3.2. The Curing Process

Curing of non-comminuted meats can be performed by one of three basic procedures. In the first, dry curing, the curing agents are rubbed onto the surface of the meat whilst in the second process, pickling, the meats are immersed in a pickle of the curing agents in water. In both these methods the meats are held at 3–4°C until the agents have penetrated into the centre of the meats. The low temperatures involved reduce the chances of putrefactive anaerobes growing but nevertheless spoilage problems can occur due to the slow penetration of the brine. These problems are largely overcome in the third procedure, stitch pumping, introduced in more recent years. In this process the pickle is injected into the deeper tissues by means of long needles, with several holes along their length, which are arranged in rows so that hundreds of separate injections are made. A variation of this technique is to pump the pickle into the vascular system which channels it to the various parts of the body. In both these injection procedures the meats are subsequently immersed in a pickling brine. To increase the shelf-life of the bacon a dry salting process has been developed whereby bacon sides are passed through a cloud of dry salt after removal from the curing brine (Gardner, 1983).

Bacon, ham, salted beef and salted pork are the principal non-comminuted meat products of the curing process and of these only the first will be considered in any detail.

3.3.3. The Microbiology and Spoilage of Bacon and Ham

3.3.3.1. Unsmoked Bacon
Bacon is conventionally cured as whole sides and when cured in this way the term 'Wiltshire bacon' is used.

Brines used in the curing of bacon usually contain 20–27% salt and this has a profound effect on the types of organisms present. The microbiological flora of a typical Wiltshire curing brine is dominated by micrococci which are able to tolerate the low a_w environment. These organisms also become predominant on the sides during the curing process so the normal heterogeneous flora of fresh meat is largely replaced by this group. As well as growing on bacteriological media containing 20% NaCl, these *Micrococcus* spp. are psychrotrophic growing at 4°C and usually at 0°C (Patterson, 1963). Another of their important characteristics is their ability to reduce nitrate to nitrite and they therefore play an important part in the curing process where sodium nitrate is included in the curing brine.

After curing, which lasts from 4 to 14 days, the sides of bacon are drained and allowed to mature for a further 5–10 days at 4°C; during these processes there is a gradual reduction in the salt concentration of the bacon down to levels typically below 10% (w/v based on the aqueous phase). In fact, nowadays, higher salt concentration bacons may be regarded as bacons with a final salt concentration of >5%; at maturation with such bacons bacterial counts range from 10^4 to 10^6 per cm^2 and whilst the predominance (>60%) of micrococci is maintained there is an increase in the proportion of Gram negative bacteria, particularly *Acinetobacter* and *Vibrio* spp., on the sides. This change in flora does not occur on the rind probably due to its lower a_w (Gardner & Patton, 1969). During subsequent storage of the bacon there is a gradual increase in bacterial numbers up to a maximum of *ca* 10^8 microorganisms per cm^2 after 2–3 weeks at 10°C. Qualitatively, the flora now consists of approximately equal proportions of members of the genera *Micrococcus, Vibrio* and *Acinetobacter* although if the bacon is held under chill conditions vibrios tend to predominate, particularly on the meat surface. The high count on the surface of a side of bacon is associated with slime formation and is commonly caused by halophilic vibrios but there should be no obvious deterioration in the quality of the bacon as changes within the meat are normally minimal (Gardner, 1971). One such change is that of bone taint which is caused mainly by vibrios and micrococci. It is characterized by an unpleasant smell noticed when deboning the product; it results from faulty curing or curing meats of too high a pH.

When spoilage does eventually occur it is commonly caused by micrococci and vibrios together with a variety of yeasts and moulds, including *Torulopsis* and *Aspergillus* spp., respectively. Off-odours and flavours are mainly associated with the fat rather than the lean meat although in the latter micrococci can induce proteolytic changes. Hydrolysis of fats is caused by bacterial and tissue lipases whilst oxidative rancidity causes yellowing of the fat.

With lower salt (2–5 % NaCl) bacons, now so popular in Europe and in the United States, the microbiological flora is more varied and less salt tolerant forms figure more extensively. With such cures and bacon, bacteria from the genera *Lactobacillus, Leuconostoc* and *Streptococcus* (i.e. the 'lactic acid bacteria') are more readily isolated and it is these organisms together with micrococci that are now responsible for the microbiological spoilage. With the so-called lactic acid bacteria dominating the spoilage flora, the characteristic spoilage is now souring resulting from the acid produced by these organisms. As the salt concentration drops there is an ever increasing incidence of Gram negative bacteria and, if the salt concentration drops too low, pseudomonads as well as vibrios and acinetobacters are able to grow.

3.3.3.2. Smoked Bacon

As well as providing a desirable flavour and colour, smoking also contributes to the preservation of bacons. The effect is both bacteriostatic (i.e. stopping bacterial growth) and bactericidal (i.e. killing bacteria) although fungi are also affected to some degree.

Smoke acts in two ways: first, by drying the surface layers it further reduces the a_w and accentuates the effects of salt; second, it impregnates the tissues with chemical preservatives such as formaldehyde and phenols which inhibit bacterial growth. In addition, large numbers of bacteria are killed on the bacon during the smoking process, the numbers depending on the time and type of smoking (Handford & Gibbs, 1964). Micrococci, yeasts and moulds are most frequently isolated on smoked bacon although lactic acid bacteria are more likely to predominate where liquid smoke is used. As these 'lactics' are responsible for a less offensive souring spoilage than that associated with micrococci, and at a later stage, the shelf-life of the product is extended.

3.3.3.3. Ham

The processes involved in the curing of hams are similar to those employed for bacon except that sugar is often added to the cures. This can be attacked by bacteria, particularly lactobacilli, and the fermentations produce souring of various types; however, it has been suggested that lactobacilli may be useful in maintaining the stability of curing brines by preventing excessive pH rise.

In general the microorganisms found on hams are similar to those on bacon and the flora consists mainly of micrococci, streptococci and lactobacilli, the proportions depending on the salt concentration and period of storage. Higher salt concentration hams also tend to support the growth of a greater proportion of yeasts and possibly moulds.

3.3.3.4. Comminuted Cured Meats

There is a wide variety of products coming under this heading including fermented raw sausages (e.g. salami and cervelat) and cooked smoked sausages (e.g. Frankfurters and Bologna-type sausages). The former group are comminuted cured meats which are prepared by adding curing salts, sugar and spices to meats which have been cut up. Mixed starter cultures of lactobacilli and micrococci or, possibly, a pure culture of a *Pediococcus* sp. are added so that the meats undergo a fermentation process during their curing and drying. These lactic acid bacteria, relatively resistant to the inhibitory effects of the curing salts grow rapidly (Egan & Roberts, 1987); this growth is accompanied by sugar fermentation causing a drop in pH to *ca* 4.8, which coupled with the high salt concentration (>6%), gives stability to the product. Spoilage is usually caused by moulds (e.g. *Aspergillus* spp.) able to tolerate the low a_w conditions.

Frankfurters and similar types of sausage are typically smoked and cooked to an internal temperature of 65–75°C, the aim being to kill vegetative bacterial cells; the cure will again contain both salt and nitrite amongst other ingredients. These products have high a_w's and it is the presence of the nitrite which prevents the germination of the spores of *Clostridium* and *Bacillus* spp. which will inevitably resist the heating process. Spoilage is usually caused by the growth of lactic acid bacteria which are usually post-cooking contaminants (Gardner, 1983).

Vacuum-packaging is widely used to prolong the shelf-life of these comminuted meat products.

3.4. SPOILAGE OF VACUUM-PACKED MEATS

3.4.1. Types of Packaging Materials

The packaging materials used by the food industry vary from the highly impermeable, required for vacuum-packaging to the highly permeable and from the opaque to the transparent. Materials may consist either of single components such as polyethylene (polythene) or polyvinyl chloride (PVC), or of multiple components where the properties required of the packaging material cannot be satisfied by a single film. Where multiple components are used the packaging material is either made up of laminates of two or more films, or are prepared by co-extruding polymers together to form a single material. An example of a multi-component packaging material is Cryovac BBI which is constructed from a triple co-extrusion of EVA/PVDC/EVA (Taylor, 1985). Alternatively coatings may be applied to one or both sides of a single component film such as cellulose.

Most films are relatively impermeable to water vapour since weight losses must be minimized during storage. Gas permeability may vary enormously (see Table 3.1) and this can have a dramatic effect on the shelf-life and spoilage characteristics of the packaged meat.

3.4.2. Influence of Packaging Materials on the Microbiological Flora

Fresh meats are normally packed in oxygen-permeable films in order to conserve the bright red colour of oxygenated myoglobin. Conversely, cured meats are packed in oxygen-impermeable films in order to prevent fading of the colour due to oxidation. However, chilled beef in particular is now widely distributed, as boneless joints, in boxed vacuum packs rather than as carcasses (Taylor, 1985). In this form there is a much improved shelf-life and, in addition, weight losses due to surface drying

Table 3.1

Permeabilities of Meat Packaging Materials to Oxygen and Carbon Dioxide[a]

Material	Permeability[b] to	
	Oxygen	Carbon dioxide
Polyethylene (low density)	8 500	44 000
Polyethylene (high density)	1 840	7 900
Polypropylene	3 000	7 900
Polyvinyl chloride (PVC)	4 200	17 000
Polyester (PET)	79	240
Polyvinylidene chloride (PVDC)	10	53
Ethylvinyl acetate (EVA)	12 000	38 000
Cryovac BBI (50 μm)	20	80
Nylon/polyethylene (15/37 μm)	90	600

[a]From Taylor (1985).
[b]Results expressed as cm^3/m^2/24 h/atm gas for film thicknesses of 25 μm except where stated otherwise.

are restricted. Vacuum-packaging is also becoming more popular in the retail trade in spite of justifiable criticisms of loss of colour, although this is less of a problem with pork. The compensations are that the beef has a considerably extended shelf-life and a rapid regeneration of the normal red colour occurs once the pack is opened or the meat is re-packed in an oxygen-permeable film.

The availability of oxygen within the pack has some effect on the microbial flora; meats have a high demand for oxygen as do many microorganisms, so oxygen levels can be readily depleted in the more impermeable packs without a vacuum being applied. At the same time, carbon dioxide (CO_2) levels tend to increase in such packs at a rate dependent upon the permeability of the film. However, packaging materials are more permeable to CO_2 than to oxygen so that a film of low permeability may exclude oxygen but still allow CO_2 to escape and maintain the vacuum in the pack. There are advantages in increasing the concentration of CO_2 within the pack as it is inhibitory to many microorganisms including moulds and pseudomonads, the latter group being the dominant flora on spoiled fresh meats. The lactic acid bacteria are much more resistant to higher CO_2 levels and so may be expected to figure in the spoilage pattern associated with packaged meats (Newton & Rigg, 1979).

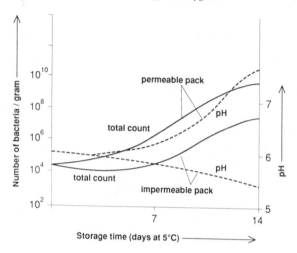

Fig. 3.3. Comparison of numbers of bacteria and pH in fresh minced meat stored at 5°C under permeable and impermeable conditions.

Another factor affecting the bacterial spoilage pattern is that of a_w which is likely to be high within the pack wrapped even in a permeable film. Since there is little or no water loss from the package, bacterial growth is not reduced by a drop in a_w but the effects of a_w are subordinate to those of carbon dioxide and oxygen within the pack.

3.4.3. Spoilage of Packed Fresh Meats

When packaged meat is stored at warm temperatures it undergoes normal putrefactive spoilage. Such meat is therefore always stored at refrigeration temperatures and only such storage will be considered.

The growth of microorganisms is retarded on vacuum-packed fresh meats stored at 0–5°C and a lag phase of *ca* 4 days to 4 weeks (at 0°C) is usually observed. Subsequent growth is slow so that a storage life of 10–12 weeks is possible at 0–1°C (Egan & Roberts, 1987); by this time the final total count of *ca* 10^7 per cm^2 (or per gram in minced meat) will be reached which is only about 1% of that obtained with permeable films (Fig. 3.3). Qualitatively the microbial flora in the impermeable pack becomes dominated by lactic acid bacteria (mainly lactobacilli and leuconostocs) which represent 50–90% of the total flora at the end of storage; *Brochothrix thermosphacta* and pseudomonads are typically only a minor part of the flora (Fig. 3.4).

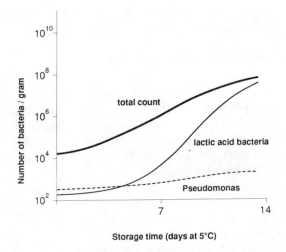

Fig. 3.4. Growth of bacteria in vacuum-packed fresh meat.

The lactic acid bacteria attack carbohydrates preferentially but due to the small amount present in meat relatively little acid can be formed and, as a consequence, only a slight fall in pH is evident (Fig. 3.3). This means that even at the maximum cell densities of these organisms relatively little spoilage is apparent; spoilage is associated with sour or cheesy odours due to the formation of fatty acids, these principally being acetic and butyric acids. Other end-products formed include the diamines cadaverine, putrescine and tyramine, the last named being produced as a direct result of the activities of specific lactobacilli (Edwards *et al.*, 1987*b*).

With increasing permeability of the package film there is a gradual change in the spoilage flora to one that comprises a high proportion of pseudomonads; thus the spoilage changes are more typical of unwrapped fresh meat (Fig. 3.3). The relative proportions of pseudomonads, lactic acid bacteria and *B. thermosphacta* are dependent principally on the concentration of carbon dioxide and oxygen within the pack. *B. thermosphacta*, like pseudomonads, decreases in numbers in impermeable packs due to its sensitivity to the higher concentrations of CO_2 which develop in such packs. If rather less impermeable films are used *B. thermosphacta* can thrive so that it often grows to high numbers with lamb and pork in particular when wrapped in intermediate permeability films which accumulate some carbon dioxide but still contain low levels of oxygen (Dainty *et al.*, 1983); in such packs *B. thermosphacta* often represents 20–30% of the total bacterial count at spoilage.

3.4.4. Spoilage of Vacuum-Packed Bacon

Mature bacon normally carries 10^4–10^6 bacteria per cm^2 which are subsequently distributed over the cut surfaces at the time of slicing before packaging. Because of this the bacon has a rather limited shelf-life but it can be stored at ambient temperatures without loss of colour.

As previously mentioned, micrococci are the main types of bacteria developing on the higher salt concentration bacons when stored at ambient temperatures (*ca* 20°C). When such bacons are stored under vacuum-packaged conditions, micrococci still predominate and rise in numbers to about 10^7 per gram after about 9 days' storage. Obvious spoilage is apparent at about 2 weeks and is characterized by a rancid odour. With lower salt concentration bacons the flora is initially more mixed but counts again reach a peak at about 9 days and spoilage follows a few days later. At this time the flora consists of approximately equal proportions of micrococci, streptococci (i.e. enterococci) and other lactic acid bacteria (i.e. lactobacilli and leuconostocs) (Tonge *et al.*, 1964); with high storage temperatures (e.g. 25°C) Gram negative forms (e.g. *Vibrio* spp. and *Proteus* spp.) may be responsible for putrefactive spoilage.

The advantage of chill storage is that it takes between 3 and 4 weeks for bacterial counts to reach a maximum and hence obvious spoilage may be delayed for up to 5 weeks. Qualitatively the changes in the microbial flora during storage reflect the influence of salt concentration and the flora is essentially similar to that for vacuum-packed bacon stored at higher temperatures (Dempster, 1972; 1973); lower temperatures tend to favour the growth of lactic acid bacteria at the expense of micrococci, a tendency that increases with smoked bacon (Gardner, 1983).

3.4.5. Modified Atmospheric Packaging

Vacuum-packaging is a form of modified atmospheric packaging (MAP) since the atmospheric gas concentrations are altered before storage. Other forms of MAP include atmospheric pressure reduction to achieve hypobaric storage conditions and atmospheric modification by the use of gas blends (Genigeorgis, 1985). The latter are now widely used in shipping and storage containers but the proportion of each gas changes with time and this may create problems. Greater control of the atmosphere can be achieved by periodically refilling the containers in order to maintain the desired blend of gases; this is known as controlled atmospheric packaging (CAP).

Because of the inhibitory effects of carbon dioxide on the growth of bacteria it is always included in gas blends but it can also be used in the pure form. In the latter case, although causing discolouration of fresh meats, carbon dioxide considerably extends the shelf-life of even retail cuts of meat. Thus Blickstad & Molin (1983) found that pork could be stored in carbon dioxide for up to 3 months at 0°C (*cf.* 2 weeks for aerobic packs) before obvious spoilage occurred, this being associated with a total dominance of lactobacilli (10^7 per cm^2). The colour changes induced in red meats in particular caused by too high a concentration of carbon dioxide can be offset by blending the gas with either oxygen (e.g. 40% CO_2 : 60% O_2) or oxygen and nitrogen, the latter being used solely as a 'filler' gas. It must be remembered that with MAP oxygen levels decrease during storage whilst carbon dioxide levels increase due to the activities of respiratory enzymes. Bartkowski *et al.* (1982) found that a minimum initial concentration of 15% carbon dioxide was critical in controlling bacterial growth. They concluded that a gas blend of 75% O_2 : 15% CO_2 : 10% N_2 was ideal since it also gave the best appearance to the raw steaks used in the experiments; in fact the recommended mixture was far more effective than conventional vacuum-packaging in containing bacterial growth.

Since MAP foods are necessarily stored at refrigeration temperatures questions have been asked about the growth of psychrotrophic pathogens such as *Clostridium botulinum* type E, *Listeria monocytogenes* and *Yersinia enterocolitica*. *C. botulinum* poses a special problem in packaged fish (see Section 3.6.7) but the other two pathogens are often isolated from meats. It would appear that both are capable of growth in certain gas mixtures but not in others, and that various other factors such as the pH and exact storage temperatures also play a significant part as growth determinants. In a detailed review Farber (1991) pointed out that MAP foods often suffer from the weakness that the growth of aerobic spoilage bacteria is inhibited so that spoilage is not apparent and yet growth of pathogens could have occurred; furthermore, the extended shelf-life of MAP foods may allow dangerous levels of pathogens to develop. There is some cause for concern, therefore, at the rapid proliferation of MAP products and Farber correctly concludes that additional research is needed to ensure their microbiological safety.

3.4.6. Sous Vide

It might be pertinent at this point to touch on 'sous vide' a form of vacuum packaged cooking which has grown enormously in popularity in the last few years. In essence sous vide involves the preparation of raw and/or partially cooked food (fish is especially popular), its packaging in impermeable plastic pouches which are then sealed under vacuum and subsequently 'lightly' cooked (i.e. 'pasteurized'), cooled, and then stored at a recommended temperature of 0–3°C for a period of up to 14 days or more depending on the food; products are reheated before consumption.

The dangers of psychrotrophic pathogens mentioned in the previous section are equally applicable again here. The aerobic spoilage bacteria are generally extremely sensitive to heat whereas the pathogens show much greater resistance, so that their growth may be possible with inadequately heated foods. Prolonged storage at 3°C or under of lightly cooked foods could prove hazardous but if slightly higher storage temperatures (e.g. 10–12°C) are maintained—unfortunately always a possibility in supermarket display cabinets and the home refrigerator—the range of pathogenic bacteria now capable of growth increases alarmingly to include, for example, salmonellas, *Staphylococcus aureus* and *Bacillus cereus*. Recommended storage temperatures must be strictly observed whilst some of the extended storage times suggested for certain sous vide products must be viewed with extreme caution.

3.5. SPOILAGE OF POULTRY

The term 'poultry' applies to a range of domestic fowls and whilst the following paragraphs deal exclusively with chickens, the general principles apply to other commonly eaten fowls such as turkeys and ducks. In addition, the discussion will be limited to commercially dressed chickens since this is now accepted as the normal method of marketing.

3.5.1. Effects of Processing on the Microbiological Flora

When live birds are brought into the processing plant they are harbouring large numbers of microorganisms of many different types on their feathers and feet and in their intestines. The various stages in processing

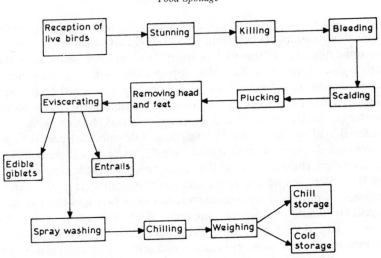

Fig. 3.5. Stages in the processing of poultry.

are outlined in Fig. 3.5 and only those that have a significant effect on the microbiological condition of the carcass will be discussed.

Scalding to loosen the feathers is performed by immersing birds for 30 s in a tank of hot water (*ca* 55°C). There is a reduction in the numbers of organisms on the carcass due to the washing effect and to the destruction of heat-sensitive bacteria including, in particular, psychrotrophic spoilage bacteria. Even where lower scalding temperatures are needed (*ca* 50°C for birds to be air chilled; see below) psychrotrophic bacteria are still destroyed.

The mechanical feather pluckers increase the bacterial load on the skin of birds and also cause cross-contamination and 'aerosol' problems, particularly with *Staphylococcus aureus*. Evisceration also increases the bacterial load on the skin by spreading faecal types onto the surface; such bacteria can be easily transferred to other carcasses again causing cross-contamination problems.

The microbial load reaches a maximum immediately prior to spray washing. This process, which produces an approximately 90% reduction of microorganisms per carcass, is followed by chilling which is performed by one of two methods (Mead, 1982). In the first of these immersion or 'spin' chillers are used. The system utilizes one, two or three units in series, each consisting of a large tank through which chlorinated water flows continuously in one direction; carcasses can either flow in the same direction as the cooling water (through-flow systems) or they can be moved mechanically in the opposite direction by using a helical screw (counter-flow systems); additional cooling is obtained by the addition of ice to one or more of the units. In addition to effectively cooling the birds, the chiller system, if properly operated, further reduces the bacterial load on the carcass by some 90%. Efficiency is dependent upon a controlled flow of chlorinated water (Mead & Thomas, 1973). Counter-flow chilling is by far the better process because carcasses leaving the system are washed in the cleanest water. In spite of this there is evidence of microbial cross-contamination in chillers and this can be of particular significance where salmonella contaminated birds are concerned. To reduce further this risk requirements have been introduced on minimum water usage per carcass and minimum water temperatures at entry and exit points; these requirements are now included in Codes of Practice operable within the EC (Mead, 1982); as a further protective measure it has been suggested that carcasses should be chilled in individual bags (Shackleford, 1988).

The second method of chilling is by air. This process, used where chickens are to be retailed chilled, tends to dry the skin and hence retard the growth of psychrotrophic spoilage bacteria. A further advantage of this method is that cross-contamination is reduced although air-chilled birds are likely to be more heavily contaminated than water-chilled poultry (Mead, 1982).

After the preliminary chilling, microbial counts on chicken skin range from 5×10^3 to 1×10^5 per cm^2 whilst counts in the visceral cavity are usually $<1 \times 10^4$ per cm^2; the lowest counts will be found in hygienically and efficiently operated processing lines. Qualitatively the microbial flora is extremely mixed at this stage and amongst the more commonly isolated bacterial groups will be micrococci, flavobacteria, various intestinal types, such as *Escherichia, Enterobacter* and *Streptococcus* spp., and *Acinetobacter* spp. (Barnes & Thornley, 1966).

3.5.2. Spoilage of Chickens Held at Chill Temperatures

When chickens are stored at chill temperatures most of the microbial growth occurs on the skin and, to a lesser extent, in the lining of the visceral cavity. Over a period of *ca* 10 days the number of bacteria increases to a maximum on the skin of between 10^9 and 10^{10} per cm^2. This increase in numbers is accompanied by off-odours (*ca* 10^7 bacteria per cm^2) and copious slime production (*ca* 10^8 per cm^2) and a rise in pH to *ca* 7·5 and so, as might be expected, there are many similarities between the spoilage of raw chicken and other meats.

Furthermore, as with chilled meats, the spoilage flora of refrigerated chickens becomes dominated by *Pseudomonas* spp. (both fluorescent and non-fluorescent types). In fact at advanced stages of refrigerated spoilage chicken skin often fluoresces when illuminated with ultraviolet light due to the presence of large numbers of fluorescent pseudomonads. At the time of spoilage pseudomonads represent 70–80% of the flora but also present in smaller numbers are *Acinetobacter* spp. (*ca* 10% of the flora) and *Alteromonas putrefaciens* (Barnes & Thornley, 1966; Barnes, 1976). The latter organism is particularly interesting as its growth is much faster on chicken leg tissue (pH 6·5) than on breast (pH 5·8); thus the predominance of pseudomonads is less pronounced on the former tissue (Barnes & Impey, 1968).

Using gas chromatography and mass spectrometry, attempts have been made to identify the off-odours produced during the refrigerated spoilage of chickens. Freeman *et al.* (1976) found that 22 volatiles were produced during spoilage under chill conditions and 15 of these were produced as a result of microbial attack on muscle tissue and were responsible for or associated with the characteristic off-odours of the later stages of spoilage. Amongst these 15 compounds were hydrogen sulphide, methyl mercaptan, dimethyl sulphide, methyl acetate, ethyl acetate, methanol, ethanol and benzaldehyde; the large variety of compounds identified clearly illustrates the complexity of this problem in terms of the role of microorganisms in the spoilage process.

3.6. SPOILAGE OF FISH AND SHELLFISH

3.6.1. Bacteriology of the Newly Caught Fish

Most of the microbiological studies dealing with fish have concentrated on marine varieties and only these types will be considered. It is generally accepted that the flesh of newly caught healthy fish is sterile but bacteria are found in variable numbers in three sites on the fish: the slime coat, the gills and the intestines. Numbers on skin have been reported to range from 10^3 to 10^5 per cm^2, on gills from 10^3 to 10^4 per g

Food Microbiology and Hygiene

tissue and in the intestines from 10^2 to 10^9 per ml contents (Hobbs, 1983). However, it has been claimed that numbers of bacteria on fish in unpolluted waters are at the lower end of the ranges quoted and that the higher numbers result from the poor hygienic standards on board ship during initial handling (Huss *et al.*, 1974).

Variations in marine environments affect the types of bacteria present on the skin and gill surfaces of newly caught fish. Thus in the cooler seas of the northern hemisphere the bacterial flora is dominated by psychrotrophic Gram negative rods. Liston (1957) found the flora of North Sea flatfish (skate and lemon sole) to comprise *Pseudomonas/Alteromonas* spp. (60%), *Acinetobacter/Moraxella* (14%) with the majority of the remainder being other types of Gram negative rods; Georgala (1958) found that the bacterial flora of North Sea cod consisted of pseudomonads (44%) and acinetobacters (32%) together with a variety of miscellaneous types. In the warmer waters off India, the east coast of South Africa, Australia and in the Adriatic the proportion of mesophiles increases and micrococci and coryneforms become more important. Thus Gillespie & Macrae (1975) found micrococci (49%) to predominate on newly caught fish from Australian waters; pseudomonads (only 18%), coryneforms (12%) and acinetobacters (only 9%) were the other main groups isolated.

3.6.2. The Effect of Initial Processing and Storage in Ice on Board Ship

Gutting of the fish on board ship tends to spread the intestinal flora over the surface of the fish. The principal organisms found in the intestines are *Vibrio* spp. although many other genera are represented. Fish are then washed in sea water and either packed in crushed ice or possibly frozen or stored in refrigerated sea water. The conventional method is to pack in ice and the effects of this form of storage will be considered in some detail.

The ice in which the fish are to be preserved is itself usually contaminated (*ca* 10^3 per ml of ice melt water) and, in addition, the holds of the fishing vessels normally have an indigenous flora composed of *Pseudomonas* and *Acinetobacter* spp. (Shewan, 1961). Thus when fish, which already contain a fairly high proportion of pseudomonads and alteromonads, are placed in ice they are likely to be further contaminated with these organisms. Partly because of the relatively high proportion of pseudomonads and alteromonads on the skin of fish initially and in the storage environment and partly because of the relatively high pH of many fish species, spoilage in ice is relatively rapid. Even under the best

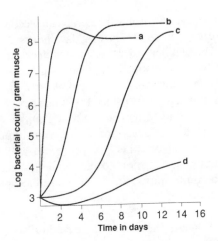

Fig. 3.6. Numbers of bacteria on fresh fish stored at: (a) 25°C, (b) 7°C, (c) 0°C, and (d) —4°C.

conditions with a temperature of 0°C total counts may reach a maximum of $>10^8$ per g after 12 to 14 days' storage (Fig. 3.6). Occasionally in the holds of vessels temperatures may reach 6 or 7°C and under these conditions maximum numbers are reached within 5 or 6 days. Spoilage of fish is thus much faster than spoilage of raw meat, maximum numbers of bacteria in the latter being reached only after 9 or 10 days' storage at 7°C.

The actual quality of fish landed at the ports is dependent upon the time it has been held in ice and the hygienic conditions on board the fishing vessels. Average bacterial counts on fish taken from trawler holds on docking are *ca* 10^6 per cm^2 of skin surface and, qualitatively, the proportion of *Pseudomonas* spp. is higher than in the newly caught fish. During prolonged storage in ice pseudomonads and alteromonads become predominant groups and represent 80–90% of the spoilage flora at peak numbers (Hobbs, 1983). Even with fish caught in warmer waters the pseudomonads and alteromonads are predominant at the time of spoilage and therefore overgrow the initial mainly Gram positive flora.

3.6.3. The Effect of Handling Ashore

After landing, the fish may be left on the quay for many hours un-iced in boxes or kits. Under such conditions the temperature of the fish will rise and growth of the psychrotrophic bacteria will become more rapid so that a 10-fold increase in numbers in a few hours can be anticipated (Shewan, 1961). The wooden fish boxes still widely used ashore harbour vast numbers of bacteria, counts often being in excess of 10^6 per cm^2 even on 'clean' boxes; however, it is probable that these bacteria have little effect on the spoilage of fish since the time spent by fish in boxes is usually limited to less than 12 h. The fish may then be re-boxed, often in ice, and transported to the processor. Here, depending on the type of fish, they may be filleted or otherwise processed. All these events, including filleting and final transport to the retailer, affect the bacteriological flora which becomes more varied and more mesophilic in character with increasing degrees of handling (Shewan, 1971).

3.6.4. Chemical Changes Induced by Bacteria in Fish

Like meat, fish may be spoilt by naturally occurring autolytic enzymes or by bacterial activity. Spoilage is principally due to the activities of enzymes produced by Gram negative rods, particularly *Pseudomonas/ Alteromonas* strains. These organisms invariably become predominant during prolonged storage of fish whether or not stored in ice.

The role of *Pseudomonas*-like organisms in the spoilage of fish has been studied in some detail. Adams *et al.* (1964) characterized potential spoilers by inoculating them as pure cultures into sterile muscle press juice from sole and measuring their ability to produce off-odours, volatile reducing substances and trimethylamine. They found that only about 10% of the initial flora were spoilers using the above criteria and that the proportion of spoilers was never more than 30% of the total flora during storage at 5°C. The spoiling bacteria were later identified as being mainly pseudomonads (many of these would now be classed as *Alteromonas* spp.), acinetobacters or vibrios; it was stressed that even though pseudomonads were the predominant group only a small proportion were active spoilers (Lerke *et al.*, 1965). Similar results were obtained with cod by Shaw & Shewan (1968) who found that the proportion of active spoilers to the total viable population did not change significantly during spoilage and always remained below 25%; again *Pseudomonas/Alteromonas* spp. were the principal group but in this work *Acinetobacter/Moraxella* spp. and *Vibrio* spp. were not implicated as spoilers.

Spoilage bacteria first utilize low molecular weight compounds such as nucleotides and amino acids present in fish muscle and it is the breakdown of these materials which is responsible for off-odours and other spoilage effects; thus, as with meats, protein plays a minor role in spoilage (Lerke *et al.*, 1967). Studies in which pure cultures of spoilage-inducing pseudomonads have been inoculated into blocks of sterile fish muscle indicate that different strains generate totally different odours. Some of the volatiles produced were 'fruity' and probably esters whilst others were 'sulphidy' (Herbert *et al.*, 1971). More definitive studies using gas-liquid chromatography have identified some of the principal volatiles in spoiling fish as methyl mercaptan, dimethyl sulphide, dimethyl disulphide, hydrogen sulphide, trimethylamine, ethyl acetate and ethanol (Miller *et al.*, 1973a, b); it should be noted that many of these volatiles have also been associated with meat spoilage (see Section 3.2.4) and that they need only be present in very low concentrations to produce obvious off-odours.

Trimethylamine formation is of great interest since it has long been used as a measure of the degree of fish spoilage (Malle *et al.*, 1986). Trimethylamine is produced by the reduction of trimethylamine oxide which is present in appreciable quantities in most marine fish but not in other animals. It is now clear that alteromonads, and *A. putrefaciens* in particular, are mainly responsible for this conversion which enables these essentially aerobic bacteria to satisfy their oxygen requirements and continue to grow in the depleted oxygen conditions developing in the fish tissues; this conversion may well account, in part, for the greater perishability of marine fish compared to other animal fresh foods (Easter *et al.*, 1982; Hobbs, 1987).

3.6.5. Salted Fish

Two basic techniques may be used in the salting of fish, either 'dry' or 'wet' (i.e. pickling) salting. In the former, used for non-fatty fish like cod, salt is spread over the surface of the fish and layers of fish are interspersed with salt layers. In wet salting, used for fatty fish like herring, the fish are pickled in salt in barrels; combined 'dry' and 'wet' salting can be used and variations in the latter can be achieved by using suitable spices and vinegar.

Irrespective of the method used, the added salt lowers the a_w of the fish and, as has already been discussed with cured meats, this has a profound effect on normal spoilage patterns. The normal predominantly Gram negative flora of fish is relatively sensitive to high salt concentrations and so bacterial numbers decline. The final flora depends on the strength of the cure. In many instances micrococci become predominant but in the highest salt concentrations a specialized group of bacteria can sometimes cause spoilage problems (Graikoski, 1973). These bacteria, represented by the genera *Halobacterium* and *Halococcus* are termed 'extreme halophiles' and tolerate salt concentrations in excess of 20% NaCl; indeed, they require 10–15% NaCl in culture media before they will grow. The spoilage produced is a red discolouration on the surface of the fish ('pink' fish) when stored at too high a temperature; it is caused by the growth of these organisms which are pigmented red (Shewan, 1971).

Another type of spoilage known as 'dun' is caused by a halophilic mould, *Sporendonema expizoum*; it is associated with the formation of peppered spots which are visible on the fleshy side of salted fish, particularly cod, the spots ranging in colour from chocolate to brown and fawn (Van Klaveren & Legendre, 1965). Lightly salted cod is also susceptible to 'sliming' which is caused by the growth of the indigenous Gram negative flora (primarily pseudomonads); this condition is characterized by a slimy beige coloured layer on the surface of the fish (Graikoski, 1973).

There are other salted fish products, popular in Scandinavian countries, in which some degree of breakdown (i.e. fermentation) of the fish tissue has taken place. This process is controlled by suitable combinations of salt, sugar and acid, and preservatives such as sodium nitrate and benzoic acid may also be included; typically, processing is carried out in barrels wth the fish immersed in brine. Examples of such products include French anchovy and Scandinavian tidbits and surstrøming. The microbial flora should be dominated by Gram positive bacteria but inadequate processing can allow a typical Gram negative spoilage flora to develop (Hobbs, 1987).

3.6.6. Smoked Fish

Before smoking, fish are gutted and given a preliminary salting treatment, the NaCl concentration depending on the level desired in the fish. The salting is often comparatively light and the preservative effect is therefore minimal as with, for example, finnan haddock, kippers and smoked cod. With certain fish such as 'red' herring and smoked salmon the salting is more extreme and may play a significant part in preserving the fish (Cutting, 1965).

Following brining the fish are either 'cold' or 'hot' smoked. In the commonly used cold smoking process the temperature of the fish should not exceed that at which protein is denatured (*ca* 30°C). The process, which includes both drying of the fish and impregnation with wood smoke, results in some reduction in the numbers of bacteria caused mainly by phenolic substances present in the smoke; overall, however, the effects of this process on the microbial flora are insubstantial (Graikoski, 1973).

In hot smoking, used for specialized canned products such as brisling (sprats) and sild (small herring) the temperature of the fish is raised to *ca* 70°C for 30 min so only the more heat-resistant bacteria should survive; the microbial flora therefore consists mainly of micrococci and *Bacillus* spp.

The microbiological changes occurring during the storage of smoked fish have not been studied in any detail but the predominant microorganisms at the time of spoilage depend largely on the processing conditions. With lightly brined cold smoked fish pseudomonads may become the major group but a slight increase in salt concentration could result in micrococci predominating. One of the most common causes of spoilage of smoked fish is due to mould contamination; both *Penicillium* and *Aspergillus* spp., which grow readily at chill temperatures, are present in the sawdust used for smoke production and may subsequently develop in the stored fish. With the higher salt concentration smoked fish storage life is prolonged for several weeks or months even at high ambient temperatures and little change in the microbial flora is likely to occur.

3.6.7. Packaged fish

Vacuum-packaging of fish has understandably been viewed with caution because of the risk of botulism (see Chaper 2). The same concern has been expressed about modified atmospheric packaging but most data available suggest that spoilage is likely to precede toxin formation when fish is stored at chill temperatures; however, a storage temperature of 8°C or above could initiate toxin formation before spoilage becomes evident (Lindroth & Genigeorgis, 1986). At these higher chill temperatures it appears that vacuum-packaging is more conducive to toxin formation than gas atmospheres (100% CO_2 or 70% CO_2 : 30% air; Garcia *et al.*, 1987). Further protection against *Clostridium botulinum* type E can be provided by ensuring residual levels of >4·5% NaCl and 50 μg/kg sodium nitrite in the fish flesh following brining (Cuppett *et al.*, 1987). Because the relationships amongst the various inhibitory factors and their efficacy are so complex, mathematical models have been developed which predict the possibility of toxin formation in different fish stored under various conditions (e.g. Baker & Genigeorgis, 1990); such models enable processors to identify more readily any potential weaknesses in the preservation protocol thus further ensuring safe fish products of this nature.

The advantage of modified atmospheric packaging is that the shelf-life of fish can be significantly extended. Thus Molin *et al.* (1983) found that spoilage of herring fillets was delayed by an additional 4 weeks when stored in carbon dioxide at 2°C rather than in air; as with raw meats the spoilage flora was dominated by *Lactobacillus* spp., the typical Gram negative bacterial flora being totally overgrown.

3.6.8. Shellfish

3.6.8.1. Crustaceans
Freshly caught shrimps are highly perishable due to bacterial and enzymatic activity. Bacterial activity is enhanced by the high content of amino acids but autolytic enzymes (proteases) cause rapid breakdown of protein providing bacteria with an ideal growth substrate. Because of their perishable nature, shrimps are preferably either frozen or boiled as soon as possible after catching but storage in ice is also common. The initial flora is similar to that of freshly caught fish (Hobbs & Hodgkiss, 1982); storage in ice results in an increase in the proportion of *Acinetobacter/Moraxella* species which account for over 80% of the flora at the time of spoilage (Fieger & Novak, 1961). This group does not appear to be responsible for bacterial spoilage, however. The active spoilers are *Alteromonas* spp. and pseudomonads play a relatively minor role in the spoilage of this food; spoilage is accompanied by increases in ammonia, trimethylamine, hypoxanthine and acetic acid (Van Spreekens, 1977).

Autolytic enzymes are particularly active in lobsters and this makes them another highly perishable food. Use is made of these enzymes in conditioning which involves storage in ice for 2 to 4 days. As with shrimps, further storage results in a pronounced increase in the proportion of acinetobacters and moraxellas which are predominant at the time of spoilage (Walker *et al.*, 1970); this is characterized by increasing concentrations of ammonia and trimethylamine. Crab meat, too, spoils rapidly and crabs are therefore cooked in boiling water immediately after capture. Studies on crabs have concentrated on the bacteriology of the cooked meat: it is likely that the flora at the time of spoilage is again dominated by acinetobacters and related species.

3.6.8.2. Molluscs

Of the bivalve shellfish eaten by humans, oysters, scallops and mussels are probably most frequently consumed. The main microbiological problem associated with these foods is the hazard of food poisoning resulting from the not infrequent pollution of their growth habitat (see Chapter 2). It is therefore necessary to cleanse these foods by depuration with chlorinated or ozonated water, with water disinfected by iodophors or, as is practised in the United States, with water disinfected by uv light (Richards, 1988). The natural flora of the bivalves may thus change extensively during treatment, and spoilage characteristics may vary depending on the efficiency of the cleansing process. An important feature of bivalves is the significant amount of carbohydrate (3–6%) present in their flesh and this can influence the type of spoilage obtained. If fermentative bacteria such as *Escherichia coli* and other coliforms are not removed during the cleansing process, spoilage is primarily one of souring, acids being formed by the dissimilation of the carbohydrate (Fieger & Novak, 1961). With properly cleansed molluscs held at chill temperatures the spoilage is totally different and is associated with increases in volatile bases and hypoxanthine, and the flora is now dominated by *Acinetobacter/Moraxella* spp. (Thomson *et al.*, 1974).

3.7. DAIRY PRODUCTS

3.7.1. Milk

3.7.1.1. Microbiology of Raw Milk

Even when drawn under aseptic conditions, milk always contains microorganisms which are derived from the milk ducts in the cow's udder. Numbers vary from quarter to quarter and from cow to cow but are roughly in the range 10^2–10^3 organisms per ml. In practice, freshly drawn milk contains *ca* 5×10^3 to 5×10^4 organisms per ml, contaminants coming from the outside of the udder, from milking equipment and from human handlers. Many different microorganisms may be present including species of *Pseudomonas, Acinetobacter/Moraxella, Flavobacterium, Micrococcus, Streptococcus, Corynebacterium, Lactobacillus* and coliforms (Cousins & Bramley, 1981). Furthermore, it must be appreciated that infected udders (see p. 46) are going to introduce potentially pathogenic bacteria into the milk. This problem is relatively widespread, it having been shown that some 30% of British dairy herds suffer from mastitis (Wilson & Richards, 1980); some control of this infection can be achieved by applying a disinfectant to the teats after milking and by antibiotic treatment of non-lactating cows.

As milk is an ideal growth medium for bacteria it is essential to cool it as rapidly as possible. The introduction of refrigerated farm bulk milk tanks over the last 25 years, coupled with the bulk collection of milk in refrigerated tankers, has markedly influenced the bacteriological quality of raw milk supplies. The major effect of this change has been to reduce the quantity of milk spoiled by souring (Murray & Stewart, 1978). Souring of milk at normal temperatures is caused by the lactic acid bacteria which grow mainly at temperatures above 10°C. These bacteria produce lactic acid from the milk sugar (lactose) which induces a sour flavour and later coagulation of the milk. Most lactic acid bacteria are killed by pasteurization but a few (e.g. *Streptococcus thermophilus*) are thermoduric and can cause post-pasteurization problems.

With rapid cooling and refrigerated storage of milk the problems are somewhat different. Nowadays it is the psychrotrophs, mainly pseudomonads, which are primarily responsible for spoilage problems. Psychrotrophic bacteria, originally derived from soil and water, are readily isolated from farm milking equipment, pipes and bulk milk tanks (Thomas *et al.*, 1971). Inefficient or delayed cooling of milk markedly increases the proportion of psychrotrophs but growth of this group continues more slowly at recommended storage temperatures for raw milk (3–7°C). Counts of bacteria in bulk tanks vary from $<10^4$ to 10^6 per ml with a mean count of 20 000 per ml of milk (Panes *et al.*, 1979); variations reflect the extent and type of contamination and storage conditions but a large majority of the bacteria will be psychrotrophs. A large proportion of these produce proteases and lipases, and many of these enzymes are unaffected by pasteurization; in fact both sets of enzymes are resistant to a heat treatment of 140°C for 5 s and 77°C for 17 s (Griffiths *et al.*, 1981). Defects due to proteases include bitterness and the principal spoilage effect of lipases is rancidity.

3.7.1.2. Pasteurization

This process involves heating the milk to a temperature high enough to destroy all pathogenic bacteria such as *Mycobacterium tuberculosis, Salmonella* spp. and *Brucella* spp; in so doing the large majority of other bacteria including spoilage bacteria are killed and the keeping quality of the milk is thus enhanced. Most of the milk produced in the UK is pasteurized by the high-temperature, short-time (HTST) method in which milk is held at 72°C for at least 15 s and then cooled rapidly to less than 10°C; the older low-temperature, long-time (LTLT) or 'holder' method (63°C for 30 min) is still occasionally employed but only on a small scale.

The term 'thermoduric' is applied to those bacteria that survive pasteurization due to their innate heat resistance although they fail to grow at pasteurization temperatures. Thermoduric bacteria consist mainly of a few species of *Streptococcus* (e.g. *S. thermophilus* and *S. faecalis*), *Micrococcus* (e.g. *M. luteus*) and *Microbacterium* (e.g. *M. lacticum*) together with the spores of certain *Bacillus* spp., particularly *B. cereus* and *B. subtilis* (Nelson, 1981). These bacteria are readily isolated from poorly cleaned dairy equipment and pipelines although numbers in refrigerated bulk milk tanks are usually small (Mackenzie, 1973). Spoilage of pasteurized milk held at normal temperatures is caused principally by the thermoduric bacteria, the predominant organism at the time of spoilage usually being *B. cereus*. This organism causes the defect termed 'bitty cream' and is responsible for the 'sweet' curdling of pasteurized milk (i.e. coagulation of milk by rennin without acid formation).

The psychrotrophic bacteria, so important in raw milk, are readily killed by pasteurization even though some of their enzymes are unaffected (see above). However, psychrotrophs can be an important cause of spoilage in pasteurized milk if post-pasteurization contamination occurs. This contamination should be negligible but with inadequate cleaning of equipment it can be very significant. Such contamination should be minimized since pasteurized milk is usually stored at *ca* 7°C, a temperature at which psychrotrophs grow well. Pasteurized milk with minimal post-pasteurization contamination should have a shelf-life of at least 7–10 days at 7°C.

3.7.1.3. UHT Milk

Ultra-high-temperature (UHT) milk is homogenized milk that is exposed to a temperature of not less than 130°C for at least 1 s, this process rendering the milk sterile. The original form of sterilization of milk was to heat the milk at *ca* 100°C for 30 min in hermetically sealed bottles; such milk was characterized by a cooked flavour and a rich and creamy texture which, together with its darker appearance, rendered it a rather unattractive product. Modern UHT processes are of two basic types: in the first (indirect heating) the milk is sterilized in a heat exchanger whilst in the second (direct heating) the milk is treated by being directly injected with steam under pressure. The sterile product is then filled aseptically in special cartons (e.g. Tetra Pak) which are then heat-sealed. UHT milk has an appearance, flavour and nutritional quality similar to that of conventionally pasteurized milk and it should remain acceptable for several months without refrigeration.

Spoilage of UHT milk can be caused occasionally by the growth of spore formers, mainly *Bacillus stearothermophilus* or *B. subtilis*, the spores of which have either survived the UHT process or are post-treatment contaminants (Murray & Stewart, 1978). More commonly, spoilage can result from the continuing activities of heat-stable proteases and lipases produced by psychrotrophs in the raw milk before processing; such spoilage can significantly reduce the theoretical shelf-life. Enhanced deactivation of these enzymes can be achieved by modification of the normal heating process; typically this involves a dual heat treatment (140°C for 5 s followed by 60°C for 5 min) which can extend the shelf-life of the product by several months (Bucky *et al.*, 1988). Gelation of stored UHT milk, which may also be caused by a chemical process, can also be attributed to proteases.

3.7.2. Butter

Butter is a comparatively stable product microbiologically as it has a low moisture content (15%) and a high fat content (80%). The water is present in the form of a fine emulsion in the fat phase and the physical conditions in the water droplets probably exert an inhibitory effect on bacterial growth. Furthermore, many butters are salted at concentrations varying between *ca* 3 and 13% NaCl and this helps in preservation.

The main source of microorganisms in butter is the cream from which it is made; this is pasteurized in the case of 'sweet' butters. The churning of cream into butter increases the numbers of organisms as they become concentrated in the buttermilk but at the end of processing numbers are low and with the highly salted butters further reductions are likely during storage. Certain 'sweet' butters utilize a 'starter' culture. These cultures of known bacteria are inoculated into milk or cream to induce souring under controlled conditions thus giving predictable and desirable characteristics to the butter; with 'sweet' butters the cream, after inoculation, is held at low temperatures to keep the acidity from increasing prior to churning.

'Sour' butters are normally made from pasteurized cream inoculated with starter organisms (*Streptococcus lactis* and *S. cremoris* are commonly used); the cream is incubated at room temperature until a low pH (4·5–5) is attained. The cream is then churned but is not salted because salt and acid react together to give undesirable flavours. Large numbers (10^7 to 10^8 per gram) are required to produce the acid. An alternative process involves the natural souring of cream which is then pasteurized before churning.

Thus the microbial content of fresh butters varies considerably depending on the manufacturing process, 'sweet' butters containing far fewer microorganisms than 'sour' butters.

Spoilage of butter can be of microbial, enzymatic or chemical origin; many undesirable flavours may stem from the cream itself but this aspect of spoilage is not considered here.

Microbial spoilage is caused principally by psychrotrophic bacteria as butter is usually stored under refrigeration. Pseudomonads and related Gram negative rods which enter the product post-pasteurization are often responsible for rancidity caused by the hydrolysis of butter fat liberating fatty acids. Putrid flavours and surface taint result from the proteolytic activities of *Alteromonas putrefaciens* growing on the surface of the butter. Moulds also grow on the surface producing discolouration; commonly implicated are members of the genera *Alternaria, Cladosporium, Aspergillus, Penicillium, Mucor* and *Rhizopus*. Rancidity may be induced by lipases present in the cream and chemical reactions include the oxidation of unsaturated fats.

3.7.3. Cheese

There are over 400 known varieties of cheese grouped into about 20 general classes. Most of these can be made from the same milk by varying the microorganisms, enzymes and salt added, and by changes in the temperature during manufacture and curing.

Cheeses are classified by their texture or the degree of hardness and two major groups of natural cheeses are recognized. The first group, the ripened cheeses, vary from the very hard, low moisture content grating cheeses (e.g. Parmesan), through the hard cheeses (e.g. Cheddar) to the higher moisture content, semi-soft cheeses (e.g. Stilton) and soft cheeses (e.g. Camembert). The second group of cheeses are the unripened soft cheeses with a high moisture content such as cottage cheese.

Most cheeses are manufactured using the same basic processes. Nowadays pasteurized milk is normally used but the ripening process occurs more slowly and, because the natural flora has been largely destroyed, 'starter' cultures of bacteria must be added to the milk. In the case of Cheddar cheese the starter culture is typically composed of a mixture of selected strains of *Streptococcus lactis* and *S. cremoris* although other lactic acid bacteria may be included (Chapman & Sharpe, 1981). The starter converts the sugar lactose into lactic acid thus inducing the first stage in cheese manufacture, i.e. souring or 'ripening' of the milk. When the milk has reached the required acidity rennet is added which helps in curd formation. Later stages in processing involve treatment of the curd which, after salting and pressing, is allowed to mature; with Cheddar cheese maturation takes about 4 months.

Microbiologically the starter culture organisms reach a peak of *ca* 10^9 per g in cheese during ripening but after some 48 h the streptococci decline in numbers. They are replaced by lactobacilli which represent 99% of the population in mature cheese. Lactobacilli slowly decompose protein during maturation and this helps to flavour the cheese; in fact lactobacilli (e.g. *L. acidophilus*) can be added deliberately to enhance flavour.

When considering the spoilage of cheese it should be emphasized that the harder cheeses with lower moisture contents have longer shelf-lives than the softer cheeses. Microbiological spoilage of mature Cheddar cheese is rare, provided that suitable starter cultures are used which produce sufficient acidity and do not give rise to bitterness or off-flavours (Chapman & Sharpe, 1981). Faults that can arise include the surface growth of moulds which produce discolouration effects although there is little penetration of or attack on the cheese. Many different moulds and yeasts have been implicated in such spoilage and include species of *Penicillium* (green discolouration), *Cladosporium* (green to black) and *Candida* (black). However, the harder cheeses have a wax coating or develop a rind and this minimizes the problem. In more recent years film wrapped and vacuum-packed cheeses have become popular and these forms of packaging should prevent fungal growth by excluding air.

Bacterial spoilage of cheeses is more common during their manufacture and ripening. If the pH is too high pseudomonads, which always contaminate the product to some extent, are able to grow rapidly and cause sliminess. 'Gassy' cheese is a common problem and is caused by coliforms fermenting the lactose with the production of carbon dioxide; this gas may also be produced by certain lactic acid bacteria and even by clostridia. The latter may be controlled by the incorporation of nisin, an antibiotic produced by certain strains of *Streptococcus lactis*, in the cheese. This antibiotic is particularly active against clostridia in cheeses with a higher pH.

Various flavour defects, of which bitterness and rancidity are the most important, can occur in cheese; many of these defects are caused by heat resistant lipases produced by pseudomonads.

The need for suitable starter cultures has already been stressed; the conversion of all the lactose to lactic acid for optimal flavour is important in the early stages of cheese production and this is dependent on reliable starters. Unfortunately these cultures are susceptible to phage attack (see p. 17) which is regarded as the major problem in cheese manufacturing today (Mabbitt *et al.*, 1987). Phage attack usually results in diminished acid production by one, or possibly more, of the starter strains; other enzyme functions important in cheese making may be impaired. Phages can be readily isolated from raw milk and whey and contamination of the starter may occur quickly if high standards of plant cleaning are not maintained. Inevitably, even with a careful rotation and selection of starter strains, phage infections occur. Thus the development of phage resistant cultures is of profound commercial interest (Mabbitt *et al.*, 1987). This can be achieved by incorporating plasmids (i.e. small independent fragments of DNA) into the DNA of the starter strain; these plasmids could carry phage resistance factors but, equally, other desirable properties such as optimum acidity production and flavour potential could be encoded in this way (Heap & Lawrence, 1988).

3.7.4. Yoghurt

Yoghurt is a fermented milk product made by adding a mixed starter culture (*Lactobacillus bulgaricus* and *Streptococcus thermophilus*) to milk that has been heated to *ca* 90°C for 5 min to destroy the indigenous flora. As with cheese, lactic acid is produced during incubation at 40–42°C and this reduces the pH to 4·0; traces of other products such as diacetyl and acetaldehyde contribute towards the flavour characteristics. After incubation the yoghurt is cooled to 15–20°C when fruit and/or flavours are mixed in prior to packaging. The yoghurt is then cooled to 4°C to prevent further acid development and the low storage temperature coupled with the acidity of the product should ensure that spoilage by non-acid tolerant bacteria is prevented; yeasts (e.g. *Saccharomyces cerevisiae*) can be carried in on fruits and can occasionally be responsible for 'doming' (i.e. gas production) in sweetened fruit yoghurts (Robinson & Tamine, 1981). Slow growth of the starter organisms continues at the storage temperature and this restricts the shelf-life to about 4 weeks as by that time residual proteolytic and lipolytic activity will impair flavour.

3.8. EGGS AND EGG PRODUCTS

3.8.1. The Chicken's Egg and its Spoilage

The chicken's egg is generally regarded as being sterile at the time of laying unless it has been infected congenitally, usually by certain salmonellas. Contamination of the egg occurs after laying and access of microorganisms into the egg is most common through cracks in the shell. The shell, covered with a water-repelling cuticle, acts as a mechanical barrier if intact, but an alternative means of entry for microorganisms is through pores which perforate the egg shell (see Fig. 3.7). These pores should be filled with plugs but in the largest pores the plugs may become dislodged. Penetration is aided by moisture setting up capillary effects in the pores. Beneath the shell are two membranes which further retard invasion by bacteria for limited periods but probably offer no barrier to the infiltrating hyphae of moulds (Board & Fuller, 1974).

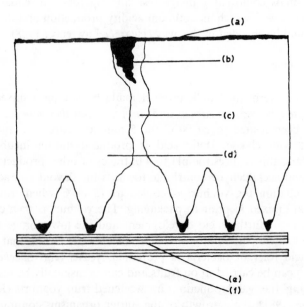

Fig. 3.7. Simplified diagram of a section through an egg shell showing: (a) cuticle, (b) cuticular plug, (c) pore, (d) matrix, (e) outer shell membrane, and (f) inner shell membrane (after Board, 1966).

The white (albumen) of eggs contains a battery of antimicrobial agents which restrict or totally inhibit the growth of invading microorganisms provided levels of contamination are low (Tranter & Board, 1982). Conalbumin (ovotransferrin) is perhaps the most effective antimicrobial agent, both against Gram positive and Gram negative bacteria; it acts as a chelating agent, removing iron which is essential for growth (Board, 1969). The yolk, however, is a rich source of nutrients and contains no inhibitory agents; thus rapid growth of invading organisms is possible if the yolk is involved. This can occur when the yolk comes to rest in the uppermost part of the egg about 10 days after laying. If penetration of the egg has occurred in the membrane area impinging on the yolk the defence mechanisms of the egg are short-circuited and heavy growth of invading bacteria is likely. Spoilage of eggs is caused principally by Gram negative bacteria which produce characteristic rots (Board, 1965); these are listed in Table 3.2.

Table 3.2
Egg Spoilage Caused by Bacteria

Type	Bacteria involved	Changes induced
Green rots	Pseudomonas fluorescens	Bright green fluorescent albumen when viewed under UV. Yolk may disintegrate. Fruity odour.
Pink rots	Pseudomonas spp.	Pink discolouration of albumen.
Colourless or white rots	Pseudomonas spp.	Watery albumen and yolk may disintegrate.
Black rots	Proteus vulgaris Aeromonas spp. Pseudomonas spp.	Watery albumen which turns brown. Yolk disintegrates and is blackened. Strong odour with gas (H_2S) present. May be green fluorescence with pseudomonads.
Red rots	Serratia marcescens	Bright red albumen and yolk may be infected. No smell.

3.8.2. Egg Products

Although the contents of fresh eggs are usually sterile, commercially produced egg products (liquid, frozen and dried) used to be heavily contaminated with bacteria. In particular, evidence accumulated that liquid whole
egg was frequently contaminated with salmonellas that were well able to
withstand the comparatively low temperatures frequently used in the baking industry. Heat treatment regulations were introduced in the UK in
1963 and since then all liquid whole egg intended for distribution in the
chilled state, for freezing or for spray drying, has been pasteurized at
64·4°C for 2½ min and then cooled immediately. The heat-resistant
salmonellas should be destroyed by this treatment and, in particular, *S.
enteritidis* phage type 4, should not survive (Humphrey *et al.*, 1990); furthermore the baking properties of the egg are not impaired (Shrimpton *et
al.*, 1962). In addition, in the chilled liquid state pasteurized whole egg
can be safely stored for at least 6 days without significant increases in
bacterial counts although post-pasteurization contamination, mainly
caused by coliforms, must be avoided.

In Britain, liquid egg albumen is now conventionally pasteurized although
heat-treatment regulations have yet to be implemented. Untreated
chilled albumen is normally spoiled by pseudomonads and related Gram
negative rods whilst faecal streptococci and lactobacilli are predominant
in pasteurized albumen (Barnes & Corry, 1969). More severe heat treatment is required to destroy salmonellas in liquid egg yolk but again
heating regulations have still to be introduced.

3.9. VEGETABLES AND FRUITS

As soon as vegetables and fruits are harvested physiological changes
occur and some of these lead to a loss in quality. Respiratory activity
involving the breakdown of carbohydrates by the plant enzymes continues and the changes induced, whether advantageous or deleterious, are
markedly influenced by the maturity of the plant when harvested; thus
plants can usually be stored for lengthy periods with little change in
quality if harvested at the right time. Many fleshy fruits such as bananas
are harvested before maturation and ripening continues thereafter but
citrus fruits only ripen satisfactorily on the tree (Duckworth, 1966).

However, although spoilage can be induced by autolytic enzymes, it is caused more usually by the activities of microorganisms and this aspect is considered in this section.

3.9.1. Spoilage by Fungi

The low pH (<4·5) of most fruits means that spoilage is caused mainly by fungi. On the other hand, the pH range of most vegetables varies between 5·0 and 7·0 and thus spoilage may be caused by either fungi or bacteria although the former are again the more important group. In terms of their spoilage characteristics fungi are often somewhat arbitrarily divided into two groups: the plant pathogens which infect the plant before harvesting and the saprophytic fungi which attack the commodity after harvesting. An important property of most spoilage organisms, both fungal and bacterial, is their ability to secrete pectolytic enzymes which soften and disintegrate plant tissues. Thus the growth of fungi on fruits and vegetables usually results in severe tissue breakdown causing mushy areas; this spoilage is termed a 'rot'. The names given to the different rots indicate the appearance of the food when spoiled. Many of the more important forms of fungal spoilage are listed in Table 3.3.

An important cause of spoilage is *Penicillium*, many species of which are able to attack fruit; perhaps as much as 30% of all fruit decay can be attributed to this genus. Many fruit-vegetables such as tomatoes and cucumbers, and vegetables such as potatoes and beetroots are also susceptible. Another important disease is *Rhizopus* soft rot which affects a wide range of fruits and vegetables particularly during transit under poor refrigeration. Harvested strawberries and potatoes are often attacked and spoilage is indicated by soft, mushy areas with greyish mycelium evident in the affected areas.

3.9.2. Spoilage by Bacteria

Although bacteria are of little importance in the spoilage of fruit some 35% of microbial spoilage losses in vegetables can be attributed to them. The bacteria responsible are chiefly members of the genera *Erwinia* and *Pseudomonas*. The most common forms of spoilage are bacterial soft rots which affect most vegetables and some fruit-vegetables such as tomatoes and cucumbers; soft rot bacteria can also cause pre-harvest

Table 3.3
Fungi Important in the Spoilage of Harvested Fruits and Vegetables[a]

Crop	Type of spoilage	Causative fungus (genus)
Tree fruits		
Apples and pears	Blue mould rot	*Penicillium*
	Grey mould rot	*Botrytis*
Peaches and plums	Rhizopus soft rot	*Rhizopus*
Soft fruits		
Raspberries and blackberries	Soft rot	*Mucor* *Rhizopus*
	Green mould rot	*Cladosporium*
Strawberries	Grey mould rot	*Botrytis*
	Soft rot	*Mucor* *Rhizopus*
Grapes	Grey mould rot	*Botrytis*
Sub-tropical fruits		
Bananas	Crown rot	*Fusarium*
Citrus fruits	Stem-end rot	*Alternaria*
	Brown rot	*Phytophthora*
	Blue and green mould rots	*Penicillium*
	Sour rot	*Geotrichium*
Vegetable fruits		
Tomatoes	Grey mould rot	*Botrytis*
	Black rot	*Alternaria*
	Soft rot	*Rhizopus*
	Sour rot	*Geotrichium*
Cucumbers	Green mould rot	*Cladosporium*
	Grey mould rot	*Botrytis*
Vegetables		
Carrots	Grey mould rot	*Botrytis*
	Watery soft rot	*Sclerotinia*
Cabbage	Black rot	*Alternaria*
	Brown rot	*Phytophthora*
Potatoes	Blight	*Phytophthora*
	Dry rot	*Fusarium*
	Soft rot	*Rhizopus*

[a]From Duckworth (1966) and Dennis (1987).

infection. Certain erwinias and pseudomonads, in particular, are important plant pathogens causing diseases such as blights, wilts, cankers and leaf spots (Garrett, 1982).

The soft rot bacteria, of which *Erwinia carotovora* is the most important, are present on the plant at harvest and they usually gain entry through damaged tissue; even the injury created by cutting lettuces and cabbages from the stem can provide a site for the bacteria causing soft rots (Dennis, 1987). Growth of such bacteria is so rapid that moulds are unable to compete and are not normally isolated from commodities infected with bacterial soft rot. Pectic enzymes are readily formed by the bacteria producing soft, mushy tissues and within a few days there is major tissue breakdown. In the case of potatoes the whole tuber may collapse, with tomatoes the outer skin may remain intact whilst the entire contents have changed to turbid liquid and with leafy vegetables slimy masses are produced.

3.9.3. Control of Microbial Spoilage

Many microorganisms gain access to or contaminate plant material during harvesting or subsequent handling. Therefore it is desirable to use equipment that is as clean as possible and to minimize mechanical damage of plant material; many external organisms can be removed from fruits and vegetables by being washed in water although washing can reduce the storage life of vegetables if inadequately drained.

A suitable storage environment is essential so that physiological and microbiological deterioration are minimized. Storage is normally under chill conditions (0–5°C) but certain commodities such as potatoes and cucumbers are best stored at 7–10°C. Pseudomonads are usually responsible for bacterial spoilage of vegetables stored at 0–2°C whilst erwinias are more important at slightly higher temperatures (Dennis, 1987). The optimum relative humidity is in the 90–95% range and storage life can be enhanced with, for example, apples and pears by controlled or modified atmospheric storage (reduced oxygen and increased carbon dioxide concentrations). Sealed plastic films encourage high humidity within the pack and increased microbial spoilage can result. Perforated plastic films largely overcome this problem but humidity can be higher than with the unwrapped product.

There is now an increased demand for prepared mixed salads which are sold in cling film packs stored at chill temperatures. Although these salads are packaged normally in air, during storage there may well be a depletion of oxygen and an accumulation of carbon dioxide due, primarily, to the respiratory activities of the vegetable enzymes. Washing of the vegetables before slicing, dicing or shredding reduces the microbial load but organisms build up rapidly on the cutting equipment so that microbial counts are often high on the packaged vegetables; in this situation microorganisms can also contribute to the gaseous exchange within the pack. Growth of bacteria can be rapid on the cut surfaces of the vegetables so that the shelf-life of this type of product is necessarily limited. In spite of the changed gaseous environment there is still sufficient oxygen present for the spoilage flora to be dominated by pseudomonads together with other Gram negative bacteria (Brocklehurst *et al.*, 1987). Where mayonnaise is included in salad packs (e.g. coleslaw) bacteria rapidly decline in numbers and are replaced by yeasts (*Saccharomyces* spp.) which will eventually spoil the product (Dennis, 1987).

Microbial spoilage of fruits and vegetables can be controlled by their treatment with chemicals, either at the pre-harvest or post-harvest stages. In the latter case common control measures include dipping or spraying with fungicides/bactericides such as borax (sodium tetraborate), sorbic acid, phenylphenates, diphenyl and iodophors, and fumigation with sulphur-containing dusts or SO_2.

3.10. CEREAL BASED PRODUCTS

The microbial flora of harvested cereal grains such as corn, wheat and oats contains up to many millions of bacteria and moulds per gram. However, the low a_w of grains effectively inhibits the growth of all microorganisms provided storage conditions are satisfactory but in moist conditions mould growth is likely.

Certain steps involved in flour manufacture reduce the microbial load and of these bleaching has the greatest effect. Mould counts remain fairly constant, in the low thousands per gram, in properly stored flours and the most commonly isolated species are members of the genera *Penicillium*, *Aspergillus* and *Rhizopus*. Bacteria decrease in numbers during storage and counts of <1000 per g are usual, with *Bacillus* spp. the predominant group. Where the moisture content is above normal mould growth is likely and at still higher a_w levels growth of *Bacillus* spp. will occur.

Commercially produced bread should be of sufficiently low moisture content to inhibit growth of most microorganisms except moulds which are the principal spoilage agents; in fact, moulds are said to be responsible for the loss of 1% of annual bread production (Seiler, 1971). Amongst the most common are *Rhizopus nigricans*, the 'bread mould' which produces characteristic black dots of sporangia, *Penicillium* and *Aspergillus* spp., which produce green conidia in abundance, and *Neurospora sitophila*, the 'red bread' mould. Mould spoilage is encouraged by slicing, wrapping the bread when too warm and storage in a warm, moist environment.

Ropiness of bread, caused by *Bacillus* spp., is now rarely seen in commercially produced bread. It is initially characterized by brownish spots accompanied by an unpleasant odour and, later, disintegration of the crumb or slices follows; the spoilage is caused by hydrolysis of flour protein and starch which produces stickiness and stringiness in the bread. Control is best achieved by low temperature storage, the addition of preservative (e.g. calcium propionate or sorbic acid) and the use of good-quality flour (Frazier & Westhoff, 1988).

Moulds are responsible for most spoilage problems in cakes although the situation is complicated by the wide variety of ingredients that may be incorporated some of which, such as dairy and imitation creams, custard and chocolate, have been implicated in bacterial food poisoning incidents (Seiler, 1978). Generally, mould growth is again controlled by low temperature storage and low a_w levels together with the use of preservatives as in bread.

Pastas are made from a stiff dough which is extruded in different shapes. There is then a warm drying stage in processing which requires careful control since both *Salmonella* spp. and *Staphylococcus aureus* could proliferate during this period; these organisms have been implicated in food poisoning outbreaks mainly involving pastas containing egg although egg-free pastas have been shown to contain these pathogens on occasions (Seiler, 1988). With careful drying and subsequent storage in dry conditions spoilage should be prevented.

3.11. BEER

The main ingredients of beer are malted barley, hops and water. The malt is ground up with water to form a mash and the enzymes naturally present convert the starch into an easily fermentable sugar, maltose. This 'wort' is then boiled together with hops which are added principally for flavouring. After cooling, yeasts (strains of *Saccharomyces cerevisiae*) are added to the cooled wort to convert maltose to alcohol and carbon dioxide. The beer is subsequently matured in storage tanks at 0°C for several weeks after which finishing processes are performed. Lager beers differ from typical ales in a number of ways including utilizing a different yeast strain, *S. carlsbergensis*, in the fermentation process which is carried out at 10–15°C rather than at 15–20°C used with ales.

Poor-quality beer is most commonly caused by non-microbial effects but only microbiological factors will be considered here. One of the most important is the need to maintain active, healthy yeasts with the required characteristics as the quality of the final product is markedly influenced by the yeast used. Microbial beer contamination, which occurs during or after the cooling of the wort, produces hazes, pellicles, acidification and undesirable flavours or consistency (Rivière *et al.*, 1977). Beer can be affected by different types of 'wild' yeast (often contaminants of the 'starter' yeast culture) which can cause cloudiness, flavour defects and pellicle formation. So-called 'killer' strains of *S. cerevisiae* can destroy the desirable yeasts by secretion of a polypeptide 'zymocin' and thus become the main fermenters; undesirable end-products of fermentation result in off-flavours and turbidity (Campbell, 1987). 'Ropiness' in beer, in which the liquid becomes viscous and pours as an 'oily' stream, is caused by *Acetobacter* or *Pediococcus* spp. Acetobacters also cause sourness in beers by oxidizing ethyl alcohol to acetic acid. Diacetyl ($CH_3COCOCH_3$) is produced by a pediococcus or *Lactobacillus* spp. and this flavour defect of lager beers, in particular, is characterized by a honey-like odour and taste.

For quality to be assured careful control from the early production stages to the point of sale is necessary. Thus clean and healthy barley should be used to reduce the risk of mould spoilage whilst the required fermentation changes are dependent upon suitable pure yeast cultures. Post-fermentation spoilage by microorganisms can be largely prevented by using clean and properly functioning equipment; particular attention should be paid to the beer filters, to filling equipment and to pipework throughout.

3.12. WINE

Wine is produced by the fermentation of the juice of crushed grapes known as 'must'. The fermentation can proceed normally, induced principally by the activities of a sequence of natural yeast populations. It is customary, however, to treat the musts with sulphur dioxide as a gas or as potassium metabisulphite in order to suppress the natural flora including undesirable yeasts and lactic acid bacteria (Goswell, 1986); a starter culture of yeast is then added and thus greater control over the fermentation process can be achieved. At the end of active fermentation the wine is transferred to storage tanks for ageing after which it is filtered and bottled or otherwise stored.

The microorganisms causing wine spoilage are principally wild yeasts and bacteria although, as with beer, defects are frequently non-microbial in origin. Important spoilage yeasts include *Candida*, *Pichia* and a number of *Saccharomyces* spp. which all produce growth films on the surface of wines. Some yeasts may be desirable in certain wines but may cause spoilage in others where residual sugar is desired. Wine spoilage bacteria are principally acetobacters and lactic acid bacteria. The former produce sourness whilst the latter, represented by the genera *Lactobacillus, Leuconostoc* and *Pediococcus*, produce lactic and acetic acids from sugars; these acids are usually accompanied by turbidity, off-flavours and possibly the evolution of carbon dioxide. Ropiness, although less common than in beers, can also be caused by leuconostocs.

3.13. SAUERKRAUT

Sauerkraut is another food produced by a fermentation process. The raw material is shredded cabbage to which salt (2–3%) is added to control the fermentation and to release sugars required by the lactic acid bacteria for growth. Incubation is at around 20°C and the lactic acid produced reduces the pH to *ca* 3·5 thus giving a preservation effect. The principal organisms involved are leuconostocs and lactobacilli which thrive in the near-anaerobic conditions prevailing.

Spoilage of the sauerkraut occurs as a result of incorrect processing conditions. Thus 'slimy kraut' is due to the growth of the wrong strains of lactobacilli as a result of too high an incubation temperature. Poor salt distribution may allow the growth of a variety of pectolytic and proteolytic organisms causing 'rotted kraut' or, with high salt densities, *Rhodotorula* spp. causing 'pink kraut'.

3.14. CANNED FOODS

Traditionally, canning is a method of food preservation in which food is closed within hermetically sealed containers. Heat is applied to the containers in such a way as to destroy or inactivate microorganisms, their toxins and enzymes, thereby rendering the food free from possible spoilage or harmful effects. From a biological viewpoint the process may fail in one or both of two ways. The first is a post-process infection involving leakage of microorganisms inwards through faulty seams and the second is the survival of organisms as a result of inadequate heat treatment.

An alternative canning process, used mainly for fluid foods (e.g. sauces, soups and fruit concentrates), involves the continuous sterilization of the product using a high-temperature, short-time (HTST) heat treatment (e.g. 130°C for 30 s). Typically the foods are passed through heat exchangers for product heating, a holding tube where the food is maintained at the required temperature/time, and finally into further heat exchangers where rapid cooling is achieved; containers and lids are sterilized separately with superheated steam and the food is added to the container which is then lidded in the sterile environment (Frazier & Westhoff, 1988). In fact this aseptic packaging principle has been used with various forms of flexible packages made with plastics and aluminium foil, the pre-sterilization of which is often done by means of hot hydrogen peroxide or uv irradiation (Mitchell, 1988). Spoilage of these aseptically processed foods is again caused by seam leakage or under processing.

Before making a more detailed review of the microbiological spoilage of canned foods it should be mentioned that spoilage may also be caused by chemical changes, the most important of which is the 'hydrogen swell'. This results from the reaction of the can metal (iron) with acidic foods when the hydrogen liberated causes the can to swell. The higher the acidity of the food, the greater the likelihood of this problem developing although suitable internal lacquers should largely eliminate this fault.

3.14.1. Leaker Spoilage

A wide variety of microorganisms may be associated with spoilage following post-process seam leakage and the main source of these organisms is the water used for cooling the cans after processing. The microbiological quality of this water thus markedly influences the frequency of reinfection and total bacterial counts of the water must be less than 100 per ml (Table 3.4). An interesting facet of this problem is that motile bacteria gain entry through seams more readily than non-motile forms. Other factors affecting can leakage are: (1) the vacuum applied which contributes by increasing the flow of minute volumes of cooling water into the can; (2) the viscosity of the can contents; (3) the

Table 3.4
The Influence of the Number of Bacteria in the Cooling
Water on the Rate of Reinfection[a]

Number of bacteria per millilitre of water	% of cans reinfected
10^2	2
10^3	8
10^4	18
10^5	30
10^6	48
10^7	62

[a]From Put *et al.* (1972).

size and shape of the hole or channel; (4) bacterial morphology (McEldowney & Fletcher, 1990). Bacterial numbers must be reduced to a minimum by adequate chlorination of the cooling water.

Many other factors contribute to post-process reinfection and defective cans are amongst the most important. The most likely point of entry for microorganisms used to be at the junction of the soldered side seam with the double seams of the can lid or base. The widespread introduction in the 1980s of cans with welded side seams which only have a minimum overlap has undoubtedly helped to reduce these sites of seam leakage. The junction of the side seam of the can with the base and lid is now far less bulky than with the older flanged side seam so that the chances of bacteria gaining entry at these points are substantially reduced. Another change in can construction which has been introduced for specialized commodities (e.g. baby foods) is that the base and side of the can are now being formed continuously from the same sheet of metal; these 'drawn' or 'two-piece' cans thus have no side seam or base-side junction so that seam leakage in such cans is only possible at the junction between the body and lid unless, of course, there are blemishes in the tin or aluminium plate.

A further leakage point is the double seam itself (Figs 3.8 and 3.9), particularly that formed by the food canner after filling and lidding. It is essential to ensure optimum seam thickness and overlap of cover and body hooks; the quality and amount of lining compound in the double seam area is also of great importance (Put *et al.*, 1972). A third possible although least likely entry point is directly through a small hole or cut in the metal from which the can was formed.

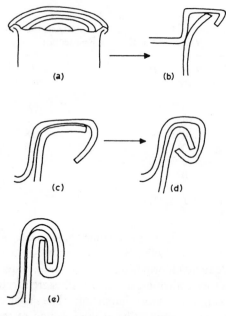

Fig. 3.8. Steps in can closure after filling: (a) positioning of lid, (b) to (e) progressive operations in the formation of the double seam.

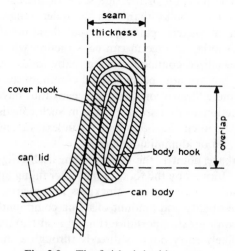

Fig. 3.9. The finished double seam.

In a detailed review of microbiological leaker spoilage Put *et al.* (1972) listed certain guiding principles which, if adhered to, would ensure that canned foods remain sterile, wholesome and safe for the consumer. These principles were:

(1) ensure that the construction of the double seam, the lap and the side seam are in accordance with accepted quality standards; (2) avoid rough handling of cans; (3) avoid excessive deformation of can ends during sterilization and cooling due to sudden pressure changes; (4) correctly chlorinate cooling water to a residual level of 1–2 mg/litre of free chlorine measured in the water drained from the seam after the cooling operation. Also ensure that the cooling water conforms to the chemical and bacteriological standards laid down for drinking water; (5) wash and disinfect at frequent intervals all surfaces of mechanical handling equipment which might come in contact with the double seam; (6) dry cans immediately after cooling and transport them on clean dry surfaces; (7) check cannery hygiene by regular microbiological surveys; (8) insist on and supervise high hygienic standards amongst employees.

Even with the advent of more modern types of cans these principles still hold true to-day.

Where spoilage results from seam leakage the bacteria implicated usually have relatively low optimum growth temperatures (25–35°C) and would be readily killed by the heat processing. In addition, more than one type of bacterium is usually isolated; these types commonly include members of the genera *Pseudomonas, Alcaligenes* and *Flavobacterium*, together with coliforms and micrococci. The above types are common contaminants where chlorination levels in the cooling water are minimal or non-existent; where chlorination is only slightly sub-standard, *Bacillus* and *Clostridium* spp. may be involved since their spores are highly resistant to chlorination. Clostridial spores may also gain access to cans post-cooling from dirty tracks and other items of equipment thus again stressing the need for good cannery hygiene (Lake *et al.*, 1985). When it is remembered that nearly two-thirds of all forms of spoilage in canned foods are due to reinfection after heating, the importance of control measures, especially adequate chlorination of cooling water, cannot be overstressed.

3.14.2. Spoilage Due to Inadequate Heat Treatment

The aim of the heat processing is to destroy or inactivate microorganisms and their products, that is, to ensure canned foods are 'commercially sterile'. Such foods may not be sterile in the absolute sense of the term but any surviving spores or microorganisms are no longer capable of growth.

The heat process necessary for commercial sterilization is determined to a considerable extent by the pH of the food. Above pH 4·5 *Clostridium botulinum* spores, which are markedly heat-resistant, can germinate and growth of vegetative cells with consequent toxin production is thus a distinct possibility. Foods with pH values above 4·5 are therefore given severe heat treatments at temperatures in the 110–121°C range. Conversely foods with pH values of below 4·5 are given relatively mild heat treatments, temperatures in excess of 100°C being usually unnecessary. However, whilst it is generally assumed that foods with pHs of 4·5 and below inhibit the growth of *C. botulinum* it has been shown that spore germination and growth is apparently possible in certain high protein foods with pH values as low as 4·2 (Smelt *et al.*, 1982); such growth is facilitated by the presence of other bacteria denaturing substances toxic to clostridia in the food. It

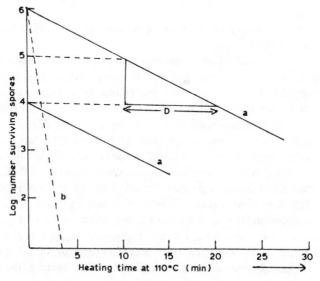

Fig. 3.10. Hypothetical survivor curves showing: (a) curves for heat-resistant spore, and (b) curve for heat-sensitive spore.

has been suggested that this growth may result from localized regions in the food having slightly higher pHs than those recorded.

The rate of killing bacterial spores (or vegetative cells for that matter, although they are much less heat-resistant) is a function of temperature and time; the higher the temperature, the greater the rate of destruction for any given time. Bacterial death is said to be logarithmic which means that equal percentages of surviving cells are killed in each successive unit of time. As can be seen in Fig. 3.10, the *D* value, represented by the slope of the survival curve is defined as the time at any given temperature for a 90% reduction in viability to be effected (thus for the heat-resistant spore the *D* value is *ca* 10 min whilst for the heat sensitive spore the value is less than 1 min). This means that the greater the number of spores that are present in any given food, the longer the time necessary for their destruction. From a practical standpoint it is therefore important to ensure that foods being canned have a predictable number of spores in them which can be destroyed by the normal heat treatment applied; thus the microbiological quality of raw materials and process line sanitation standards must be carefully monitored.

3.14.2.1. Spoilage of Low Acid Foods (pH >4.5)

Foods in this category include canned meats, poultry, fish, vegetables, soups, baked beans, spaghetti and milk puddings, and spoilage of these foods is caused mainly by *Bacillus* and *Clostridium* spp. due to the heat resistance of their spores.

B. stearothermophilus is the only species of *Bacillus* of commercial importance and is responsible for 'flat sour spoilage', i.e. it produces acids from carbohydrates to sour the food but no gas is formed so the ends of the can retain their normal shape. *B. stearothermophilus* is an obligate thermophile, all strains grow at 65°C and none below about 35°C. Extremely heat-resistant spores are produced which are some ten times more resistant than the spores of *C. botulinum*. Thus foods that are processed to destroy *C. botulinum* spores may well contain viable *B. stearothermophilus* spores. However, because of its high minimum growth temperature, cans held at temperatures below 35°C will not spoil even though the spores may be present (Gillespy & Thorpe, 1968). This is a good example of the concept of a commercially sterile pack. Canned foods susceptible to this type of spoilage include peas and similar vegetables which receive a fairly lengthy heat process. Although *B. coagulans* has also been implicated in flat sour spoilage, the spores of this organism are not very heat-resistant and it is of little importance nowadays.

C. thermosaccharolyticum is another important thermophilic organism producing spores that are not as heat-resistant as those of *B. stearothermophilus*; it produces 'swell' or 'TA' (thermophilic anaerobe) spoilage in canned foods. Such spoilage is caused by the fermentation of carbohydrates to form acids and large quantities of gas, carbon dioxide and hydrogen; if sufficient pressure builds up, the ends of the can become distended and eventually the side seam may split and the can contents burst out. Canned beans in tomato sauce are amongst the foods most commonly affected although rates of spoilage by the organism in the UK are negligible (Gillespy & Thorpe, 1968).

C. nigrificans, or more correctly *Desulfotomaculum nigrificans*, produces 'sulphur stinker' spoilage. *D. nigrificans* is essentially a Gram negative obligate anaerobe producing spores with a heat resistance intermediate between *B. stearothermophilus* and *C. thermosaccharolyticum* spores (Speck, 1981). Hydrogen sulphide, produced by the breakdown of protein, is soluble in the product and reacts with the iron of the container to produce iron sulphide causing blackening of the food. As excess hydrogen sulphide is formed, the aroma of the characteristic spoilage is very unpleasant. Fortunately this type of spoilage is rare due to the extremely low incidence of the spores of the organism and to its high minimum growth temperature. Sulphur stinker spoilage has been associated with both canned mushrooms and milk puddings in recent years.

Control of thermophilic spore formers which survive the heat treatment is best achieved by ensuring that cans are cooled as rapidly as possible after processing. In particular, 'flat sour' thermophiles multiply very rapidly in the 50–70°C range and therefore failure to cool cans immediately after processing to a temperature of *ca* 35°C could allow considerable multiplication and lead to serious spoilage. Control is also aided by: (a) ingredient (sugar, starch, dried milk, etc.) selection to assure freedom from or only low numbers of thermophilic spores; (b) thorough cleaning of raw materials and (c) sound line sanitation.

3.14.2.2. Canned Hams

The microbiology of canned hams is rather different from other low acid foods and merits separate discussion. The preservation of canned hams is partly based on a mild heat treatment which kills many of the more heat-sensitive bacteria; for this to be achieved an internal temperature of *ca* 70°C should be attained. Due to their relative heat stability, certain streptococci often cause spoilage problems such as gelatin liquefaction and souring. Clostridia and bacilli are also implicated and in recent years the potential danger of *Clostridium botulinum* in this food has aroused much interest. A number of interrelated factors, i.e. salt, nitrite, pH and storage temperature, play an important part in preservation by inhibiting the outgrowth of germinated spores. In a review, Ingram (1976) pointed out that spores damaged by heating in meats are more readily inhibited by nitrite than unheated spores. The indications are that nitrite inhibits *C. botulinum* by several different mechanisms although inactivation of specific enzymes would appear to be a major cause (Carpenter *et al.*, 1987); inhibitory effects are further stimulated by increasing the salt concentration and/or decreasing the pH. Misgivings have been expressed about the possible toxicity of nitrites to humans. It has been suggested that the nitrites may react with some of the amines naturally present in meat to form nitrosamines, some of which have been shown to be carcinogenic under certain conditions of use (Crosby & Sawyer, 1976). Thus since nitrite has a key role in inhibiting spore outgrowth we are faced with a dilemma and much research has been undertaken to elucidate the problems. Alternative methods of preservation have been suggested including the use of nisin, an antibiotic produced by lactic streptococci (see p. 142), which is permitted in foods (Hurst, 1981). Nisin is particularly effective if combined with low levels of nitrite (Rayman *et al.*, 1981). Sorbic acid or potassium sorbate can also be used in combination with nitrite at reduced levels; their effectiveness is enhanced by the inclusion of phosphoric acid which is again a permitted additive (Huhtanen *et al.*, 1983). Whichever curing system is to be used in the future it would seem that nitrite levels are inevitably going to be reduced; nitrite may be retained at low levels because of its useful role in stabilizing meat colour.

3.14.2.3. Spoilage of High Acid Foods (pH <4·5)

Foods with pH values of <4·5 include tomatoes, pears, peaches, pineapples and other fruits and their juices, together with pickles and sauces. Microorganisms causing spoilage of these products are much more varied and a number of the more important ones and the spoilage effects they induce are outlined below.

The thermophilic 'flat sour' organism *B. coagulans* can grow in pH values in the 4·0–4·5 range and is sometimes involved in the spoilage of tomatoes and tomato juice, usually as a result of seam leakage. *B. coagulans* spores are heat-sensitive and should be destroyed by the mild heat treatments used, although as an additional safety measure processing temperatures in excess of 100°C have been recommended (Thompson, 1981). Certain clostridia, particularly *C. pasteurianum* and *C. butyricum*, also grow well in this pH range and cause spoilage, with gas production, of canned tomatoes, pears and other fruits. The spores produced by these clostridia are even more heat labile and since they are destroyed in 15 min at 100°C it is unlikely that processing is inadequate.

Amongst the non-spore forming bacteria the lactic acid bacteria are most frequently incriminated in the spoilage of these high acid foods (Dennis, 1987). *Lactobacillus brevis* commonly causes fermentation of tomato ketchup, Worcester sauce, pickles and salad dressings. Other lactobacilli and leuconostocs occasionally cause spoilage of a range of canned fruits and fruit juices. These lactic acid bacteria (i.e. some *Lactobacillus* spp. and all *Leuconostoc* spp.) produce gas as well as acid from the syrup sugar so that spoilage is accompanied by can distension; other end products include acetic acid and ethyl alcohol so that with these diverse products such lactic acid bacteria are termed 'heterofermentative'. With 'homofermentative' lactic acid bacteria (i.e. the remaining lactobacilli and all *Streptococcus* spp.) only lactic acid is produced by the fermentation of sugar.

Yeasts are extremely heat-sensitive and are therefore rarely involved in spoilage of canned foods. Certain *Torulopsis* spp. may occasionally cause gaseous spoilage of sweetened condensed milk, which relies upon the high sugar content rather than upon a substantial heat treatment as the method of preservation. Other yeasts, *Saccharomyces* spp., have produced spoilage in citrus juices and pickles. Like yeasts, moulds rarely cause spoilage but there are two notable exceptions in *Byssochlamys fulva* and *B. nivea* the ascospores of which are unusually heat-resistant tolerating 85°C for 30min. The foods attacked are mainly strawberries and raspberries which may totally disintegrate in spoilage due to the action of the pectolytic enzymes produced by the organisms (Put & Kruiswijk, 1964). Control of this type of spoilage is best achieved by pre-treating the infected fruit with gaseous methyl bromide or peracetic acid and by careful cleaning of both the raw material and the processing equipment.

3.15.　FROZEN FOODS

3.15.1.　Influence of Sub-Zero Temperatures on Microorganisms

Freezing normally commences in foods at −1 to −3°C and as the temperature is further reduced more of the water in food becomes frozen. Thus at temperatures slightly below 0°C unfrozen water is available for the growth of microorganisms and growth down to $-7°C$ is possible for a few specialized bacteria and even down to $-10°C$ for certain moulds (Ingram & Mackey, 1976). As the temperature falls from $0°C$ a series of eutectics (i.e. ice: solute mixtures) is formed which is accompanied by an increasing concentration of dissolved solids in the unfrozen water. As well as lowering the freezing point of the remaining unfrozen water these increasing solute concentrations also progressively lower the a_w and this has an increasingly deleterious effect on the microbial population; thus organisms capable of growth in foods at sub-zero temperatures must also tolerate lowered a_w values. A small percentage of water remains unfrozen at temperatures well below $-100°C$; however, for practical purposes the 'freezable' water in meat and fish is totally frozen at -50 to $-70°C$ whilst for fruits and vegetables the corresponding figures are -16 to $-20°C$.

3.15.2. Factors Affecting Viability of Microorganisms During Freezing

Although some microorganisms are killed by freezing, approximately 50% may survive although this figure is influenced by a number of factors including the type of organism, the rate of freezing and the composition of substrate being frozen (MacLeod & Calcott, 1976).

Bacterial spores are unaffected by freezing and, in general, Gram positive rods and cocci are more resistant than Gram negative bacteria. At conventional freezing rates varying from the 'slow' process used in the home freezer unit to the 'quick' processes used by the food industry it has long been established that viability of organisms is enhanced as the freezing rate increases (Fig. 3.11, curve (a)). This increase of survival is probably mainly due to the diminishing contact time of the susceptible organisms with harmful high solute concentrations in the unfrozen water. When freezing is more rapid, viability decreases probably due to the formation of internal ice crystals causing destruction of the cell membranes (curve (b)). With extremely fast freezing rates, as, for example, when liquid nitrogen is used, ice crystal formation is reduced and is replaced by 'vitrification' (curve (c)). When foods are frozen commercially the bacterial viabilities obtained will be predominantly as in curve (a). There are certain substances such as glucose, milk solids, fats and sodium glutamate which are known to be 'protective' and improved viabilities are obtained in their presence; the mechanisms by which these cryoprotectants prevent freezing damage have yet to be elucidated (Mackey, 1984).

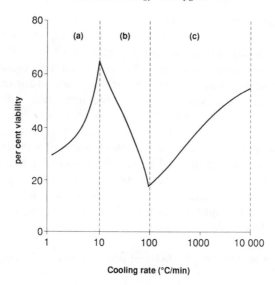

Fig. 3.11. Effect of freezing on viability of typical Gram negative rod (after MacLeod & Calcott, 1976).

3.15.3. Effect of Cold Storage

Whilst the main losses in viability occur during initial freezing, further kill-off of bacteria occurs during frozen storage. Provided the storage temperature is low enough, death rates are minimal but at normal frozen food storage temperatures (−20°C) some loss in viability is evident, particularly in the early days of storage. Foods show a far greater reduction in viable counts when held at −5 to −10°C than at −20°C but whilst the higher storage temperatures may be an effective method of reducing counts they contribute to an increased rate of deterioration of the food resulting from other causes. Even when microbial growth is completely inhibited the product quality can still deteriorate due to the continued activity of released microbial enzymes or to the indigenous enzymes present in the food; in the case of vegetables these enzymes must be destroyed by blanching. Other harmful physico-chemical and biochemical changes can occur during freezing and cold storage (Boegh-Soerensen & Jul, 1985).

3.15.4. Freezing Injury to Cells

When bacteria are frozen and subsequently thawed three categories of cells can be distinguished: uninjured, injured and killed. Uninjured cells are capable of growth on minimal nutritive media or on the selective media normally used in their isolation; conversely, killed cells would be unable to grow on any medium. Injured cells are more demanding nutritionally whilst repair of the freezing-induced injury is being effected; they only grow on media which provide certain energy-requiring factors necessary for the repair of the injury. This repair is rapid being completed in under 2 h; it can also be performed by cells in the thawed-out food provided the required nutrients are available (Ray *et al.*, 1972). This finding has important applications when using selective media for the enumeration of bacteria from frozen foods; recoveries may be substantially reduced, giving a false picture, if injury repair is precluded. Repair is best effected by pre-incubating the samples in a nutritionally complex medium for *ca* 2 h before enumerating bacteria on selective media but many modifications to this protocol have been suggested (Ray, 1986). It is equally possible for repair to be effected in most frozen foods after they have been thawed out but thaw out times for bulkier food items could be lengthy thus delaying analysis.

3.15.5. Thawed Foods and Their Spoilage

When foods are very rapidly frozen the number of microorganisms surviving the freeze-thaw cycle is partly dependent on the rate of thawing, somewhat lower recoveries being obtained with slower thawing; this is due to the growth of the very small ice crystals within the microbial cell causing increased cellular damage (Mackey, 1984). The survivors start multiplying, as in the normal growth cycle, after a lag period (see pp. 17–18) but this period is extended by the inherently low temperature of the food so that the log phase of growth may take 3–6 h to become established. When frozen foods are allowed to defrost over a long period at, say, 3–10°C psychrotrophs may dominate the flora and subsequently cause spoilage. In other cases the types of organisms growing will depend on the temperature at which the thawed food is held but with most foods the organisms predominating would be similar to those in the equivalent unfrozen product. Particular problems arise with larger packs, e.g. frozen turkeys, where a temperature gradient is established between the warm surface and the cold interior. If such packs are defrosted at too high a temperature growth of bacteria on the surface could become quite rapid. In general, however, if thawing is reasonably fast and the food is used within a few hours no problems should arise. Within this time it is perfectly safe to re-freeze thawed food, although to maintain textural, flavour and nutritional qualities it may be unwise.

There are a number of cases where the spoilage flora of the thawed frozen food is different from that of the harvested counterpart and of these the spoilage of defrosted peas is a good example. During processing, leuconostocs and streptococci build up on the production lines and it is these organisms which dominate the spoilage flora; they attack the sugars (mainly sucrose) present in the peas with a consequent drop in pH accompanied by a yellow appearance. Other changes induced include the production of copious amounts of slime on the surface of the peas together with typical 'vinegary' or 'buttery' odours (Cavett *et al.*, 1965). The spoilage profiles in other frozen green vegetables are essentially similar (Dennis, 1987).

3.16. DEHYDRATED FOODS

3.16.1. Methods of Drying

Drying is the oldest method of food preservation and can be performed in a number of different ways. Reference has already been made to salting and smoking which act indirectly by lowering the a_w of the food. Direct removal of water can be effected by three basic methods, i.e. sun drying, mechanical drying and freeze drying.

Sun drying, restricted to hot, dry climates, is employed with fruits such as raisins, prunes and figs which are spread out on trays and occasionally turned whilst drying. Conventional mechanical drying, performed in kilns or tunnels involves the passage of heated air over foods, particularly vegetables, which must be diced or shredded to increase the surface area:volume ratio thus enhancing drying efficiency. Dehydration today widely utilizes the 'fluidized-bed' principle in which warm air is blown up between the food particles creating an agitated state thereby further enhancing drying efficiency. Liquid foods such as milk and egg products and coffee can be spray-dried by atomizing the liquid into a current of hot air; this technique has largely superseded drum drying previously extensively used for milk products. Freeze drying is basically the dehydration of frozen material under a high vacuum and with heating limited so as not to prevent the ice being converted directly to water vapour (i.e. sublimation). This technique, commonly used for meats and fish which are denatured by normal drying methods, gives the highest quality product obtainable by any drying method; in particular, cell damage is minimized as are various degradative reactions which often occur during conventional drying such as protein denaturation and enzymatic and non-enzymatic browning reactions.

3.16.2. Influence of Drying and Freeze Drying on Microorganisms

In conventional drying used for vegetables the microbial flora will be modified by initial processing operations such as dicing; increases in counts can be anticipated unless the equipment is scrupulously clean. However, as with frozen vegetables, blanching must be carried out on most vegetables and this substantially reduces microbial counts. Although temperatures of up to 90°C are used during the first drying stage, the rapid moisture loss from foods during this period induces a cooling effect and helps to maintain the temperature between 40 and 50°C; hence only a small reduction in microbial counts occurs. In the secondary stage of drying food temperatures are higher (60–70°C) and yeasts and many bacteria are killed at this time; counts of 10^3–10^4 per gram should be achievable (Dennis, 1987). The residual flora after drying is composed principally of spore formers (*Bacillus* spp.), enterococci and a variety of moulds (e.g. *Aspergillus*, *Penicillium*, *Alternaria* and *Cladosporium* spp.). With spray-dried milk and egg the temperatures reached are not so high and in consequence a much more varied flora may persist although it is again likely to be dominated by spore formers and thermoduric streptococci; *Salmonella* spp. may survive the spray-drying process so that milk so used is now pasteurized prior to drying (Mossel & Shennan, 1976).

Since freeze drying was originally devised as a means of preserving biological material it is not surprising that microbial counts in freeze-dried foods are often high. Foods must be frozen before being dried and the effects of freezing, discussed earlier, are manifested; in fact more bacteria are killed by freezing than by dehydration during the freeze drying process (Frazier & Westhoff, 1988). Although drying is performed from the frozen state under vacuum, heat can still be applied to sublimate the ice. As the frozen interface recedes, temperatures near the surface will increase to those of the heating plate (40–50°C) and in consequence some destruction of heat-sensitive bacteria in this region is now possible. However, it is only in the very last stages of drying that the core temperature increases above 0°C so no kill-off can be anticipated in this zone. These conditions mean that something like 30% of the original flora is present after freeze drying and counts on foods dried in this way often exceed 100000 per gram (Saleh *et al.*, 1966).

3.16.3. Storage Stability of Dried Foods

Many changes can occur during the storage of dried foods and most are non-microbial in origin. The commonest and most important change is non-enzymic (Maillard) browning which involves a complex series of chemical reactions between reducing sugars and amino acids or proteins. The a_w levels that must be achieved during drying to stop this browning are far lower than those required to inhibit microbial growth so that microbial spoilage should not arise. In fact during storage there is a decrease in numbers of viable organisms although the spores of bacteria and moulds remain unaffected. If dried foods are incorrectly packaged or are stored under moist conditions, sufficient water may be re-absorbed to enable moulds to grow but the water uptake should not be such as to permit bacterial growth.

3.16.4. Rehydration

When dried foods are rehydrated similar responses are shown by the contained microorganisms as with thawed frozen foods; there is a lag phase of growth and many organisms exhibit metabolic injury (Gibbs, 1986). Clearly the temperature of the water used for rehydration may have a marked effect on the flora and on the subsequent rate of spoilage of the food. If boiling water is used *Bacillus* spp. will predominate and cause spoilage and at progressively lower rehydration temperatures the flora will become more varied and contain more heat-labile organisms. When refrigerated storage is used the storage life of most rehydrated foods is restricted to 1 or 2 days but storage at room temperature should clearly be limited to no more than a few hours.

3.16.5. Intermediate Moisture Foods

The term 'intermediate moisture foods' is applied to a heterogeneous collection of foods with a_w values in the range 0·60–0·85, equivalent to moisture contents of 20–40%, and which do not require refrigeration for stability. Foods in this group include dried fruits, certain bakery products and salted meats and fish all of which have already been discussed, together with jams, syrups and honey. Spoilage of the latter group is attributed to osmophiles, microorganisms growing in high sugar concentrations (65–70%) and tolerating low pH values (<4·0). The most common spoilage agents are osmophilic yeasts (*Saccharomyces* and *Torulopsis* spp.) which ferment the sucrose with the production of alcohol. Certain moulds may develop on the surface of jams, the most common being species of *Aspergillus* and *Penicillium*. Since osmophiles are heat-sensitive and therefore readily killed during heat processing, spoilage of these products is only possible after recontamination which may occur through faulty sealing or after opening the containers. Furthermore, it is probable that moisture re-absorption is necessary before growth can be initiated.

3.17. IRRADIATED FOODS

3.17.1. Types of Radiation

The types of radiation potentially available for the preservation of foods include alpha, beta and gamma radiation, sources of α-, β- and γ-particles respectively. These radiations are of a high frequency with a high energy content and they have the power to break molecules into oppositely charged units termed ions; these radiations are therefore called ionizing radiations. Ultraviolet (uv) radiations have a lower frequency and energy but are also lethal to microorganisms.

Alpha radiation is of no value since α-particles have very poor penetration characteristics and are even stopped by a sheet of paper! Beta particles, which are identical to electrons, also suffer from relatively poor penetration characteristics, the maximum depth of penetration into foods being only 1–2 cm; their use is therefore restricted to thin food products such as beefburgers and to the surface treatment of foods to reduce microbial loads. Gamma particles, short wavelength electro-magnetic radiations, have excellent powers of penetration; foods of up to 30–40 cm thick can be treated. Since particle sources are readily available in the form of radioisotope cobalt 60, this is the most suitable γ-particle emitter for industrial use; in fact there were nearly 100 cobalt 60 plants in operation worldwide in the early 1980s (Ley, 1983).

Because of consumer resistance to gamma radiation, uv radiation has been suggested as an alternative means of treatment of foods. Unfortunately, powers of penetration are very low so that its use is restricted to treatment of food surfaces such as raw meats where substantial reductions (>99%) in bacterial numbers can be achieved (Stermer *et al.*, 1987); it is also useful in the continual treatment of bread wrapping machines, moulds being removed from the packaging material.

3.17.2. Effect of Radiation on Microorganisms

The sensitivity of microorganisms to radiation varies enormously (Table 3.5). Gram negative bacteria are generally more sensitive than Gram positive forms with bacterial spores being strongly resistant. The comparative sensitivity of Gram negative bacteria is of considerable significance since many common food spoilage organisms (e.g. pseudomonads) and food poisoning bacteria (e.g. salmonellas) are readily inactivated by relatively low irradiation doses. Unfortunately these advantages can be counterbalanced by the radiation resistance exhibited by bacterial spores including those of *Clostridium botulinum*; their presence in irradiated foods can be a cause of concern where the background bacterial flora has been eliminated. As in the case of inactivation by heat (see p. 159) bacterial death is again essentially logarithmic, so the more bacteria (or spores) that are present in the food the greater the radiation dose required to eliminate them.

In general, the radiation resistance of moulds is similar to vegetative bacteria whilst yeasts tend to be rather more resistant (ICMSF, 1980). Viruses are unaffected except at doses equivalent to or greater than those required to inactivate spores; some of the smallest viruses require doses of >200 kGy to achieve a million-fold reduction in their numbers.

Inactivation of bacteria, yeasts and moulds mainly results from damage to DNA and the differing sensitivities of these microorganisms is a reflection of the efficiency of their DNA repair mechanisms (Moseley, 1984). Ultraviolet light-induced damage is mainly associated with the formation of cross-linkages in DNA causing distortion of the molecule and hence suppression of DNA replication. Ionizing radiation primarily causes breaks in the DNA strand with the same consequence.

For the organism to survive it is imperative that it is able to repair the damage quickly. Thus there are a number of different repair mechanisms, some involving a whole battery of enzymes; repair is effected swiftly, from fractions of seconds to a few minutes. If there are too

Table 3.5
Approximate Radiation Doses to Effect a Reduction of a Million in Viable
Numbers of Different Microorganisms[a]

Microorganisms	Dose (kGy)
Gram negative bacteria	
Escherichia coli	2
Salmonella enteritidis	4
Vibrio parahaemolyticus	<1
Moraxella sp	7
Pseudomonas fluorescens	<1
Gram positive bacteria	
Micrococcus sp	4
Staphylococcus aureus	5–10
Streptococcus faecalis	5
Leuconostoc sp	3
Bacillus sp (vegetative cells)	3
Bacillus cereus (spores)	25
Clostridium perfringens (spores)	25
Clostridium botulinum A (spores)	25
Moulds and Yeasts	
Aspergillus flavus	3
Candida sp	4
Saccharomyces cerevisiae	10
Viruses	>30

[a]From ICMSF (1980).

many lesions on the DNA for the organism to cope with, replication
will cease. Some of the more complex enzyme repair systems are found
in all organisms including humans although the repair mechanisms are
far less efficient in higher animals; thus a human would certainly be
killed by exposure to a radiation dose of 0.005 kGy.

3.17.3. High Dose Applications

Following favourable reports from various committees it was agreed by
the Board of the International Committee on Food Microbiology and
Hygiene that the irradiation of food up to a dose of 10 kGy introduced
no special microbiological problems (ICFMH, 1983). High dose irradiation treatments have been somewhat arbitrarily defined as applications

of >10 kGy so that, following the ICFMH recommendations, their use is likely to be extremely limited. In fact, high dose applications are generally concerned with the sterilization of foods and, from a microbiological standpoint, have much in common with the thermal sterilization of canned foods. Thus in foods with a pH of >4·5 the sterilizing dose must again be based on calculations assuring the absolute elimination of *Clostridium botulinum* spores; this means in practice that doses of 45 kGy are necessary. Such high doses result in undesirable side effects such as the formation of offensive odours, texture breakdown and changes in colour (Urbain, 1978).

In spite of these constraints there are still specialized niches for radiation sterilized foods. Patients with reduced immune responses such as those with AIDS, certain types of cancer, and heart or kidney transplant patients can be given diets based on radiation sterilized foods; similar foods are supplied to astronauts. Slightly lower doses (e.g. 25 kGy) have been proposed for the treatment of canned hams which would allow a substantial reduction in nitrite levels but still ensure safety from *C. botulinum* (see p. 161). However, the fear of botulism understandably remains so that radiation sterilized foods must have a limited market. This is stressed by the fact that toxins produced by food poisoning bacteria are remarkably radiation resistant (Rose *et al.*, 1988). Thus *C. botulinum* type A toxin retains 15% of its activity with a radiation dose of *ca* 25 kGy whilst *Staphylococcus aureus* enterotoxin A retains slight activity even at >50 kGy. High radiation doses in the 10–20 kGy range have found a wider use in the treatment of spices. In this application the objective is to eliminate bacterial cells and substantially reduce the numbers of bacterial spores which are often present in enormous numbers on natural spices (Urbain, 1978).

3.17.4. Low Dose Applications and the Spoilage of Foods so Treated

Radiation processing involving treatments of up to 10 kGy is often referred to as 'radiation pasteurization' although this category has now been sub-divided based on the objective in mind. Thus if extended shelf-life is the aim the process is termed 'radurization'; if it is to remove particular pathogens, 'radicidation' (ICMSF, 1980). In practical terms, however, essentially similar kill-offs are achieved with both processes since similar radiation doses are employed.

Radurization, the removal of food spoilage organisms in order to extend the shelf-life of foods, has been applied mainly to meats and fish.

Relatively low doses of 1–2 kGy are sufficient to destroy *Pseudomonas* spp. as well as much of the remainder of the indigenous Gram negative flora on raw meats (Tarkowski *et al.*, 1984); a dose of 2·5 kGy will eliminate pseudomonads (>10^5 per g) from raw meat whilst Enterobacteriaceae (i.e. *Escherichia coli, Salmonella* spp. etc.) and *Brochothrix thermosphacta* show slightly greater resistance (Niemand *et al.*, 1983). The more radiation resistant species remaining on meats, poultry and fish will typically include moraxellas, lactic acid bacteria and, at the lower end of the dose range, *B. thermosphacta*. Such organisms will cause spoilage eventually although an extension of the shelf-life by a factor of 3 or 4 can be anticipated since their growth rates and spoilage potential are much lower than that of pseudomonads (ICMSF, 1980).

Mention has already been made of the comparative sensitivity of salmonellas to radiation so that the practicability of radicidation has been explored, particularly in relation to poultry. Salmonellas are more radiation resistant than pseudomonads but numbers are normally low on meats and poultry so that a lower dose will suffice. Mulder *et al.* (1977) have suggested that 2·5 kGy is sufficent to eliminate salmonellas from chicken carcasses since they are normally present at <1000 per carcass. If exceptionally high numbers (e.g. 10^6 per g) were to be present doses of 10 kGy may be required; unfortunately such a dose will partially destroy sensitive vitamins (e.g. thiamine) and produce off-odours but these problems can be overcome by irradiating the poultry in the frozen state which has little effect on the killing power (Hanis *et al.*, 1989).

As well as using irradiation as a means of inactivating salmonellas, its use for the elimination of *Listeria monocytogenes* has also been suggested. Patterson (1989) found that the radiation sensitivity of *L. monocytogenes* on chicken meat was similar to that of salmonellas and concluded that doses of 2·5–7 kGy would be sufficient to remove the organism. Similar findings have been reported by Huhtanen *et al.* (1989) who concluded that *L. monocytogenes* could be eliminated from meats with doses of <10 kGy and that a dose of 2 kGy was sufficient to destroy 10^4 cells of the organism. Other food poisoning and pathogenic bacteria such as *Vibrio parahaemolyticus, Yersinia enterocolitica, Campylobacter* spp. and *Shigella* spp. are equally susceptible and could effectively be inactivated from foods by low irradiation doses (Farkas, 1989). Finally, parasites such as *Taenia solium, T. saginata* and *Trichinella spiralis* can be successfully treated in meats at dose levels of up to 7 kGy (Urbain, 1978).

There are, however, certain meats such as bacon where radiation appears to be of questionable value. The indigenous flora including micrococci, lactic acid bacteria and acinetobacters is inherently more resistant to the effects of irradiation and doses of 10 kGy could prove inadequate; more seriously foreign odours persist in these higher lipid containing foods after treatment even though the appearance is unaffected (Dempster *et al.*, 1986).

In conclusion, radiation as a means of food processing produces no unique microbiological problems—the problem of the resistant organism(s) is common to many food processing techniques. Thus the same degree of control must be exercised with radiation processing as with any other type of process so that the quality of raw materials and the hygiene standards applied on the processing line, and to the finished product, are as important here as with any other process.

BIBLIOGRAPHY

Ayres, J. C., Mundt, J. O. & Sandine, W. E. (1980). *Microbiology of Foods*. San Francisco, Freeman.

Bean, P. G. (1983). Developments in heat treatment processes for shelf-stable products. In: *Food Microbiology: Advances and Prospects*. Eds T. A. Roberts and F. A. Skinner, London, Academic Press.

Davies, F. L. & Law, B. A. (1984). *Advances in the Microbiology and Biochemistry of Cheese and Fermented Milk*. London, Elsevier Applied Science Publishers.

Dempster, J. F. (1985). Radiation preservation of meat and meat products: a review. *Meat Science*, **12**, 61–89.

Dixon, N. M. & Kell, D. B. (1989). A review: the inhibition by CO_2 of the growth and metabolism of micro-organisms. *Journal of Applied Bacteriology*, **67**, 109–36.

Gardner, G. A. (1982). Microbiology of processing: bacon and ham. In: *Meat Microbiology*. Ed. M. H. Brown, London, Applied Science Publishers Ltd.

Gill, C. O. (1982). Microbial interactions with meats. In: *Meat Microbiology*. Ed. M. H. Brown, London, Applied Science Publishers Ltd.

Gray, J. I. & Pearson, A. M. (1984). Cured meat flavour. *Advances in Food Research*, **29**, 1–86.

Hauschild, A. H. W. & Simonsen, B. (1985). Safety of shelf-stable canned cured meats. *Journal of Food Protection*, **48**, 997–1009.

Herbert, D. A. & Bettison, J. (1987). Packaging for thermally sterilised foods. In: *Developments in Food Preservation—4*. Ed. S. Thorne, London, Elsevier Applied Science.

Hersom, A. C. & Hulland, E. D. (1980). *Canned Foods-Thermal Processing and Microbiology*, 7th edn, Edinburgh, Churchill Livingstone.

International Commission on Microbiological Specifications for Foods (ICMSF) (1980). *Microbial Ecology of Foods*, Vol. 2 (Food Commodities). New York, Academic Press.

Lund, B. M. (1982). The effect of bacteria on post-harvest quality of vegetables and fruits, with particular reference to spoilage. In: *Bacteria and Plants*. Eds M. E. Rhodes-Roberts and F. A. Skinner, London, Academic Press.

McMeekin, T. A. (1982). Microbial spoilage of meats. In: *Developments in Food Microbiology—1*. Ed. R. Davies, London, Applied Science Publishers Ltd.

Mayes, F. J. & Takeballi, M. A. (1983). Microbial contamination of the hen's egg: a review. *Journal of Food Protection*, **46**, 1092–8.

Mead, G. C. (1983). Effect of packaging and gaseous environment on the microbiology and shelf life of processed poultry products. In: *Food Microbiology: Advances and Prospects*. Eds T. A. Roberts and F. A. Skinner, London, Academic Press.

Murphy, M. F. (1981). Microbiology of butter. In: *Dairy Microbiology*, Vol. 2 (The Microbiology of Milk Products). Ed. R. K. Robinson, London , Applied Science Publishers.

Nielsen, H-J. S. (1985). Refrigerated storage of packaged meat. In: *Developments in Food Preservation—3*. Ed. S. Thorne, London, Elsevier Applied Science Publishers.

Nursten, H. E. (1986). Maillard browning reactions in dried foods. In: *Concentration and Drying of Foods*. Ed. D. MacCarthy, London, Elsevier Applied Science Publishers.

Patterson, M. F. (1990). A review. The potential for food irradiation. *Letters in Applied Microbiology*, **11**, 55–61.

Rainbow, C. (1981). Beer spoilage microorganisms. In: *Brewing Science*, Vol. 2. Ed. J. R. A. Pollock, London, Academic Press.

Van Arsdel, W. B., Copley, M. J. & Morgan, A. I. (1973). *Food Dehydration*, 2nd edn. Westport, Avi Publishing Co.

REFERENCES

Adams, R., Farber, L. & Lerke, P. (1964). Bacteriology of spoilage of fish muscle. II. Incidence of spoilers during spoilage. *Applied Microbiology*, **12**, 277–9.

Ayres, J. C. (1960). Temperature relationships and some other characteristics of the microbial flora developing on refrigerated beef. *Food Research*, **25**, 1–18.

Baker, D. A. & Genigeorgis, C. (1990). Predicting the safe storage of fresh fish under modified atmospheres with respect to *Clostridium botulinum* toxigenesis by modeling length of the lag phase of growth. *Journal of Food Protection*, **53**, 131–40.

Barnes, E. M. (1976). Microbiological problems of poultry at refrigerator temperatures — a review. *Journal of the Science of Food and Agriculture*, **27**, 777–82.

Barnes, E. M. & Corry, J. E. L. (1969). Microbial flora of raw and pasteurized egg albumen. *Journal of Applied Bacteriology*, **32**, 193–205.

Barnes, E. M. & Impey, C. S. (1968). Psychrophilic spoilage bacteria of poultry. *Journal of Applied Bacteriology*, **31**, 97–107.

Barnes, E. M. & Thornley, M. J. (1966). The spoilage flora of eviscerated chickens stored at different temperatures. *Journal of Food Technology*, **1**, 113–9.

Bartkowski, L., Dryden, F. D. & Marchello, J. A. (1982). Quality changes of beef steaks stored in controlled gas atmospheres containing high or low levels of oxygen. *Journal of Food Protection*, **45**, 41–5.

Blickstad, E. & Molin, G. (1983). Carbon dioxide as a controller of the spoilage flora of pork, with special reference to temperature and sodium chloride. *Journal of Food Protection*, **46**, 756–63.

Board, R. G. (1965). The properties and classification of the predominant bacteria in rotten eggs. *Journal of Applied Bacteriology*, **28**, 437–53.

Board, R. G. (1966). Review article: the course of microbial infection of the hen's egg. *Journal of Applied Bacteriology*, **29**, 319–41.

Board, R. G. (1969). The microbiology of the hen's egg. *Advances in Applied Microbiology*, **11**, 245–81.

Board, R. G. & Fuller, R. (1974). Non-specific antimicrobial defences of the avian egg, embryo and neonate. *Biological Reviews*, **49**, 15–49.

Boegh-Soerensen, L. & Jul, M. (1985). Effects of freezing/thawing on foods. In: *Microbiology of Frozen Foods*. Ed. R. K. Robinson, London, Elsevier Applied Science Publishers.

Brocklehurst, T. F., Zamon-Wong, C. M. & Lund, B. M. (1987). A note on the microbiology of retail packs of prepared salad vegetables. *Journal of Applied Bacteriology*, **63**, 409–15.

Bucky, A. R., Hayes, P. R. & Robinson, D. S. (1988). Enhanced inactivation of bacterial lipases and proteinases in whole milk by a modified ultra high temperature treatment. *Journal of Dairy Research*, **55**, 373–80.

Campbell, I. (1987). Microbiology of brewing: beer and lager. In: *Essays in Agricultural and Food Microbiology*. Eds J. R. Norris and G. L. Pettipher, Chichester, John Wiley and Sons.

Carpenter, C. E., Reddy, D. S. A. & Cornforth, D. P. (1987). Inactivation of clostridial ferredoxin and pyruvate-ferredoxin oxidoreductase by sodium nitrite. *Applied and Environmental Microbiology*, **53**, 549–52.

Cavett, J. J., Dring, G. J. & Knight, A. W. (1965). Bacterial spoilage of thawed frozen peas. *Journal of Applied Bacteriology*, **28**, 241–51.

Chapman, H. R. & Sharpe, M. E. (1981). Microbiology of cheese. In: *Dairy Microbiology*, Vol. 2 (The Microbiology of Milk Products). Ed. R. K Robinson, London, Applied Science Publishers.

Cousins, C. M. & Bramley, A. J. (1981). The microbiology of raw milk. In: *Dairy Microbiology*, Vol. 1 (The Microbiology of Milk). Ed. R. K. Robinson, London, Applied Science Publishers.

Crosby, N. T. & Sawyer, R. (1976). N-nitrosamines: a review of chemical and biological properties and their estimation in foodstuffs. *Advances in Food Research*, **22**, 1–71.

Crouse, J. D., Anderson, M. E. & Naumann, H. D. (1988). Microbial decontamination and weight of carcass beef as affected by automated washing

pressure and length of time in spray. *Journal of Food Protection*, **51**, 471–4.
Cuppett, S. L. *et al.* (1987). Effect of salt level and nitrite on toxin production by *Clostridium botulinum* Type E spores in smoked Great Lakes Whitefish. *Journal of Food Protection*, **50**, 212–7.
Cutting, C. L. (1965). Smoking. In: *Fish as Food*, Vol. 3 (Processing: Part 1). Ed. G. Borgstrom, New York, Academic Press.
Dainty, R. H. (1971). The control and evaluation of spoilage. *Journal of Food Technology*, **6**, 209–24.
Dainty, R. H., Shaw, B. G., De Boer, K. A. & Scheps, E. S. J. (1975). Protein changes caused by bacterial growth on beef. *Journal of Applied Bacteriology*, **39**, 73–81.
Dainty, R. H., Shaw, B. G. & Roberts, T. A. (1983). Microbial and chemical changes in chill-stored red meats. In: *Food Microbiology: Advances and Prospects*. Eds T. A. Roberts and F. A. Skinner, London, Academic Press.
Dempster, J. F. (1972). Vacuum packed bacon; the effects of processing and storage temperature on shelf life. *Journal of Food Technology*, **7**, 271–9.
Dempster, J. F. (1973). Microbial progression in sliced vacuum packaged bacon at refrigeration temperatures. *Journal of Applied Bacteriology*, **36**, 543–52.
Dempster, J. F., McGuire, L. & Halls, N. A. (1986). Effect of gamma radiation on the quality of bacon. *Food Microbiology*, **3**, 13–7.
Dennis, C. (1987). Microbiology of fruits and vegetables. In: *Essays in Agricultural and Food Microbiology*. Eds J. R. Norris and G. L. Pettipher, Chichester, John Wiley and Sons.
Duckworth, R. B. (1966). *Fruit and Vegetables*. Oxford, Pergamon.
Easter, M. C., Gibson, D. M. & Ward, F. B. (1982). A conductance method for the assay and study of bacterial trimethylamine oxide reduction. *Journal of Applied Bacteriology*, **52**, 357–65.
Edwards, R. A., Dainty, R. H. & Hibbard, C. M. (1983). The relationship of bacterial numbers and types to diamine concentration in fresh and aerobically stored beef, pork and lamb. *Journal of Food Technology*, **18**, 777–88.
Edwards, R. A., Dainty, R. H. & Hibbard, C. M. (1987a). Volatile compounds produced by meat pseudomonads and related reference strains during growth on beef stored in air at chill temperatures. *Journal of Applied Bacteriology*, **62**, 403–12.
Edwards, R. A., Dainty, R. H., Hibbard, C. M. & Ramantanis, S. V. (1987b). Amines in fresh beef of normal pH and the role of bacteria in changes in concentration observed during storage in vacuum packs at chill temperatures. *Journal of Applied Bacteriology*, **63**, 427–34.
Egan, A. F. & Roberts, T. A. (1987). Microbiology of meat and meat products. In: *Essays in Agricultural and Food Microbiology*. Eds J. R. Norris and G. L. Pettipher, Chichester, John Wiley and Sons.
Farber, J. M. (1991). Microbiological aspects of modified-atmosphere packaging technology—a review. *Journal of Food Protection*, **54**, 58–70.
Farkas, J. (1989). Microbiological safety of irradiated foods. *International Journal of Food Microbiology*, **9**, 1–15.
Fieger, E. & Novak, A. F. (1961). Microbiology of shellfish deterioration. In: *Fish as Food*, Vol. 1 (Production, Biochemistry and Microbiology). Ed. G. Borgstrom, New York, Academic Press.

Frazier, W. C. & Westhoff, D. C. (1988). *Food Microbiology*, 4th edn. New York, McGraw-Hill.

Freeman, L. R. *et al.* (1976). Volatiles produced by microorganisms isolated from refrigerated chicken at spoilage. *Applied and Environmental Microbiology*, **32**, 222–31.

Garcia, G. W., Genigeorgis, C. & Lindroth, S. (1987). Risk of growth and toxin production by *Clostridium botulinum* nonproteolytic types B, E and F in salmon fillets stored under modified atmospheres at low and abused temperatures. *Journal of Food Protection*, **50**, 330–6.

Gardner, G. A. (1965). The aerobic flora of stored meat with particular reference to the use of selective media. *Journal of Applied Bacteriology*, **28**, 252–64.

Gardner, G. A. (1971). Microbiological and chemical changes in lean Wiltshire bacon during aerobic spoilage. *Journal of Applied Bacteriology*, **34**, 645–54.

Gardner, G. A. (1983). Microbial spoilage of cured meats. In: *Food Microbiology: Advances and Prospects*. Eds T. A. Roberts and F. A. Skinner, London, Academic Press.

Gardner, G. A. & Patton, J. (1969). Variations in the composition of the flora on a Wiltshire cured bacon side. *Journal of Food Technology*, **4**, 125–31.

Garrett, C. M. E. (1982). Bacterial diseases of food plants—an overview. In: *Bacteria and Plants*. Eds M. E. Rhodes-Roberts and F. A. Skinner, London, Academic Press.

Genigeorgis, C.A. (1985). Microbial and safety implications of the use of modified atmospheres to extend the storage life of fresh meat and fish. *International Journal of Food Microbiology*, **1**, 237–51.

Georgala, D. L. (1958). The bacterial flora of the skin of North Sea cod. *Journal of General Microbiology*, **18**, 84–91.

Gibbs, P. A. (1986). Microbiological quality of dried foods. In: *Concentration and Drying of Foods*. Ed. D. MacCarthy, London, Elsevier Applied Science Publishers.

Gill, C. O. (1976). Substrate limitation of bacterial growth at meat surfaces. *Journal of Applied Bacteriology*, **41**, 401–10.

Gill, C. O. & Newton, K. G. (1977). The development of aerobic spoilage flora on meat stored at chill temperatures. *Journal of Applied Bacteriology*, **43**, 189–95.

Gill, C. O. & Newton, K. G. (1980). Development of bacterial spoilage at adipose tissue surface of fresh meat. *Applied and Environmental Microbiology*, **39**, 1076–7.

Gillespie, N. C. & Macrae, I. C. (1975). The bacterial flora of some Queensland fish and its ability to cause spoilage. *Journal of Applied Bacteriology*, **39**, 91–100.

Gillespy, T. G. & Thorpe, R. H. (1968). Occurrence and significance of thermophiles in canned foods. *Journal of Applied Bacteriology*, **31**, 59–65.

Goswell, R. W. (1986). Microbiology of table wines. In: *Developments in Food Microbiology—2*. Ed. R. K. Robinson, London, Elsevier Applied Science Publishers.

Graikoski, J. T. (1973). Microbiology of cured and fermented fish. In: *Microbiological Safety of Fishery Products*. Eds C. O. Chichester and H. D. Graham, New York, Academic Press.

Griffiths, M. W., Phillips, J. D. & Muir, D. D. (1981). Thermostability of proteases and lipases from a number of species of psychrotrophic bacteria of
dairy origin. *Journal of Applied Bacteriology*, **50**, 289–303.

Handford, P. M. & Gibbs, B. M. (1964). Antimicrobial effects of smoke constituents on bacteria isolated from bacon.In: *Microbial Inhibitors in Food*.
Ed. N. Molin, Stockholm, Almqvist and Wiksell.

Hanis, T. *et al.* (1989). Poultry meat irradiation—effect of temperature on chemical changes and inactivation of microorganisms. *Journal of Food
Protection*, **52**, 26–9.

Heap, H. A. & Lawrence, R. C. (1988). Culture systems for the dairy industry.
In *Developments in Food Microbiology—4*. Ed. R. K. Robinson, London,
Elsevier Applied Science.

Herbert, R. A., Hendrie, M. S., Gibson, D. M. & Shewan, J. M. (1971).
Bacteria active in the spoilage of certain sea foods. *Journal of Applied
Bacteriology*, **34**, 41–50.

Hobbs, G. (1983). Microbial spoilage of fish. In: *Food Microbiology : Advances
and Prospects*. Eds T. A. Roberts and F. A. Skinner, London, Academic
Press.

Hobbs, G. (1987). Microbiology of fish. In: *Essays in Agricultural and Food
Microbiology*. Eds J. R. Norris and G. L. Pettipher, Chichester, John Wiley
and Sons.

Hobbs, G. & Hodgkiss, W. (1982). The bacteriology of fish handling and processing. In: *Developments in Food Microbiology—1*. Ed. R. Davies, London,
Applied Science Publishers Ltd.

Huhtanen, C. N., Feinberg, J. I., Trenchard, H. & Phillips, J. G. (1983). Acid enrichment of *Clostridium botulinum* inhibition in ham and bacon prepared with
potassium sorbate and sorbic acid. *Journal of Food Protection*, **46**, 807–10.

Huhtanen, C. N., Jenkins, R. K. & Thayer, D. W. (1989). Gamma radiation sensitivity of *Listeria monocytogenes*. *Journal of Food Protection*, **52**, 610–13.

Humphrey, T. J., Chapman, P. A., Rowe, B. & Gilbert, R. J. (1990). A comparative study of the heat resistance of salmonellas in homogenized whole egg,
egg yolk or albumen. *Epidemiology and Infection*, **104**, 237–41.

Hurst, A. (1981). Nisin. *Advances in Applied Microbiology*, **27**, 85–123.

Huss, H. H. *et al.* (1974). The influence of hygiene in catch handling on the
storage life of iced cod and plaice. *Journal of Food Technology*, **9**, 213–21.

Ingram, M. (1971). Microbial changes in foods-general considerations. *Journal
of Applied Bacteriology*, **34**, 1–8.

Ingram, M. (1976). The microbiological role of nitrite in meat products. In:
Microbiology in Agriculture, Fisheries and Food. Eds F. A. Skinner and J.
G. Carr, London, Academic Press.

Ingram, M. & Dainty, R. H. (1971). Changes caused by microbes in spoilage of
meats. *Journal of Applied Bacteriology*, **34**, 21–39.

Ingram, M. & Mackey, B. M. (1976). Inactivation by cold. In: *Inhibition and
Inactivation of Vegetative Microbes*. Eds F. A. Skinner and W. B. Hugo,
London, Academic Press.

International Committee on Food Microbiology and Hygiene (ICFMH) (1983).
The Microbiological Safety of Irradiated Food. Codex Alimentarius
Commission, Document CX/FH83/9.

International Commission on Microbiological Specifications for Foods (ICMSF) (1980). *Microbial Ecology of Foods*, Vol. 1 (Factors Affecting Life and Death of Microorganisms). New York, Academic Press.

Kelly, C. A., Dempster, J. F. & McLoughlin, A. J. (1981). The effect of temperature, pressure and chlorine concentration of spray washing water on numbers of bacteria on lamb carcases. *Journal of Applied Bacteriology*, **51**, 415–24.

Lake, D. E. *et al.* (1985). Enumeration and isolation of mesophilic anaerobic sporeformers from cannery post-processing equipment. *Journal of Food Protection*, **48**, 794–8.

Lerke, P., Adams, R. & Farber, L. (1965). Bacteriology of spoilage of fish muscle. III. Characterization of spoilers. *Applied Microbiology*, **13**, 625–30.

Lerke, P., Farber, L. & Adams, R. (1967). Bacteriology of spoilage of fish muscle. IV. Role of protein. *Applied Microbiology*, **15**, 770–6.

Ley, F. J. (1983). New interest in the use of irradiation in the food industry. In: *Food Microbiology: Advances and Prospects*. Eds T. A. Roberts and F. A. Skinner, London, Academic Press.

Lindroth, S. E. & Genigeorgis, C. A. (1986). Probability of growth and toxin production by nonproteolytic *Clostridium botulinum* in rockfish stored under modified atmospheres. *International Journal of Food Microbiology*, **3**, 167–81.

Liston, J. (1957). The occurrence and distribution of bacterial types on flatfish. *Journal of General Microbiology*, **16**, 205–16.

Mabbitt, L. A., Davies, F. L., Law, B. A. & Marshall, V. M. (1987). Microbiology of milk and milk products. In: *Essays in Agricultural and Food Microbiology*. Eds J. R. Norris and G. L. Pettipher, Chichester, John Wiley and Sons.

Macaskie, L. E., Sheard, A. G., Dainty, R. H. & Henderson, P. J. F. (1984). Glycerol utilization by *Brochothrix thermosphacta*. *Journal of Applied Bacteriology*, **56**, 137–43.

Mackenzie, E. (1973). Thermoduric and psychrotrophic organisms on poorly cleansed milking plants and farm bulk milk tanks. *Journal of Applied Bacteriology*, **36**, 457–63.

Mackey, B. M. (1984). Lethal and sublethal effects of refrigeration, freezing and freeze-drying on micro-organisms. In: *The Revival of Injured Microbes*. Eds M. H. E. Andrew and A. D. Russell, London, Academic Press.

MacLeod, R. A. & Calcott, P. H. (1976). Cold shock and freezing damage to microbes. In: *The Survival of Vegetative Microbes*. Eds T. R. G. Gray and J. R. Postgate, Cambridge, University Press.

Malle, P., Eb, P. & Tailliez, R.(1986). Determination of the quality of fish by measuring trimethylamine oxide reduction. *International Journal of Food Microbiology*, **3**, 225–35.

McEldowney, S. & Fletcher, M. (1990). The effect of physical and microbiological factors on food container leakage. *Journal of Applied Bacteriology*, **69**, 190–205.

Mead, G. C. (1982). Microbiology of poultry and game birds. In: *Meat Microbiology*. Ed. M. H. Brown, London, Applied Science Publishers.

Mead, G. C. & Thomas, N. L. (1973). The bacteriological condition of eviscerated chickens processed under controlled conditions in a spin-chilling system and sampled by two different methods. *British Poultry Science*, **14**, 413–19.

Miller, A., Scanlan, R. A., Lee, J. S. & Libbey, L. M. (1973a). Identification of the volatile compounds produced in sterile fish muscle (*Sebastes melanops*) by *Pseudomonas fragi*. *Applied Microbiology*, **25**, 952–5.

Miller, A., Scanlan, R. A., Lee, J. S. & Libbey, L. M. (1973b). Volatile compounds produced in sterile fish muscle (*Sebastes melanops*) by *Pseudomonas putrefaciens, Pseudomonas fluorescens* and an *Achromobacter* species. *Applied Microbiology*, **26**, 18–21.

Mitchell, E. L. (1988). A review of aseptic processing. *Advances in Food Research*, **32**, 1–37.

Molin, G. & Ternström, A. (1986). Phenotypically based taxonomy of psychrotrophic *Pseudomonas* isolated from spoiled meat, water and soil. *International Journal of Systematic Bacteriology*, **36**, 257–74.

Molin, G., Stenström, I-M. & Ternström, A. (1983). The microbial flora of herring fillets after storage in carbon dioxide, nitrogen or air at 2°C. *Journal of Applied Bacteriology*, **55**, 49–56.

Moseley, B. E. B. (1984). Radiation damage and its repair in non-sporulating bacteria. In: *The Revival of Injured Microbes*. Eds M. H. E. Andrew and A. D. Russell, London, Academic Press.

Mossel, D. A. A. & Shennan, J. L. (1976). Microorganisms in dried foods: their significance, limitation and enumeration. *Journal of Food Technology*, **11**, 205–20.

Mulder, R. W. A. W., Notermans, S. & Kampelmacher, E. H. (1977). Inactivation of salmonellae on chilled and deep frozen broiler carcasses by irradiation. *Journal of Applied Bacteriology*, **42**, 179–85.

Murray, J. G. & Stewart, D. B. (1978). Advances in the microbiology of milk and dairy products. *Journal of the Society of Dairy Technology*, **31**, 28–35.

Nelson, F. E. (1981). The microbiology of market milk. In: *Dairy Microbiology*, Vol. 1 (The Microbiology of Milk). Ed. R. K. Robinson, London, Applied Science Publishers.

Newton, K. G. & Rigg, W. J. (1979). The effect of film permeability on the storage life and microbiology of vacuum-packed meat. *Journal of Applied Bacteriology*, **47**, 433–41.

Niemand, J. G., van der Linde, H. J. & Holzapfel, W. H. (1983). Shelf-life extension of minced beef through combined treatments involving radurization. *Journal of Food Protection*, **46**, 791–6.

Nottingham, P. M. (1982). Microbiology of carcass meats. In: *Meat Microbiology*. Ed. M. H. Brown, London, Applied Science Publishers.

Panes, J. J., Parry, D. R. & Leech, F. B. (1979). *Report of a survey of the quality of farm milk in England and Wales in relation to EEC proposals*. London, MAFF.

Patterson, J. T. (1963). Salt tolerance and nitrate reduction by micrococci from fresh pork, curing pickles and bacon. *Journal of Applied Bacteriology*, **26**, 80–5.

Patterson, M. (1989). Sensitivity of *Listeria monocytogenes* to irradiation on poultry meat and in phosphate-buffered saline. *Letters in Applied Microbiology*, **8**, 181–4.

Put, H. M. C. & Kruiswijk, J. T. (1964). Disintegration and organoleptic deterioration of processed strawberries caused by the mould *Byssochlamys nivea*. *Journal of Applied Bacteriology*, **27**, 53–8.

Put, H. M. C., Van Doren, H., Warner, W. R. & Kruiswijk, J. T. (1972). The mechanism of microbiological leaker spoilage of canned foods: a review. *Journal of Applied Bacteriology*, **35**, 7–27.

Ray, B. (1986). Impact of bacterial injury and repair in food microbiology: its past, present and future. *Journal of Food Protection*, **49**, 651–5.

Ray, B., Janssen, D. W. & Busta, F. F. (1972). Characterisation of the repair of injury induced by freezing *Salmonella anatum*. *Applied Microbiology*, **23**, 803–9.

Rayman, M. K., Aris, B. & Hurst, A. (1981). Nisin: a possible alternative or adjunct to nitrite in the preservation of meats. *Applied and Environmental Microbiology*, **41**, 375–80.

Richards, G. P. (1988). Microbial purification of shellfish: a review of depuration and relaying. *Journal of Food Protection*, **51**, 218–51.

Rivière, J., Moss, M. O. & Smith, J. E. (1977). *Industrial Applications of Microbiology*, London, Surrey University Press.

Robinson, R. K. & Tamine, A. Y. (1981). Microbiology of fermented milks. In: *Dairy Microbiology*, Vol. 2 (The Microbiology of Milk Products). Ed. R. K. Robinson, London, Applied Science Publishers.

Rose, S. A. *et al.* (1988). Studies on the irradiation of toxins of *Clostridium botulinum* and *Staphylococcus aureus*. *Journal of Applied Bacteriology*, **65**, 223–9.

Saleh, B. A., Silverman, G. J. & Goldblith, S. A. (1966). Microbial evaluation of commercial freeze-dried foods. *Food Technology*, **20**, 671–4.

Seiler, D. A. L. (1971). The hygienic production of bread and flour confectionery. In: *Hygiene and Food Production*. Ed. A. Fox, Edinburgh, Livingstone.

Seiler, D. A. L. (1978). The microbiology of cake and its ingredients. *Food Trade Review*, **48**, 339–44.

Seiler, D. A. L. (1988). Microbiological problems associated with cereal based foods. *Food Science and Technology Today*, **2**, 37–40.

Shackleford, A. D. (1988). Modifications of processing methods to control *Salmonella* in poultry. *Poultry Science*, **67**, 933–5.

Shaw, B. G. & Latty, J. B. (1984). A study of the relative incidence of different *Pseudomonas* groups on meat using a computer-assisted identification technique employing only carbon source tests. *Journal of Applied Bacteriology*, **57**, 59–67.

Shaw, B. G. & Shewan, J. M. (1968). Psychrophilic spoilage bacteria of fish. *Journal of Applied Bacteriology*, **31**, 89–96.

Shewan, J. M. (1961). The microbiology of sea-water fish. In: *Fish as Food*, Vol. 1 (Production, Biochemistry and Microbiology). Ed. G. Borgstrom, New York, Academic Press.

Shewan, J. M. (1971). The microbiology of fish and fishery products—a progress report. *Journal of Applied Bacteriology*, **34**, 299–315.

Shrimpton, D. H., Monsey, J. B., Hobbs, B. C. & Smith, M. E. (1962). A laboratory determination of the destruction of α-amylase and salmonellae in whole egg by heat pasteurization. *Journal of Hygiene, Cambridge*, **60**, 153–62.

Skovgaard, N. (1985). *Brochothrix thermosphacta:* comments on its taxonomy, ecology and isolation. *International Journal of Food Microbiology*, **2**, 71–9.

Smelt, J. P. P. M., Raatjes, G. J. M., Crowther, J. S. & Verrips, C. T. (1982). Growth and toxin formation by *Clostridium botulinum* at low pH values. *Journal of Applied Bacteriology*, **52**, 75–82.

Smulders, F. J. M. *et al.* (1986). Review: lactic acid: considerations in favour of its acceptance as a meat decontaminant. *Journal of Food Technology*, **21**, 419–36.

Speck, R. V. (1981). Thermophilic organisms in food spoilage: sulfide spoilage anaerobes. *Journal of Food Protection*, **44**, 149–53.

Stermer, R. A., Lasater-Smith, M. & Brasington, C. F. (1987). Ultraviolet radiation—an effective bactericide for fresh meat. *Journal of Food Protection*, **50**, 108–11.

Tarkowski, J. A., Beumer, R. R. & Kampelmacher, E. H. (1984). Low gamma irradiation of raw meat. II. Bacteriological effects on samples from butcheries. *International Journal of Food Microbiology*, **1**, 25–31.

Tarrant, P. J. V., Pearson, A. M., Price, J. F. & Lechowich, R. V. (1971). Action of *Pseudomonas fragi* on the proteins of pig muscle. *Applied Microbiology*, **22**, 224–8.

Taylor, A. A. (1985). Packaging fresh meat. In: *Developments in Meat Science— 3*. Ed. R. Lawrie, London, Elsevier Applied Science Publishers.

Thomas, S. B., Druce, R. G. & Jones, M. (1971). Influence of production conditions on the bacteriological quality of refrigerated farm bulk tank milk—a review. *Journal of Applied Bacteriology*, **34**, 659–77.

Thompson, P. J. (1981). Thermophilic organisms involved in food spoilage: aciduric flat-sour sporeforming aerobes. *Journal of Food Protection*, **44**, 154–6.

Thomson, A. B., Davis, H. K., Early, J. C. & Burt, J. R. (1974). Spoilage and spoilage indicators in queen scallops. *Journal of Food Technology*, **9**, 381–90.

Tonge, R. J., Baird-Parker, A. C. & Cavett, J. J. (1964). Chemical and microbiological changes during storage of vacuum packed sliced bacon. *Journal of Applied Bacteriology*, **27**, 252–64.

Tranter, H. S. & Board, R. G. (1982). The antimicrobial defense of avian eggs: biological perspective and chemical basis. *Journal of Applied Biochemistry*, **4**, 295–338.

Urbain, W. M. (1978). Food irradiation. *Advances in Food Research*, **24**, 155–227.

Van Klaveren, F. W. & Legendre, R. (1965). Salted cod. In: *Fish as Food*, Vol. 3 (Processing: Part 1). Ed. G. Borgstrom, New York, Academic Press.

Van Spreekens, K. J. A. (1977). Characterization of some fish and shrimp spoiling bacteria. *Antonie van Leeuwenhoek*, **43**, 283–303.

Walker, P., Cann, D. & Shewan, J. M. (1970). The bacteriology of 'scampi' (*Nephrops norwegicus*). 1. Preliminary bacteriological, chemical and sensory studies. *Journal of Food Technology*, **5**, 375–85.

Wilson, C. D. & Richards, M. S. (1980). A survey of mastitis in the British Dairy Herd. *Veterinary Record*, **106**, 431–5.

Chapter 4

MICROBIOLOGICAL EXAMINING METHODS

4.1. THE RATIONALE OF MICROBIOLOGICAL TESTING

There may be a variety of reasons why it is necessary to examine foods qualitatively and/or quantitatively for microorganisms. The principal objectives of microbiological testing are to ensure: (1) that the food meets certain statutory standards; (2) that the food meets internal standards set by the processing company or external standards required by the purchaser; (3) that food materials entering the factory for processing are of the required standard and/or meet a standard agreed with a supplier; (4) that process control and line sanitation are being maintained.

The microbiological test methods used to monitor food quality are themselves also varied and are largely dependent on the food being analysed. It is convenient to list tests under the following general headings:

1. Estimations of total numbers of microorganisms.
2. Estimations of numbers of indicator organisms.
3. Examining for, or estimating numbers of, food spoilage organisms or other selected groups.
4. Examining for, or estimating numbers of, food poisoning or food-borne pathogenic microorganisms.
5. Examinations of the metabolic products of microorganisms; this often requires more sophisticated techniques only available in the larger laboratories.

4.2. SAMPLING

4.2.1. Sampling Rate

One of the biggest problems associated with the microbiological testing of foods is the question of the number of samples that should be analysed from a batch or a day's production to ensure that the product or food material is of the required standard. Obviously the greater the number of samples taken, the greater the confidence placed in the results obtained but, since food used for analytical purposes represents a non-recoverable expense, a compromise must be achieved between the accuracy and the economics of testing. Sampling schemes in current use include testing five or ten samples per batch, or the square root of the number of packs per batch; however, it could be argued that such a high level of sampling is totally unjustified where consistently good results are obtained. Thus the available microbiological resources should be used prudently and sampling rates should be highest with foods that are known to be hazardous or which give erratic results; sampling rates should also be high for foods which have been subjected to a process change or after a process line failure. It should also be appreciated that the level of testing is affected by the test rationale; thus fewer tests are necessary where confirmation of a satisfactory standard is the object than where elimination of an unsatisfactory product is the aim.

Any worthwhile sampling scheme should be statistically based and readers requiring information on such schemes are strongly recommended to consult *Microorganisms in Foods*, Vol. 2 (ICMSF, 1986) where much other pertinent information is also available.

4.2.2. The Representative Sample

As defined by the aforementioned international commission (ICMSF, 1986), 'a representative sample is one which reflects, as far as is possible, the composition of the lot (or batch) from which it is drawn'. It is therefore necessary to avoid any form of bias and to ensure that sufficient samples are taken. This is best achieved by some form of random sampling by the use of random number tables as recommended by the Commission (ICMSF, 1986). However, account should be taken of the possible build up of microorganisms on equipment, and hence in food, during a production run; a product may contain far fewer microorganisms at the start of a shift than towards the end and this should be allowed for in any sampling scheme. Alternatively, if equipment has been inadequately cleaned foods may remove large numbers of bacteria at the start of a production run. It may be advantageous, therefore, to remove products on a regular time basis; this method does not undermine random sampling and is frequently employed.

Other difficulties in sampling can arise if the product has a heterogeneous consistency and more frequent sampling is required if the food is not homogeneous. At one extreme only a small sample of well-mixed milk would be required to give a representative sample whereas with some of today's multi-component products it may prove impossible to get the required quantity of each component into a single representative sample and therefore a separate analysis of each component may have to be made.

Sampling must sometimes be deliberately biased when microorganisms are confined to specific zones of the food, e.g. foods such as meats, fish and fruits carry the majority of their organisms on the outer surfaces and these must therefore constitute the principal sampling regions.

4.2.3. Sampling Techniques

Liquid foods present no problem and are normally sampled by means of sterile pipettes after thorough mixing. Solid foods often present problems for the sampler who uses a variety of techniques to overcome the difficulties. With larger food units such as poultry and fish the whole carcass may be rinsed by agitation in a suitable diluent. Although only a part of the flora is recovered by such a treatment it does ensure that organisms from the visceral cavity are included, this being the region most likely to contain salmonellas; higher counts are obtained if abrasives such as sterile sand are included in the rinse solution. Moistened cotton-wool swabs have been widely used for many years for surface sampling and have the advantage of being easy to handle and, when used in conjunction with a metal template, of presenting a known surface area for analysis; after swabbing, the head of the swab is broken off into diluent which is then shaken to release the organisms. Recoveries obtained by these swab techniques are poor due to the adherence of microorganisms to the food surface and to their retention in the swab but recoveries can be improved if the sampled area is re-swabbed with a second dry swab. Agar contact methods, in which a sterile agar-based medium is pressed against the surface to be sampled, give even lower counts than swabbing techniques but they are the simplest sampling methods to use since the medium can be incubated directly (ten Cate, 1965).

With meats, fish and poultry the highest counts are obtained with so-called 'destructive sampling'. Samples can be taken with pre-sterilized cork borers which have the advantage of analysing a known surface area in conjunction with a core sample. Scalpels and knives are also popular but with these instruments the surface area:weight ratio is impossible to control. Known areas can also be sampled by using a skin scraper together with a sterile metal cylinder into which the diluent is poured (Barnes *et al.*, 1973).

Comminuted meats and particulate foods such as peas and flour are easier to handle as a known weight can be taken. Known weights can also be obtained from blocks of frozen meat or egg by drilling with a sterile bit; the drillings can easily be collected for sampling. More detailed advice on the collecting and handling of samples is again given in *Microorganisms in Foods*, Vol. 2 (ICMSF, 1986).

Certain of the microbiological sampling techniques discussed above are also used for the examination of food processing equipment: viz. rinsing with a known volume of water or diluent, swabbing, possibly with template, agar contact methods.

4.2.4. Treatment of Sample

A known weight of food sample (e.g. 10 or 25 g) is taken except where swabs or rinses have been used. The food is added to a suitable sterile diluent such as ¼ strength Ringer's solution or 0·1% peptone water and then treated so as to release into the diluent microorganisms on or in the food. The treatment typically involves mechanical blending or 'stomaching'. In the latter technique, which is now very popular, the sample together with a known volume of diluent is put in a sterile plastic bag which is pounded by paddles inside the 'stomacher'. The volume of diluent used is usually nine times the weight of sample (e.g. 25 g peas in 225 ml diluent) so that a 10^{-1} homogenate is prepared. From this, suitable dilutions can be prepared (10^{-2}, 10^{-3}, 10^{-4}, etc.) depending upon the microbiological quality of the food or surface under test (Fig. 4.1).

Further information on the microbiological assessment of surfaces and on the preparation and dilution of food homogenates can be found in appropriate titles listed in the Bibliography section.

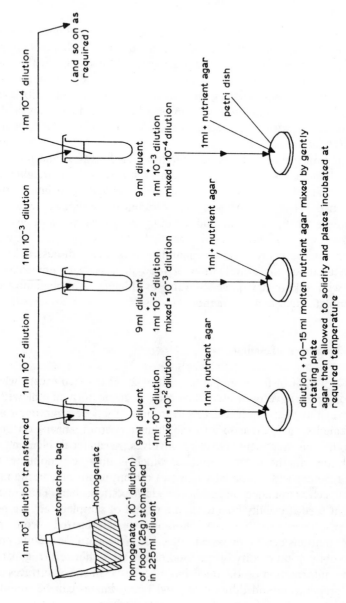

Fig. 4.1. Schematic representation of 'pour plate' method used in the enumeration of bacteria.

4.3. MICROBIOLOGICAL TEST PROCEDURES IN COMMON USAGE

4.3.1. Total Viable Count

One of the most common microbiological tests carried out on foods is the total viable count which is also known as the standard plate count or aerobic plate count. In this, suitable dilutions of the food sample are plated on or in agar-based media containing complex nutrients which support the growth of as wide a range of microorganisms as possible. Nutrients included in, for example, Nutrient Agar (Oxoid) are beef extract, yeast extract and peptone (a proteolytic enzyme digest of fresh meat containing a variety of inorganic salts, growth factors and peptides). The pH of the medium is usually adjusted to 7·0–7·4 so that bacteria rather than yeasts or moulds are recovered.

The individual bacterial cells transferred to the plate in the diluent divide in the normal way during incubation. Thus an estimate of the total number of *viable* cells (i.e. cells capable of growth in the recovery medium) in the dilution plated out can be calculated by counting the total number of bacterial colonies which develop following incubation; clearly, under normal aerobic conditions obligate aerobes and facultative anaerobes will grow.

The incubation temperatures selected depend on the food being examined. Commonly used temperatures are 55°C for thermophiles, 35–37°C for mesophiles and 20–30°C for many spoilage bacteria. Whilst the latter temperatures are suitable for most psychrotrophic bacteria as well as for many mesophiles, lower temperatures (e.g. 5°C) are sometimes used for more accurate estimates of psychrotrophs; it must always be borne in mind that no one incubation temperature completely excludes all organisms from another group.

Many plating techniques are employed for enumerating total numbers of viable bacteria and five are described briefly. Enumeration of colonies is traditionally performed manually using an illuminated colony counter with the operator counting each individual colony (see below). This can be a tedious operation and, unless a suitable number of colonies has developed in the growth medium (ideally below 300, although for statistical reasons a minimum of 30 colonies is also required), it can also be inaccurate. In recent years automatic colony counting devices have been developed which enable accurate counts to be obtained in a few seconds.

4.3.1.1. The 'Pour Plate' Method

In this method a set (or preferably duplicate sets) of Petri dishes is inoculated with 1 ml aliquots from appropriate dilutions of the food (Fig. 4.1). Some 10–15 ml of melted nutrient agar or other suitable medium, cooled to 45°C, is then added to each of the Petri dishes and mixed carefully with its aliquot. After the agar has solidified, the plates are incubated at the required temperature for a period of time depending upon the incubation conditions (e.g. 1–2 days at 37°C, 3–4 days at 20°C and 7–10 days at 5°C). After incubation, plates containing 30-300 colonies should be counted from which the number of viable cells per gram (or per cm^2) of food can be readily calculated: if an average count of 112 colonies is obtained for the 10^{-4} dilution the total count per gram of food $= 112 \times 10^4 = 1 \cdot 1 \times 10^6$.

4.3.1.2. The 'Spread Plate' Method

Here the medium is pre-poured and allowed to solidify in the Petri dishes; 0·1 ml quantities of the dilutions are spread evenly over the whole surface of the medium using sterile L-shaped glass rods. Plates are incubated as above. Advantages of this technique are, first, that heat-sensitive cells (i.e. psychrotrophs) are not killed by the molten agar which may occur to an extent in the 'pour plate' method if the temperature of the agar is too high; second, all the colonies develop on the surface of the agar and can be easily observed and picked off if necessary, whereas many colonies develop embedded in the agar in the 'pour plate' method, and these may be restricted in size and may be more difficult to subculture.

4.3.1.3. The 'Drop Plate' Method

Solidified medium is again used with this technique. Specially calibrated pipettes delivering 0·02 ml per drop are used and five separate drops (i.e. 0·1 ml) are delivered onto the surface of the plate, the drops being dried before incubation. Dilutions giving under 20 colonies per drop should be counted.

4.3.1.4. The 'Agar Droplet' Method

Since the above methods are both time-consuming and expensive in materials, a more rapid technique using far smaller quantities of materials has been developed; this is the 'agar droplet' method (Sharpe & Kilsby, 1971). In this technique the dilutions are prepared in molten agar and colonies develop in the solidified droplets (0·1 ml) during incubation.

Counting of the colonies is facilitated by using a projection viewer which magnifies the droplets approximately ten-fold.

4.3.1.5. The 'Spiral Plate' Method

Another semi-automatic method which has become popular is the 'spiral plate' (Gilchrist *et al.*, 1973) in which a machine continuously plates a known volume of sample on the surface of a rotating agar plate. The amount of sample deposited decreases as the dispensing stylus is moved from the centre to the perimeter of the rotating plate; thus the microbial colonies develop along a spiral track during incubation. Counting can be performed manually using a counting grid but laser-based automatic colony counters have been developed specifically for use with this technique. Comparisons with the more traditional methods have shown no significant differences in counts obtained (Jarvis *et al.*, 1977).

4.3.2. Counting Using Electrical Impedance Measurements

As microorganisms multiply in a growth medium minute changes in impedance occur which can be measured by passing a small electric current through the medium. Such changes can be monitored continuously and at a particular concentration of microorganisms there is a marked change in impedance; clearly with higher initial numbers this threshold concentration is reached more rapidly so that in such conditions results can be obtained within 2 h. Thus estimates can be made of the numbers of microorganisms initially present in a food by recording the time taken (detection time) to reach the threshold using samples of the food diluted in a growth medium.

There are many other advantages with this technique so it is not surprising that impedance monitoring instruments, such as that illustrated in Fig. 4.2, are becoming an integral part of the equipment used in larger food microbiology laboratories. Conventionally impedimetry has been used to estimate total viable counts and it has been shown that there is a good agreement between standard viable counts and detection times in foods such as frozen vegetables (Hardy *et al.*, 1977), meats (Bülte & Reuter, 1984), fish (Gibson *et al.*, 1984), and milk (Nieuwenhof & Hoolwerf, 1987). Another advantage is that large numbers of samples can be monitored simultaneously by means of a computer and suitable recording devices; furthermore, the times required for analysis, even for foods with relatively low total counts, are normally only about 5–8 h so that rapid quality assessments are made possible.

Fig. 4.2. Impedance measuring instrumentation. (By courtesy of Don Whitley Scientific Ltd.)

The development of impedimetry using selective media for specific microorganisms (see p. 198 *et seq.*) has been hampered by inhibitory agents used in the media preventing adequate growth of the required organisms and hence delaying detection times. However, impedance methods for salmonellas and coliforms (see p. 195) have been successfully introduced (Jarvis & Easter, 1987); by the careful development of selective media it is possible to reduce both the detection times and selectivity for salmonellas in foods (Bullock & Frodsham, 1989). Progress in the detection of other pathogenic organisms has been even more limited although suitable protocols for listerias have been suggested (Phillips & Griffiths, 1989).

4.3.3. Counting by Measurement of Adenosine Triphosphate (ATP)

A bioluminescent technique has been developed based on a reaction involving ATP and the luciferase enzyme derived from fireflies. The reaction in much simplified form can be expressed as:

$$ATP + luciferase \rightarrow light$$

The total amount of light produced is directly proportional to the quantity of ATP present and since all bacteria contain roughly the same amount of ATP per cell (approximately 10^{-15} g) it should be possible to measure the number of bacterial cells in any sample. Unfortunately, foods (e.g. meats) contain large amounts of non-microbial ATP and it is necessary to remove this ATP or separate the bacteria from the remainder of the food before bacterial ATP can be estimated. Relatively simple separation techniques have been reported for meats (Stannard & Wood, 1983; Kennedy & Oblinger, 1985) which enable results to be obtained in less than 1 h, the results correlating with the bacterial counts obtained using more conventional counting methods. Similarly promising procedures have been developed for estimating bacterial numbers in raw milks (Webster *et al.*, 1988) but sensitivity remains poor where low numbers of bacteria are encountered. Views have been expressed that the method, whilst promising, requires further development due to unreliability, expense and lack of automation (Jarvis & Easter, 1987). Hopefully newer instruments awaited will answer these criticisms.

It is also possible to estimate numbers of a specific bacterial type present on a surface or in a food sample by using phages (see p. 17). These viruses show a high degree of specificity to the bacteria they infect. In this technique luciferase genes are manipulated into the phage DNA, the phages, in turn, attaching to their specific bacteria if present. Thus the extent of luminescence gives an estimate of the numbers of that bacterium present.

4.3.4. Counting Using the Direct Epifluorescent Filter Technique

This technique, commonly abbreviated to DEFT, was originally developed for the rapid enumeration (15–20 min) of bacteria in raw milk but it has found wider use more recently (Pettipher, 1983). In this method, pre-filtered suspensions of the food (the pre-filtering is necessary with foods to remove debris after 'stomaching') are subsequently passed through a fine polycarbonate membrane filter. Bacteria retained on the surface of the filter are then stained with a fluorescent material, acridine orange, and enumerated by means of an epifluorescence microscope. The acridine orange absorbs energy from a special source of illumination provided on the microscope but the absorbed energy is lost immediately in the form of fluorescent light which is transmitted towards the eyepiece; thus individual cells fluorescing in the preparation can be counted readily by the viewer, although a fully automated counting system can be employed (Pettipher, 1986).

The DEFT is used primarily for estimating total numbers of viable bacteria in raw milks and on meat and poultry. With the latter foods there can be situations where DEFT tends to slightly overestimate counts, but the rapidity with which results can be obtained (35–45 min) still makes this an attractive technique to use (Shaw *et al.*, 1987). Difficulties in separating bacteria from food material can be overcome by using suitable degradative enzymes (e.g. proteases with minced beef) which leave the organisms in suspension leading to highly accurate counts (Walls *et al.*, 1990). The DEFT is less reliable for food products containing heat-stressed cells because certain of these non-viable bacteria may fluoresce to give false positive results (Jarvis & Easter, 1987). A development of the DEFT which is of some interest is direct epifluorescence microscopy (DEM) where bacteria trapped on the filters are incubated on media containing optical brighteners, the microcolonies produced on the filters being then examined by fluorescence microscopy. Using this technique counts of specific organisms or groups of organisms can be estimated by employing suitable selective media (Rodrigues & Kroll, 1988).

4.3.5. Microcalorimetry

During growth microorganisms degrade nutrients and the small amount of heat produced can be measured accurately using temperature-sensitive instruments (Forrest, 1972); the amount of heat produced is an indirect measure of the numbers, types and activity of the organisms involved. Developments in microcalorimetry have been restricted by high cost and the lack of sensitivity where low numbers of bacteria are found. However the technique has been used to estimate bacterial counts in raw milks (Cliffe *et al.*, 1973) and minced meat (Gram & Søgaard, 1985) the results being obtained within 24 h provided counts are above 10^5 per ml (or per g).

4.3.6. Direct Microscopic Count

It may sometimes be necessary to obtain the total numbers of microorganisms (i.e. viable and non-viable cells) present in food samples as in, for example, canned foods or pasteurized milk. Smears of the food sample, possibly diluted, are prepared on glass microscope slides and stained with a suitable dye. The total number of microbial cells in a given number of microscopic fields is counted and from this the total number of organisms per gram of food can be roughly calculated.

4.3.7. Indicator Organisms

The routine examination of foods for a wide range of pathogenic bacteria is impracticable in most laboratories either because they are inadequately equipped or because the sample size would be impractical to handle. Thus it has become normal practice to examine foods for bacteria whose presence indicates the possibility of food poisoning or other pathogenic bacteria being present. These bacteria are thus termed 'indicator organisms' and they are often regarded as being of great significance when assessing the microbiological safety and quality of foods. The principal bacteria employed as indicators are the coliforms, enterococci and, more recently, Enterobacteriaceae. Total viable counts at 37°C are sometimes a useful rough guide although their relationship with the safety of foods is a rather more tenuous one.

In more recent years this view of indicator organisms has been modified by certain food microbiologists who recognize two separate groups of such organisms (Mossel, 1983). These are : (1) 'index' organisms, whose presence in foods suggests the possibility of pathogens being present, i.e. they are concerned with safety; and (2) 'indicator' organisms, whose presence relates to the general microbiological condition of the food, i.e. they are concerned with hygienic quality. This distinction may be rather confusing since many of the old indicator organisms are now recast as index organisms although some, the Enterobacteriaceae, can be regarded as indicator organisms primarily but as index organisms with certain foods where pathogenic Enterobacteriaceae (e.g. salmonellas) may be anticipated. For simplicity the older definition will be used.

4.3.7.1. Coliforms

The principal coliform bacteria are *Escherichia coli* and *Enterobacter aerogenes*. The former is normally found in the gastrointestinal tract of man and other animals and is rarely found elsewhere whereas *E. aerogenes* is normally associated with vegetation and is only occasionally found in the intestine.

In water testing *E. coli* is the classical indicator for the possible presence of enteric pathogens. Here there is a direct relationship between the numbers of *E. coli* present and the extent of faecal pollution, the higher the numbers the greater the pollution; this is because the organism cannot multiply in water and, in fact, numbers slowly decline unless renewed pollution occurs. In food, the presence and concentration of

E. coli is of less significance and the occurrence of this organism, even in large numbers, does not necessarily imply recent heavy faecal pollution. The numbers can be influenced by many factors such as natural contamination, actual growth in the food, poorly cleaned equipment and contamination from personnel. Thus all that can be concluded with many foods is that faecal contamination, indirect or direct, occurred at some stage and that the safety of the food is possibly in question.

The detection and enumeration of *E. coli* are performed in three stages (ICMSF, 1978). The first stage, the presumptive coliform count, involves the enumeration of coliforms, both faecal and non-faecal, by use of selective media; these can be described briefly as nutrient media containing selective agents such as antibiotics, dyes or chemicals which suppress the growth of many organisms but allow the selected organisms, resistant to the agents, to grow. With coliforms bile salts are commonly incorporated in the culture media as the selective agent. Examples of selective media used for this purpose are MacConkey agar and Violet Red Bile agar. The former additionally contains a nutrient source (peptone), lactose, which is dissimilated by coliforms with the production of acid (and gas) and a pH indicator (neutral red); the latter medium contains crystal violet as an additional selective agent but is otherwise based on similar principles.

Enumeration techniques include direct plating of dilutions of homogenized food samples on the above or similar media employing the 'pour' or 'spread' plate method; incubation is for 24 h at 37°C, coliforms then appearing as red colonies. Alternatively, dilutions can be added to MacConkey or similar broths such as lauryl sulphate tryptose broth which should contain inverted fermentation (Durham) tubes to collect gas produced by the breakdown of lactose by coliforms (Fig. 4.3).

When broths are used in the enumeration of coliforms, etc., the Most Probable Number (MPM) technique is usually employed: in this, five tubes of broth are inoculated with each dilution. The presumptive MPM is based on the number of broths showing acid and gas production after 24–48 h incubation and is calculated using statistical tables (Oblinger & Koburger, 1975).

The second stage is to confirm the presence of coliform organisms although, in practice, this step is frequently missed out as the majority of presumptive coliforms are confirmed. This confirmation is made by subculturing all the broths showing acid and gas production or suspected colonies from agar into tubes of Brilliant Green Bile Broth (BGBB) which

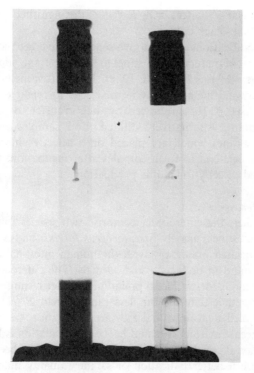

Fig. 4.3. Lauryl sulphate tryptose broth showing (1) no change and (2) indicator change and gas collected in the inverted Durham tube.

is then incubated at 37°C for 48 h. The inclusion of brilliant green renders BGBB more selective than the two media employed earlier and the production of gas in this medium confirms the presence of coliforms.

The final stage is to confirm the presence of *E. coli*. This can be done directly from stage 1 using BGBB as above but with an incubation temperature of 44°C for a 24 h period. At the same time a tube of peptone water is inoculated, again incubated at 44°C for 24 h and subsequently tested for the presence of indole. *E. coli* is virtually alone in being able to produce gas from lactose and indole from peptone at this elevated temperature so that positive reactions confirm *E. coli*. As an alternative to BGBB a solid medium (Eosin Methylene Blue agar) can be employed. Presumptive or confirmed coliforms from stages 1 and 2 are inoculated onto the surface of pre-poured plates of this medium which are then incubated at 37°C for 24 h; *E. coli* is distinguished by the metallic sheen of the blue-black colonies.

Various modifications to the above schemes have been suggested with the prime aim of reducing the time to complete the identification of *E. coli* (Hofstra & Huis In't Veld, 1988). These often include a resuscitation stage allowing metabolically injured cells to repair on a non-selective medium (see p. 165), followed by their transfer to an appropriate selective medium. Alternatively diluted food samples can be filtered through membranes which are placed on a selective medium allowing the entrapped bacteria to form colonies after incubation; these colonies are then tested directly for indole production.

4.3.7.2. Enterococci

As stated earlier, the enterococci comprise two species found in human and animal intestines, namely *Streptococcus faecalis* and *S. faecium*. The former is associated principally with the human intestinal canal whereas the latter is found in both man and animals. The enterococci are sometimes used as indicators of faecal pollution in water testing, one of the advantages being that they die out less rapidly than *E. coli*. However, a disadvantage of this group is that they are found more frequently than *E. coli* in non-faecal environments and hence their isolation is less conclusive evidence of faecal contamination. In foods it has often been argued that enterococci are a better indication of sanitary quality than *E. coli*; they are generally more resilient than coliforms, particularly in frozen and dried foods and in foods given a moderate heat treatment. However, this resilience further undermines their value as indicator organisms as their presence in, for example, heat-treated foods may be of little value if less-stable pathogens such as salmonellas have been killed by the process.

Many techniques are available for the isolation and enumeration of enterococci but they usually rely on either sodium azide, thallous acetate or antibiotics as the selective agents (Reuter, 1985) and often employ a high incubation temperature (45°C). An example of a commonly used medium is KF Streptococcus agar (Difco) which, as well as the usual nutrient sources, also contains tetrazolium chloride, an ingredient which imparts a red colour to the colonies; incubation is at 37°C for 48 h. Alternatively, glucose azide broth incubated at 45°C may be used with assessments of numbers being made by the MPM technique in conjunction with probability tables. Tubes showing acid production are regarded as positive (N.B. enterococci do not produce gas from glucose). Further identification to species or strain level is normally not required.

4.3.7.3. Enterobacteriaceae

The family Enterobacteriaceae comprises many genera including those characterized by lactose fermentation (e.g. *Escherichia* and *Enterobacter*) and those not fermenting lactose (e.g. the non-enteropathogenic *Proteus* and *Serratia* as well as *Salmonella* and *Shigella*). Mossel *et al.* (1963) claimed that a close correlation existed between counts obtained for total numbers of the family Enterobacteriaceae and the extent of faecal pollution, particularly in relation to contamination by salmonellas; because many discrepancies were found when the more conventional tests for coliforms were employed, they suggested that a test for total Enterobacteriaceae would be more reliable. The discrepancies noted when coliform tests were used were: (1) the product may contain only *Salmonella* spp. and therefore falsely reassuring results may be obtained; (2) non-lactose fermenting strains of *E. coli* may occasionally predominate; (3) occasionally strains may fail to produce gas even though they dissimilate lactose; (4) coliforms are poorly defined anyway and on these grounds the new test was justified.

The test procedure suggested by Mossel *et al.* (1963) involves enriching samples of homogenized food in BGBB (the rationale of the enriching technique is given in the next section); the BGBB used in this test is modified from that used in the confirmation of coliforms by the substitution of glucose for lactose since all Enterobacteriaceae members, by definition, dissimilate glucose with acid production. Incubation is for 24 h at 37°C and this is normally followed by plating out broths showing growth (i.e. turbid) onto modified MacConkey or Violet Red Bile agar; again modification is by sugar substitution (Mossel, 1985). All colonies developing with a characteristic red or purple appearance can be regarded as Enterobacteriaceae.

It is important to reiterate that the presence of even substantial numbers of indicator organisms does not, in itself, indicate *with certainty* that direct faecal contamination of processed foods has occurred; it may equally well suggest inadequate processing, post-process contamination or insanitary processing conditions especially where foods have subsequently been stored at temperatures permitting microbial growth.

4.3.8. Food Poisoning Bacteria

Only the more important food poisoning bacteria will be considered in this section and test procedures for salmonellas, *Clostridium perfringens* and *C. botulinum, Staphylococcus aureus, Bacillus cereus, Vibrio para-*

haemolyticus, *Listeria monocytogenes* and *Campylobacter* spp. are briefly discussed in turn.

4.3.8.1. Salmonellas

The isolation and identification of *Salmonella* spp. is conventionally a complex process involving a number of stages and enumeration is not normally attempted (ICMSF, 1978). After the sample of food (25 or 50 g) has been taken it is blended as described previously (see Section 4.2.4) except that the process is normally performed in a pre-enrichment medium such as nutrient broth. The broth is then incubated at 37°C for a few hours in order to facilitate the recovery of injured salmonellas (see Section 3.15.4). The second stage involves the inoculation of a sample of the broth into an enrichment medium which encourages the growth of salmonellas but restricts or completely inhibits growth of competitive organisms such as coliforms. Selective enrichment is a general term used to indicate that one or a group of organisms is being allowed to grow whilst competitors are being inhibited; in essence this stage sees an increase in the ratio of the organism(s) required to all other competitors.

Two commonly used enrichment media for salmonellas are selenite broth and tetrathionate broth. The former contains sodium selenite as the inhibitory agent although the dissimilation of lactose, which is incorporated in the medium, by many of the enteric bacteria present causes a pH drop which favours salmonella growth. Tetrathionate broth contains a number of inhibitory agents including bile salts, brilliant green and tetrathionate which, in combination, permit the salmonellas to grow whilst effectively inhibiting most competitors. Incubation of these broths can be at either 37 or 43°C for *ca* 24 h.

After enrichment the broths are usually plated out on selective media and both MacConkey agar and Violet Red Bile agar have been used for this purpose. However, media such as Brilliant Green Phenol Red agar and Desoxycholate Citrate agar, both having more marked selective properties, are now more generally used (Blood, 1985); both these media contain a variety of selective agents, the usual nutrients and lactose plus a pH indicator. Salmonellas as non-lactose fermenters form essentially colourless colonies on these media after incubation at 37°C for 24 h and are thus readily distinguished from the red colonies produced by the lactose fermenters. Unfortunately many other organisms produce colonies indistinguishable from those of *Salmonella* spp. on these media and therefore further tests are necessary before isolation can be confirmed. Confirmation is made by picking off suspect colonies and, after checking for purity, reinoculating them into media to test for the dissimilation of further carbohydrates, production of hydrogen sulphide, decarboxylation of lysine and motility. Provided these screening tests give the correct pattern of results, the isolate may be regarded as a salmonella and a final confirmation is made on the basis of serological tests (ICMSF, 1978). Characterization to specific serotype is possible using appropriate H and O antisera which contain specific antibodies raised against the salmonella although, in general, confirmation using polyvalent H and O antisera (representative of all the serotypes likely to be encountered) is sufficient.

From the foregoing it will be apparent that conventional methods for the detection of salmonellas in foods are tedious and labour intensive—it can take up to 5 days before a negative result is confirmed. Hence there has been a search for more rapid methods and a number of techniques have been introduced including impedimetry (see p. 191) amongst many others which have been suggested (Ibrahim & Fleet, 1985). Fluorescent-antibody staining in which a fluorescent dye is attached to the antibody can reduce the period for positive identification to 1 or 2 days but non-specific staining remains a weakness of this technique. Enzymes can also be attached to antibodies in the technique known as enzyme-linked immunosorbent assay (ELISA) and this has proved a useful tool in the detection of salmonellas. In essence the ability of the antibody to bind to a specific site on the cell is measured by adding a substrate which any enzyme retained, after suitable washing procedures, will attack (Chantler & McIllmurray, 1987). The use of monoclonal antibodies has further increased the specificity of this method (Ibrahim, 1986); commercially available kits utilizing these principles have given more rapid results (within 48 h) which correlate well with traditional methods (e.g. Todd *et al.*, 1987).

A novel technique, DNA–DNA hybridization, has been developed in recent years for the detection of a variety of pathogens including salmonellas (Hill & Keasler, 1991). As well as organisms possessing specific antigenic regions attracting complementary antibodies, all species contain portions of genetic material, DNA or RNA, which are unique in terms of their nucleotide sequences. Such sequences can be readily detected using gene probes and diagnostic kits using DNA probes are now available (Flowers *et al.*, 1987; Izat *et al.*, 1989). With *Salmonella* spp. these kits detect one or more sequences which are unique to members of the genus as a whole so that the presence of salmonellas generally, rather than specific serotypes, can be confirmed.

Overnight incubation of a food sample in a recovery medium is necessary to increase numbers of salmonellas above a threshold level (*ca* 10^5 per ml), the cultures then being collected on membrane filters and assayed by hybridization (Fitts *et al.*, 1983). This process takes anything from 1 to 24 h depending on the probe used and readers requiring more information on probes and assay systems should consult Walker & Dougan (1989).

4.3.8.2. *Clostridium perfringens and C. botulinum*

Unlike salmonellas, the isolation of low numbers of *C. perfringens* from food does not necessarily mean that a food poisoning danger exists. Only when large numbers are present is there a definite hazard and therefore enumeration techniques are essential with this organism. Pour or spread plate techniques are used with dilutions of food homogenates in conjunction with selective media. Many such media have been devised for *C. perfringens* but the majority are agar-based, and contain suitable nutrients, indicator systems and selective agents.

The ICMSF has carried out a comparative study on methods for the enumeration of *C. perfringens* in foods and concluded that Sulphite Cycloserine agar gives highest recoveries of this organism together with the lowest number of false positives (Hauschild *et al.*, 1977). This medium contains the antibiotic cycloserine and, in addition, there is an indicator system that utilizes the fact that *C. perfringens*, like a number of other clostridia, reduces sulphite to sulphide giving black colonies in the presence of an iron salt. After anaerobic incubation of pour plates at 37°C for 24 h, suspect colonies are inoculated into a confirmatory medium checking for motility (*C. perfringens* is non-motile) and the ability to reduce nitrate to nitrite (Hauschild & Hilsheimer, 1974). An alternative confirmatory test is to plate out suspect colonies on an egg yolk agar to one half of which has been added *C. perfringens* antitoxin. After incubation, colonies which have developed in the half containing no antitoxin are surrounded by a zone of opacity (the Nagler reaction) whereas colonies in the other part show no change due to the specific neutralization of the reaction with the antitoxin (Cruickshank *et al.*, 1975).

C. botulinum is not normally enumerated in foods and tests generally involve an examination of the food for botulinal toxins and the isolation of *C. botulinum* followed by toxin assays. Extreme care should be taken when examining suspect foods and it is essential to seek expert advice before contemplating such analyses (ICMSF, 1978). Where facilities are suitable, food homogenates can be examined directly for the presence of intact cells and spores using fluorescent staining methods. Food samples, unheated and heated, should also be streaked out on Blood agar, preferably containing egg yolk so that the typical *C. botulinum* colony reaction (i.e. a zone of opacity around the colony and a 'pearly layer' on the surface) can be obtained after anaerobic incubation at 30°C for 3 days. Suspect colonies are then cultured in a cooked meat broth and the supernatant liquid is then tested for botulinum toxin after a suitable incubation period (N.B. toxin assay is sometimes performed on meat broths inoculated directly with suspect food samples). Finally, a toxin assay can be performed directly on extracts of the original food.

Assays essentially involve the inoculation of the food extracts or culture supernatants into mice some of which have been protected with A, B or E antitoxin whilst the others remain unprotected. The inoculated mice are observed over a number of days and if the unprotected mice die with typical symptoms but the mice protected with a specific antitoxin do not then the test is positive for that *C. botulinum* type (ICMSF, 1978). Whilst bioassays using mice remain the most sensitive and widely used method for toxin detection, ELISA techniques employing monoclonal antibodies are being introduced and are likely to replace bioassays in the future; sensitivities to low toxin concentrations in food are now comparable with the two methods (Gibson *et al.*, 1988).

4.3.8.3.   *Staphylococcus aureus*

Examining methods for *S. aureus* can be grouped into two distinct areas: first, the enumeration of staphylococci and, second, testing for the presence of the enterotoxin in the food. As only about 50% of *S. aureus* strains are enterotoxin producers it will be appreciated that the latter test is more conclusive when investigating food poisoning outbreaks; furthermore, small numbers of *S. aureus* are commonly isolated from certain foods and this cannot be regarded as a danger.

The range of agar-based selective media which have been used for the enumeration of *S. aureus* from foods is a particularly extensive one and, as with *C. perfringens*, the ICMSF has carried out a comparative study of some of those in common use (Rayman *et al.*, 1978). It was concluded that Baird-Parker agar performed most satisfactorily and more recent surveys (e.g. Tham & Hajdu, 1987) have confirmed the status of this medium. However minor reservations have been expressed about its selectivity and supplements, mainly antibiotics, have been included in the medium to answer these criticisms (White *et al.*, 1988). The medium

(Baird-Parker, 1962) contains potassium tellurite and lithium chloride as inhibitory agents and egg yolk emulsion as an indicator system. After 24–48 h incubation at 37°C *S. aureus* produces black colonies surrounded by a clear zone (2–5 mm wide) in the medium; within this zone there is a smaller opaque region which only develops in the later stages of incubation. These reactions are highly specific for *S. aureus* but a confirmatory test (coagulase reaction) must be made (see Section 2.3.3.1).

The detection of enterotoxin from food extracts or from filtrates of cultures isolated from suspect foods has involved a variety of serological techniques such as those employing microslides, Ouchterlony plates and fluorescent antibodies (ICMSF, 1978); these techniques have proved more sensitive and less expensive than those involving animals which were used previously. As levels of enterotoxin in foods are extremely low, extraction and concentration methods have had to be employed; since such techniques are rather time-consuming, more rapid methods have been introduced in recent years (Tranter & Brehm, 1990). Assays involving the detection of specific enterotoxins in foods by latex agglutination, in which visible cross-linking of antibody-coated latex particles occurs in the presence of enterotoxin, can now be performed using commercially available kits (Bankes & Rose, 1989); results are available within a few hours and concentration of food extracts may not be necessary due to its extreme sensitivity. However the most popular method now used for the detection of staphylococcal enterotoxin is ELISA (see p. 201). The enzyme can be attached either to the enterotoxin (competitive ELISA) or to the antibody (sandwich ELISA); the latter may be the preferred choice (Notermans, 1982). Commercial kits usually employ the sandwich–ELISA principle and comparative studies performed with them have shown high sensitivities to the enterotoxin (Wieneke & Gilbert, 1987; Ewald, 1988).

4.3.8.4. *Bacillus cereus*

Whilst techniques for the detection of *B. cereus* toxins in foods are now being introduced (Shinagawa, 1990), conventionally only quantitative determination of *B. cereus* in suspect foods has been regarded as necessary and large numbers must be isolated to be of any significance. The selective media used for enumerating this organism (e.g. Holbrook & Anderson, 1980) often include polymyxin since *Bacillus* spp. are largely unaffected by this antibiotic. An indicator system involving egg yolk and possibly mannitol in conjunction with a pH indicator is often incorporated; *B. cereus* can be presumptively identified by the opaque zone surrounding the colony after incubation at 37°C for 24 h, the zone being similar to that seen with *S. aureus* in egg yolk containing media. *B. cereus* does not dissimilate mannitol so that the colonies are pale with a purple colouration of the surrounding agar whilst mannitol fermenters produce yellow colonies.

4.3.8.5. Vibrio parahaemolyticus

V. parahaemolyticus is halophilic (tolerates 8·5 % NaCl) and thrives at a high pH (8·6) and these characteristics are utilized in selective media devised for the isolation and enumeration of this organism. Enrichment in liquid media such as salt polymyxin broth or glucose salt teepol broth is normally performed before subculture onto solid selective media (Sakazaki *et al.*, 1986). One commonly used medium is Thiosulphate Citrate Bile Salt Sucrose agar (Kobayashi *et al.*, 1963) which contains bile salts as additional selective agents. Two indicator systems are incorporated in the medium, one involving thiosulphate and ferric citrate (see Sulphite Cycloserine agar, p. 202) and the other sucrose with pH indicator. Vibrios, unlike many enteric bacteria, fail to produce hydrogen sulphide whilst most *V. parahaemolyticus* strains are unusual amongst vibrios in not dissimilating sucrose. Thus, after overnight incubation at 37°C, *V. parahaemolyticus* produces large dark green colonies whilst sucrose-fermenting organisms form yellow colonies in the presence of bromothymol blue indicator. Further biochemical tests are necessary to confirm the identity of the isolates and, in particular, the Kanagawa test for potential pathogenicity must be performed. This involves plating out suspect cultures on an agar-based medium containing a 20% suspension of washed human blood cells which are haemolysed after overnight incubation (Kanagawa +ve); this confirms the potential virulence of the culture (Barrow & Miller, 1976).

4.3.8.6. Listeria monocytogenes

Since low numbers of *L. monocytogenes* are normally encountered in foods an enrichment stage (selective only) is employed. Two enrichment broths in common use are those of Lovett *et al.* (1987) and McClain & Lee (1988); the latter uses a two-stage enrichment process. Selective agents employed in these media include nalidixic acid and acriflavin; incubation is typically 30°C for 24 h although longer incubation for up to 7 days, often at lower temperatures, has been suggested as a means of improving recoveries (Lammerding & Doyle, 1989).

Following enrichment, broths are subcultured on one or more of the numerous selective media which have been formulated. Widely used is Oxford agar (Curtis *et al.*, 1989) which contains a battery of selective agents; after incubation at 37°C for 24–48 h, *L. monocytogenes* appears as black colonies with surrounding black zones due to an indicator system involving aesculin and ferric ammonium citrate producing black iron compounds. Other selective media commonly employed are modified McBride's agar (Lee & McClain, 1986) and ALPAMY agar (Van Netten *et al.*, 1988). The former contains a high concentration of lithium chloride together with moxalactam as selective agents whilst the latter again includes a range of selective and diagnostic agents. The above selective and diagnostic media are excellent examples of the complex formulations which are necessary to suppress the growth of unwanted organisms but at the same time permit characteristic growth of the required organism(s). Unfortunately, a few other *Listeria* spp. may still mimic *L. monocytogenes* on these selective media so that further testing is necessary to confirm the presence of *L. monocytogenes*. Typically such tests include culturing suspect isolates on blood agars and examining for a characteristic zone of haemolysis around the colonies together with testing for enhanced haemolysis in the vicinity of *Staphylococcus aureus* colonies; further testing also involves the inoculation of the suspect listeria into a range of carbohydrate-based media in which, after incubation, a particular pattern of results (i.e. acid production from specified carbohydrates) must be obtained (Lovett, 1988).

Inevitably, because of the often time-consuming nature of methods used conventionally for the detection of *L. monocytogenes*, rapid methods of identification are being sought; this search is made more urgent because of the concern now being expressed about the pathogenicity of this organism. The ELISA method suffers from the weakness that it only detects members of the genus *Listeria* rather than *L. monocytogenes* specifically. Monoclonal antibodies have been used with ELISA but enrichment of food samples for 40–48 h appears necessary in order to increase cell numbers before the ELISA can be performed (Beumer & Brinkman, 1989). More promising methods involve the use of gene probes. Datta *et al.* (1988) and Chenevert *et al.* (1989) have developed DNA probes which are claimed to be specific for *L. monocytogenes*. In the former method, following enrichment culturing, food samples are plated out onto a selective medium which is incubated overnight. A replica of the colonies is prepared by pressing filter paper onto the surface of the medium, the colonies are lysed and, after further stages, DNA hybridization is performed using radioactive or enzyme-labelled gene probes. Hybridization is performed more conventionally on filters with the latter method; it would seem that DNA probe and ELISA kits now available commercially have sensitivities which are at least on a par with normal cultural procedures (Heisick *et al.*, 1989).

The problem with the DNA probe is that samples must contain at least 10^5 per ml cells of *L. monocytogenes* so an overnight enrichment stage is normally necessary. However, an exciting development utilizing the so-called polymerase chain reaction (PCR) is eliciting great interest at the present time. The technique makes use of an extremely thermostable DNA polymerase obtained from a thermophilic 'flavobacterium', *Thermus aquaticus*, which is found naturally in hot springs. It is necessary to extract DNA from suspect cells and then rapidly amplify specific sequences by a factor of *ca* 10^7. This is done by heating the extracted double-stranded DNA to separate its two component strands (denaturing). On cooling, primers (a sequence of nucleotides matching the ends of the DNA strand) bind to the dissociated DNA strands; the added polymerase rapidly builds new complementary strands using the original strands as templates. The double strands are again heated to separate into what is now four components and the process continues. Using the so-called *Taq* DNA-polymerase sufficient specific DNA material from a single listeria cell can be amplified within 3 h, the polymerase having the further advantage that it is unaffected by the heating cycles. The PCR-amplified products are then run on an agarose gel electrophoresis plate for a further 1½ h when results are read. By using DNA sequences unique to *Listeria* spp. or, alternatively, by using a sequence unique to *L. monocytogenes*, the presence of either listerias in general, or specifically, *L. monocytogenes* can be readily determined in samples within 6 h (Border *et al.*, 1990; Kite *et al.*, 1991). An alternative approach is to utilize the fact that all pathogenic *L. monocytogenes* strains are haemolytic; one particular haemolysin (listeriolysin) has been gene sequenced and a unique portion has been used to develop a PCR assay although the sensitivity needs improving before it can be used for food samples (Bessesen *et al.*, 1990). Problems have also been encountered by Wernars *et al.* (1991) who found that amplification was inhibited by materials present in soft cheeses. Whilst these authors concluded that PCR was not yet feasible for the detection of low numbers of *L. monocytogenes* in cheeses, there is little doubt that this technique will become widely used for different foods and microorganisms.

4.3.8.7. Campylobacter jejuni and C. coli

Where numbers of *Campylobacter jejuni* or *C. coli* are low, enrichment is necessary and commonly used enrichment broths are those of Doyle & Roman (1982) and the Preston broth of Bolton *et al.* (1982). Both rely on the use of antimicrobial agents and antibiotics common to both broths are trimethoprim, polymyxin and cyclohexamide. Because of the high optimum growth temperature of these two campylobacters incubation is at 42°C for 24–48 h, typically under microaerophilic conditions. The selective enrichment broth of Doyle & Roman is so effective that 1 cell in 10 g food can be recovered. After enrichment samples are subcultured on selective media which, in the case of campylobacters, rely exclusively on antibiotics to prevent the growth of unwanted organisms. Amongst the best of these media is Preston Campylobacter selective agar (Bolton & Robertson, 1982) because it is more inhibitory to the background flora than most other recommended media (Fricker, 1987); again it relies on the three antibiotics mentioned above plus a fourth, rifampicin, for its selective properties. *C. jejuni* and *C. coli* produce flat spreading colonies after microaerophilic incubation at 42°C and they can be differentiated by biochemical tests.

4.3.9. Food Spoilage Organisms

Using standard nutrient media devoid of any selective ingredients it is possible to encourage the growth of particular groups of organisms by varying the incubation conditions. Incubation of such media at elevated temperatures (e.g. 55°C) will enable thermophiles to grow and such organisms may cause problems where foods are processed at high temperatures. Canned foods and foods processed at lower temperatures (50–90°) are examples of foods with potential thermophile problems. By incubating plates anaerobically a further selective element can be introduced as only the obligately anaerobic and facultatively anaerobic thermophiles can now grow.

Conversely, by incubating non-selective nutrient media at low temperatures psychrotrophs can be recovered. An incubation temperature of 5°C is recommended for most foods but a temperature of 7°C is used by the dairy industry as refrigerated milk is frequently exposed to temperatures of up to 7°C. Unfortunately because of the low incubation temperature, colonies take longer to develop and counting is delayed for 7 or more days. For this reason media are often incubated at *ca* 20°C which is ideal for most psychrotrophs, although the growth of many mesophiles at this temperature means that only a rough, although more rapid, indication of psychrotroph density is obtained.

Obviously selective media could also be based on extremes of pH and salt level, and examples will be presented in the following sections of media commonly used for the detection and/or enumeration of the principal groups of spoilage bacteria.

4.3.9.1. Pseudomonads

Many media which have been devised for the detection and enumeration of pseudomonads in foods have suffered from the weakness that they have allowed the growth of acinetobacters and certain coliforms. However, Grant & Holt (1977) described a new medium containing a number of selective agents which allowed the almost exclusive growth of pseudomonads; the agents consist of the antibiotic nalidixic acid (suppresses enteric Gram negative bacteria), nitrofurantoin and triphenyl tetrazolium chloride (suppress aerobic Gram negative rods other than pseudomonads), the dye basic fuchsin (suppresses Gram positive bacteria) and cyclohexamide (suppresses fungi). More recently, Mead (1985) formulated an extremely effective selective medium containing three antimicrobial compounds, viz. cephaloridine, fucidin and cetrimide. CFC agar allows good growth of pseudomonads found in foods but, at the same time, suppresses virtually all unwanted organisms; there may be slight growth of one of two *Alteromonas putrefaciens* strains.

With both the above media incubation should be for 2–3 days in the 20–30°C range; further biochemical tests may be necessary on selected colonies to confirm them as pseudomonads and additional screening is necessary where specific identification is required. Alternatively, use of a non-selective medium incubated at 5°C in parallel with one of the selective media outlined above could be considered; high counts on both media are strongly indicative of an overwhelming predominance of *Pseudomonas* spp.

4.3.9.2. Micrococci

The medium most commonly used for the isolation and enumeration of micrococci from foods is Mannitol Salt agar (Chapman, 1945). This medium contains 7.5 % NaCl as a selective agent and suppresses the growth of all organisms with the exception of micrococci and staphylococci. To differentiate between them mannitol is incorporated in the medium with a pH indicator (phenol red), it being assumed that only staphylococci will attack this carbohydrate with acid production.

Unfortunately this assumption is incorrect as there are a number of mannitol positive *Micrococcus* strains and mannitol negative staphylococci. More recently, Curry & Borovian (1976) devised a selective medium totally inhibiting the growth of staphylococci by the incorporation of furazolidone and this medium has been successfully used with foods for the detection of micrococci.

4.3.9.3. *Lactobacilli and Leuconostocs*
As lactic acid bacteria (i.e. members of the genera *Lactobacillus, Leuconostoc, Pediococcus* and some *Streptococcus* spp.) have complex growth requirements, any medium used for their enumeration must contain a wide variety of nutrients. No universal selective medium exists for the cultivation of all lactic acid bacteria so that it is often necessary to choose media depending on the food being examined (Reuter, 1985). A suitable and widely used medium is Rogosa's Acetate agar (Rogosa *et al.*, 1951) which in addition to the nutrients contains Tween '80' (polyoxyethylene sorbitan mono-oleate) as a specific growth stimulant. Selectivity is achieved by the low pH (5·4) together with a 0·2 M concentration of sodium acetate. There is no indicator system incorporated so that organisms capable of growing on the medium are presumptively identified as lactobacilli. As lactobacilli, in particular, are microaerophilic, best growth is obtained by incubating inoculated plates in an atmosphere of 95% hydrogen and 5% carbon dioxide (3 days at 30°C). By raising the pH to 6·2 leuconostocs (and pediococci) will also grow on the medium. Another medium of interest is Lactobacillus Sorbic Acid agar (Reuter, 1985). Sodium acetate, sodium citrate and sorbic acid together with a low pH (5·0) suppress other organisms including most lactic acid bacteria but lactobacilli generally grow well.

Preliminary screening of isolates on the above media involves Gram staining, catalase reaction (should be negative) and fermentation of carbohydrates (homofermentative or heterofermentative; see p. 162). The latter test is used for further identification to species level.

4.3.9.4. *Streptococci*
Methods for the specific enumeration of enterococci have already been discussed (see Section 4.3.7.2) but streptococci have many different characteristics, and can be isolated from a variety of environments by various means. A medium that is frequently used for enumerating streptococci in foods is the Thallous Acetate Tetrazolium Glucose agar of Barnes (1956). Thallous acetate is included as a selective agent against

various Gram positive and Gram negative bacteria whilst the triphenyl tetrazolium chloride is included as an indicator; this latter salt is reduced by certain streptococci giving characteristic red colonies.

The lactic streptococci, so important in the dairy industry, can be enumerated on the β-glycerophosphate medium of Terzaghi & Sandine (1975). This medium is again nutritionally complex to satisfy the requirements of the streptococci and the glycerophosphate is incorporated as a buffer to minimize the pH drop resulting from the formation of lactic acid by the dissimilation of lactose present in the medium. Incubation of the above media is at 30–37°C for 1 or 2 days.

4.3.9.5. Brochothrix thermosphacta

Brochothrix thermosphacta, now recognized as having a significant role in the spoilage of meat and certain meat products, is readily isolated on Streptomycin Thallous Acetate Actidione agar (Gardner, 1966). *B. thermosphacta* is markedly resistant to streptomycin and thallous acetate which prevent the growth of the majority of bacteria whilst actidione is included to suppress the growth of yeasts. A few pseudomonads may develop on the medium but they can be easily distinguished by their colony appearance. Incubation is typically at 25°C for 48 h.

4.3.9.6. Spore Formers

Methods for the isolation of *Clostridium botulinum*, *C. perfringens* and *Bacillus cereus* have been described earlier but it may be necessary to enumerate the total numbers of spore formers present in foods. As a preliminary, samples or dilutions containing them are heat treated at 80°C for 10 min to destroy all vegetative cells and then cooled and plated. The heating encourages spores to germinate (heat shocking), a process that is often difficult to initiate without a suitable stimulus. Standard nutrient media are employed when enumerating *Bacillus* spp., the plates being incubated aerobically. Obviously members of this group present as vegetative forms in food do not figure in any enumeration of this type.

Anaerobic incubation in a sealed jar is necessary for clostridia and this involves evacuation of the air and its replacement by hydrogen. The more extreme obligately anaerobic clostridia require the inclusion of a catalyst such as palladium in the jar; this converts any residual oxygen to water by combination with the hydrogen. Preliminary enrichment of the diluted sample in a cooked meat broth medium is sometimes practised but direct plating onto nutritionally complex solid media is preferable where enumeration is required. Differential Reinforced Clostridial medium (Gibbs & Freame, 1965) is frequently employed and, although a liquid medium, has many advantages. Recoveries are higher than on solid media and anaerobic jars are unnecessary. Accurate counts are not possible but if decimal dilutions are inoculated figures with a range of a times ten factor are available (e.g. >1000 but <10 000 per g). The medium contains a sulphur source/iron salt indicator system so that blackening of the medium indicates growth of clostridia almost all of which produce hydrogen sulphide.

4.3.9.7. Yeasts and Moulds

Yeasts and moulds have generally been cultivated on media with a low pH (3·5–5·5) and at a temperature of 20–30°C but many bacteria are able to grow under these conditions. To inhibit bacteria, 'broad-spectrum' antibiotics are now incorporated which suppress their growth whilst pHs of media are now often above 5·5. Typical widely used media are Oxytetracycline Glucose Yeast Extract (OGY) agar (Mossel *et al.*, 1970) and Dichloran Rose Bengal Chloramphenicol (DRBC) agar (King *et al.*, 1979). OGY utilizes the antibiotic oxytetracycline and nutrients including glucose, and is adjusted to a relatively high pH of *ca* 6·5; DRBC includes the antibiotic chloramphenicol and has a reduced pH (5·6) to prevent bacterial growth whilst dichloran and rose bengal are added to restrict the size and height of mould colonies.

Assay methods for mycotoxins differ somewhat depending on the food. Generally the food is comminuted to reduce particle size and then extracted with a suitable solvent such as chloroform after which the extract is purified. The next stage of analysis and detection typically involves some form of chromatography, usually thin-layer chromatography. Using suitable solvents which clearly separate the toxins it is possible to compare any suspicious fluorescent spots under uv with appropriate controls. Alternatively, cytotoxicity tests can be employed in the preliminary screening of many mycotoxins. These bioassays typically involve challenging tissue cultures derived from animal cells with an extract obtained from the suspect food (Buckle & Sanders, 1990); if a toxin is present cell growth will stop and this indicates that a more detailed chemical analysis is justified.

High-performance liquid chromatography (HPLC) has been used increasingly in recent years, either to clean up or for final assay of the sample. ELISA methods have also been developed and kits are now becoming readily available for the assay of different mycotoxins (Scott,

1988). In addition, monoclonal antibodies are now being employed in highly sensitive and rapid assay procedures for the routine screening of samples (Candlish *et al.*, 1988).

4.3.10. Canned Foods

Because the heat treatment of canned foods should destroy or inactivate all microorganisms, their presence in the food indicates a process failure. Thus microbiological examining techniques are restricted only to the isolation and identification of organisms. Media used fall into two main groups depending on the pH of the food. For low acid foods (pH < 4·5) the principal media are Dextrose Tryptone agar and Reinforced Clostridial medium while for high acid foods (pH <4·5) Tomato Juice agar and Malt Extract agar are commonly used.

4.3.10.1. Dextrose Tryptone Agar

This medium is used for the cultivation of 'flat-sour' thermophiles, particularly *Bacillus stearothermophilus*, and incubation is therefore at 55°C and aerobic. The organism produces acid from dextrose and a colour change from purple to yellow ensues in the presence of bromocresol purple indicator. This medium can also be used for the isolation of mesophiles and in this case incubation will be at 37°C.

4.3.10.2. Reinforced Clostridial Medium

This medium is similar to the Differential Reinforced Clostridial medium used in the enumeration of spore formers except that it is a solid medium (i.e. contains agar) and does not include the sulphur/iron salt indicator system.

4.3.10.3. Tomato Juice Agar

This medium is used for the cultivation of lactobacilli, *Bacillus coagulans* and clostridia growing in low pH canned foods. The pH of the medium can be adjusted by the addition of lactic acid to correspond with that of the food under investigation.

4.3.10.4. Malt Extract Agar

An example of a low pH medium (*ca* 4·5) used for the cultivation of yeasts and moulds from canned foods is Malt Extract agar; other examples are given in Section 4.3.9.7.

It is beyond the scope of this book to describe other essential procedures in the examination of canned foods. Readers requiring such information and further details of the media just described should consult appropriate literature in the Bibliography section.

4.3.11. Frozen and Dehydrated Foods

When examining frozen or dehydrated foods the standard microbiological methods described in this chapter are employed. However, many of the cells may be metabolically injured and thus require an adequate thawing out or rehydration period during which repair may be effected; an alternative procedure is to preincubate food samples in nutritionally complex media. Failure to carry out one of these steps will result in reduced counts on selective media (see p. 165).

4.3.12. Miscellaneous Tests

4.3.12.1. Methylene Blue Test

This test, used in the examination of milk, is an example of a dye reduction test, the time taken for decolorization to occur being related to the number of organisms in the sample. This relationship is very approximate, however, as milk heavily contaminated with inert bacteria may give an extended reduction time whilst short reduction times may be due to natural reducing systems present in the milk.

In the test 1 ml of a standard methylene blue solution is added to 10 ml of milk in a test-tube, mixed and then incubated in a water bath at 37°C. The milk and suitable controls are observed at 30 min intervals until decolorization of the milk is complete: the longer the time, the better the quality of the milk. Raw milk is considered satisfactory if it fails to decolorize methylene blue in 30 min.

4.3.12.2. The Limulus Lysate Test

Gram negative rods are important in the spoilage of proteinaceous foods and they produce lipopolysaccharide endotoxins which are actually a layer of their cell walls. Only minute (picogram) quantities of these endotoxins are required to gel a lysate protein obtained from the blood of the horseshoe crab (*Limulus*) and this reaction forms the basis of the test (Jay, 1977). Dilutions of the food sample are reacted with the lysate for 1 h at 37°C after which the end of a capillary tube is placed in the reaction mixture; the height to which the mixture is drawn up the tube determines whether or not gelling has occurred.

This test has been developed particularly as an indicator of the microbiological quality of raw minced meats but it has also been found useful for the assessment of raw milk quality. With the former food it has been found that a close correlation exists between a positive test result and the numbers of bacteria present, the titre (i.e. highest dilution of food sample gelling lysate) increasing 1000-fold as the food spoils.

4.4. COMPILATION OF SPECIFICATIONS

Employing the microbiological test methods described in this chapter it should be possible to maintain a measure of control of food processing operations. The application of these methods in food processing is discussed in later chapters but one aspect that can best be reviewed here is that of the compilation of specifications.

Specifications are, in essence, microbiological standards that are agreed within or between companies and generally have no direct legal implications. Thus specifications can be prepared for raw materials supplied to a food processor, for foods at various stages of preparation and for final products. In the last case the microbiological standards may be those agreed as reasonable and attainable by the company or they may be standards imposed by or jointly agreed with an external agency. From a microbiological viewpoint and depending on the food, specifications may include standards for total numbers of microorganisms and/or for food poisoning, indicator or spoilage organisms.

When compiling specifications for raw materials and final products it is desirable to start with as wide a range of relevant test methods as is practicable so that comprehensive data on the background microbiology can be built up. However, from the outset it is essential to obtain agreement with the other companies involved on the best methods of sampling and analysis. With the plethora of techniques available it is no wonder that different laboratories have their preferred methods but the aim should be to use methods that give the highest recovery of organisms, that give reproducible results and that are easy to use. Ideally test samples should be exchanged between laboratories and analysed by the agreed methods to ensure that similar results are being obtained.

As was mentioned at the beginning of this chapter, greater attention should be paid to raw materials and foods where erratic or unexpected results are obtained than to foods giving a consistent picture; attempts should be made to ascertain the reasons for these variations so that they may be eliminated. Where a clear pattern of results is difficult to establish, microbiological standards must be regarded as tentative in the short term although with food poisoning organisms additional constraints clearly apply. Specifications should reflect what is attainable under good manufacturing practice but should include tolerances to allow for sampling inaccuracies and for foods that are marginally outside the agreed standards. For example, a specification could include 'total viable count (30°C incubation)—not more than 5000 per g', but what about the occasional sample with a count of 5500, 6000 or even 8000 per g? To cover for these contingencies the specification might be better written thus: 'total viable count . . . 5000 per g but counts of <10000 per g accepted in no more than 10% of samples examined'. Such tolerances would not be permissible for salmonellas in, for example, frozen liquid whole egg; here the specification would include 'salmonella absent from 50 g sample' and that requirement would be absolute.

Perhaps before getting too enthusiastic about the need for a massive microbiological test regime for foods it would be as well to remind ourselves of the often quoted remark of the eminent bacteriologist, Sir Graham Wilson, who said 'Bacteriologists are better employed in devising means to prevent or overcome contamination than in examining more and more samples.' (Wilson, 1970). There is much in this! However, Goldenberg & Elliott (1973) do appraise the types and numerical values of non-legal standards that might be incorporated in specifications and for those requiring such information this article is recommended.

BIBLIOGRAPHY

Baillie, A., Crowther, J. S. & Baird-Parker, A. C. (1973). The use of fluorescent antibodies and other techniques for the detection of *Clostridium botulinum*. In: *The Microbiological Safety of Food*. Eds B. C. Hobbs and J. H. B. Christian, London, Academic Press.

Baird, R. M., Corry, J. E. L. & Curtis, G. D. W. (1987). Pharmacopoeia of culture media for food microbiology. *International Journal of Food Microbiology*, **5**, 187–299.

Cassidy, P. K. & Brackett, R. E. (1989). Methods and media to isolate and enumerate *Listeria monocytogenes*: a review. *Journal of Food Protection*, **52**, 207–14.

Corry, J. E. L., Roberts, D. & Skinner, F. A. (1982). *Isolation and Identification Methods for Food Poisoning Organisms*. London, Academic Press.

Defigueiredo, M. P. & Jay, J. M. (1976). Coliforms, enterococci and other microbial indicators. In: *Food Microbiology: Public Health and Spoilage Aspects.* Eds M. P. Defigueiredo and D. F. Splittstoesser, Westport, Avi Publishing Co.

Heathcote, J. G. & Hibbert, J. R. (1978). *Aflatoxins: Chemical and Biological Aspects.* Amsterdam, Elsevier.

Hersom, A. C. & Hulland, E. D. (1980). *Canned Foods—Thermal Processing and Microbiology,* 7th edn. Edinburgh, Churchill Livingstone.

Jarvis, B., Seiler, D. A. L., Ould, A. J. L. & Williams, A. P. (1983). Observations on the enumeration of moulds in food and feedingstuffs. *Journal of Applied Bacteriology,* **55,** 325–36.

King, A. D., Pitt, J. I., Beuchart, L. R. & Corry, J. E. L. (1986). *Methods for the Mycological Examination of Foods.* New York, Plenum Press.

Mead, G. C. (1985). Selective and differential media for *Clostridium perfringens. International Journal of Food Microbiology,* **2,** 89–98.

O'Toole, D. K. (1983). A review: methods for the direct and indirect assessment of the bacterial content of milk. *Journal of Applied Bacteriology,* **55,** 187–201.

Parry, R. T., Haysom, L., Thomas, N. L. & Davis, R. (1982). *A Manual of Recommended Methods for the Microbiological Examination of Poultry and Poultry Products.* London, British Poultry Meat Association.

Sharpe, M. E. (1979). Identification of the lactic acid bacteria. In: *Identification Methods for Microbiologists,* 2nd edn. Eds F. A. Skinner and D. W. Lovelock, London, Academic Press.

Skinner, F. A. & Quesnel, L. B. (1978). *Streptococci,* Society for Applied Bacteriology Symposium Series No. 7. London, Academic Press.

Wood, J. M. & Gibbs, P. A. (1982). New developments in the rapid estimation of microbial populations in foods. In: *Developments in Food Microbiology – 1.* Ed. R. Davies, London, Applied Science Publishers Ltd.

REFERENCES

Baird-Parker, A. C. (1962). An improved diagnostic and selective medium for isolating coagulase-positive staphylococci. *Journal of Applied Bacteriology,* **25,** 12–19.

Bankes, P. & Rose, S. A. (1989). Rapid detection of staphylococcal enterotoxins in foods with a modification of the reversed passive latex agglutination assay. *Journal of Applied Bacteriology,* **67,** 395–9.

Barnes, E. M. (1956). Methods for the isolation of faecal streptococci (Lancefield Group D) from bacon factories. *Journal of Applied Bacteriology,* **19,** 193–203.

Barnes, E. M., Impey, C. S. & Parry, R. T. (1973). The sampling of chickens, turkeys, ducks and game birds. In: *Sampling Microbiological Monitoring of Environments.* Eds R. G. Board and D. W. Lovelock, London, Academic Press.

Barrow, G. I. & Miller D. C. (1976). *Vibrio parahaemolyticus* and seafoods. In: *Microbiology in Agriculture, Fisheries and Food*. Eds F. A. Skinner and J. G. Carr, London, Academic Press.

Bessesen, M. T. *et al.* (1990). Detection of *Listeria monocytogenes* by using the polymerase chain reaction. *Applied and Environmental Microbiology*, **56**, 2930–2.

Beumer, R. R. & Brinkman, E. (1989). Detection of *Listeria* spp. with a monoclonal antibody-based enzyme-linked immunosorbent assay (ELISA). *Food Microbiology*, **6**, 171–7.

Blood, R. M. (1985). Techniques for the isolation of salmonellae and the quality control of enrichment and selective media. *International Journal of Food Microbiology*, **2**, 33–9.

Bolton, F. J. & Robertson, L. (1982). A selective medium for isolating *Campylobacter jejuni/coli*. *Journal of Clinical Pathology*, **35**, 462–7.

Bolton, F. J., Coates, D., Hinchliffe, P. M. & Robertson, L. (1982). A most probable number method for estimating small numbers of campylobacters in water. *Journal of Hygiene, Cambridge*, **89**, 185–90.

Border, P. M., Howard, J. J., Plastow, G. S. & Siggens, K. W. (1990). Detection of *Listeria* species and *Listeria monocytogenes* using polymerase chain reaction. *Letters in Applied Microbiology*, **11**, 158–62.

Buckle, A. E. & Sanders, M. F. (1990). An appraisal of bioassay methods for the detection of mycotoxins—a review. *Letters in Applied Microbiology*, **10**, 155–60.

Bullock, R. D. & Frodsham, D. (1989). Rapid impedance detection of salmonellas in confectionery using modified LICNR broth. *Journal of Applied Bacteriology*, **66**, 385–91.

Bülte, M. & Reuter, G. (1984). Impedance measurement as a rapid method for the determination of microbial contamination of meat surfaces, testing two different instruments. *International Journal of Food Microbiology*, **1**, 113–25.

Candlish, A. A. G., Haynes, C. A. & Stimson, W. M. (1988). Detection and determination of aflatoxins using affinity chromatography. *International Journal of Food Science and Technology*, **23**, 479–85.

ten Cate, L. (1965). A note on a simple and rapid method of bacteriological sampling by means of agar sausages. *Journal of Applied Bacteriology*, **28**, 221–3.

Chantler, S. M. & McIllmurray, M. B. (1987). Labelled-antibody methods for detection and identification of microorganisms. In: *Methods in Microbiology*, Vol 19 (Current Methods for Classification and Identification of Microorganisms). Eds R. R. Colwell and R. Grigorova, London, Academic Press.

Chapman, G. H. (1945). The significance of sodium chloride in studies of staphylococci. *Journal of Bacteriology*, **50**, 201–3.

Chenevert, J., Mengaud, J., Gormley, E. & Cossart, P. (1989). A DNA probe specific for *Listeria monocytogenes* in the genus *Listeria*. *International Journal of Food Microbiology*, **8**, 317–9.

Cliffe, A. J., McKinnon, C. H. & Berridge, N. J. (1973). Microcalorimetric estimation of bacteria in milk. *Journal of the Society of Dairy Technology*, **26**, 209–10

Cruickshank, R., Duguid, J. P., Marmion, B. P. & Swain, R. H. A. (1975). *Medical Microbiology*, Vol. 2, 2nd edn. Edinburgh, Churchill Livingstone.

Curry, J. C. & Borovian, G. E. (1976). Selective medium for distinguishing micro-cocci from staphylococci in the clinical laboratory. *Journal of Clinical Microbiology*, **4**, 455–7.

Curtis, G. D. W., Mitchell, R. G., King, A. F. & Griffin, E. J. (1989). A selective differential medium for the isolation of *Listeria monocytogenes*. *Letters in Applied Microbiology*, **8**, 95–8.

Datta, A. R., Wentz, B. A. & Hill, W. E. (1988). Identification and enumeration of beta-haemolytic *Listeria monocytogenes* in naturally contaminated dairy products. *Journal of the Association of Official Analytical Chemists*, **71**, 673–5.

Doyle, M. P. & Roman, D. J. (1982). Recovery of *Campylobacter jejuni* and *Campylobacter coli* from inoculated foods by selective enrichment. *Applied and Environmental Microbiology*, **43**, 1343–53.

Ewald, S.(1988). Evaluation of enzyme-linked immunosorbent assay (ELISA) for detection of staphylococcal enterotoxin in foods. *International Journal of Food Microbiology*, **6**, 141–53.

Fitts, R., Diamond, M., Hamilton, C. & Neri, M. (1983). DNA-DNA hybridization assay for detection of *Salmonella* spp. in foods. *Applied and Environmental Microbiology*, **46**, 1146–51.

Flowers, R. S. *et al.* (1987). Comparative study of a DNA hybridization method and the conventional culture procedure for detection of *Salmonella* in foods. *Journal of Food Science*, **52**, 781–5.

Forrest, W. W. (1972). Microcalorimetry. In: *Methods in Microbiology*, Vol. 6B. Eds J. R. Norris and D. W. Ribbons, London, Academic Press.

Fricker, C. R. (1987). A review. The isolation of salmonellas and campylobacters. *Journal of Applied Bacteriology*, **63**, 99–116.

Gardner, G. A. (1966). A selective medium for the enumeration of *Microbacterium thermosphactum* in meat and meat products. *Journal of Applied Bacteriology*, **29**, 455–60.

Gibbs, B. M. & Freame, B. (1965). Methods for the recovery of clostridia from foods. *Journal of Applied Bacteriology*, **28**, 95–111.

Gibson, A. M. *et al.* (1988). Evaluation of a monoclonal antibody-based immunoassay for detecting type B *Clostridium botulinum* toxin produced in pure culture and an inoculated model cured meat system. *Journal of Applied Bacteriology*, **64**, 285–91.

Gibson, D. M., Ogden, I. D. & Hobbs, G. (1984). Estimation of the bacteriological quality of fish by automated conductance measurements. *International Journal of Food Microbiology*, **1**, 127–34.

Gilchrist, J. E. *et al.* (1973). Spiral plate method for bacterial determination. *Applied Microbiology*, **25**, 244–52.

Goldenberg, N. & Elliott, D. W. (1973). The value of agreed non-legal specifications. In: *The Microbiological Safety of Food*. Eds B. C. Hobbs and J. H. B. Christian, London, Academic Press.

Gram, L. & Søgaard, H. (1985). Microcalorimetry as a rapid method for estimation of bacterial levels in ground beef. *Journal of Food Protection*, **48**, 341–5.

Grant, M. A. & Holt, J. G. (1977). Medium for the selective isolation of members of the genus *Pseudomonas* from natural habitats. *Applied and Environmental Microbiology*, **33**, 1222–4.

Hardy, D., Kraeger, S. J., Dufour, S. W. & Cady, P. (1977). Rapid detection of microbial contamination in frozen vegetables by automated impedance measurements. *Applied and Environmental Microbiology*, **34**, 14–17.

Hauschild, A. H. W. & Hilsheimer, R. (1974). Evaluation and modifications of media for enumeration of *Clostridium perfringens*. *Applied Microbiology*, **27**, 78–82.

Hauschild, A. H. W. *et al.* (1977). ICMSF methods studies. VIII. Comparative study for the enumeration of *Clostridium perfringens* in foods. *Canadian Journal of Microbiology*, **23**, 884–92.

Heisick, J. E. *et al.* (1989). Comparison of four procedures to detect *Listeria* spp. in foods. *Journal of Food Protection*, **52**, 154–7.

Hill, W. E. & Keasler, S. P. (1991). Identification of foodborne pathogens by nucleic acid hybridization. *International Journal of Food Microbiology*, **12**, 67–76

Hofstra, H. & Huis In't Veld, J. H. J. (1988). Methods for the detection and isolation of *Escherichia coli* including pathogenic strains. In: *Enterobacteriaceae in the Environment and as Pathogens*. Eds B.M. Lund, M. Sussman, D. Jones and M. F. Stringer. Journal of Applied Bacteriology Symposium Supplement No. 17, 197S–212S.

Holbrook, R. & Anderson, J. M. (1980). An improved selective and diagnostic medium for the isolation and enumeration of *Bacillus cereus* in foods. *Canadian Journal of Microbiology*, **26**, 753–9.

Ibrahim, G. F. (1986). A review of immunoassays and their application to salmonellae detection in foods. *Journal of Food Protection*, **49**, 299–310.

Ibrahim, G. F. & Fleet, G. H. (1985). Detection of salmonellae using accelerated methods. *International Journal of Food Microbiology*, **2**, 259–72.

International Commission on Microbiological Specifications for Foods (ICMSF) (1978). *Microorganisms in Foods*, Vol.1 (Their Significance and Methods of Enumeration). 2nd edn. Toronto, University of Toronto Press.

International Commission on Microbiological Specifications for Foods (ICMSF) (1986). *Microorganisms in Foods*, Vol. 2 (Sampling for Microbiological Analysis: Principles and Specific Applications). Toronto, University of Toronto Press.

Izat, A. L. *et al.* (1989). Comparison of the DNA probe to culture methods for the detection of *Salmonella* on poultry carcasses and processing waters. *Journal of Food Protection*, **52**, 564–70.

Jarvis, B. & Easter, M. C. (1987). Rapid methods in the assessment of microbiological quality; experiences and needs. In: *Changing Perspectives in Applied Microbiology*. Eds C. S. Gutteridge and J. R. Norris. Journal of Applied Bacteriology Symposium Supplement No. 16, 115S–26S.

Jarvis, B., Lach, V. H. & Wood, J. M. (1977). Evaluation of the spiral plate maker for the enumeration of micro-organisms in foods. *Journal of Applied Bacteriology*, **43**, 149–57.

Jay, J. M. (1977). The *Limulus* lysate endotoxin assay as a test of microbial quality of ground beef. *Journal of Applied Bacteriology*, **43**, 99–109.

Kennedy, J. E. & Oblinger, J. L. (1985). Application of bioluminescence to rapid determination of microbial levels in ground beef. *Journal of Food Protection*, **48**, 334–40.

King, A. D., Hocking, A. D. & Pitt, J. I. (1979). Dichloran-rose bengal medium for enumeration and isolation of molds from foods. *Applied and Environmental Microbiology*, **37**, 959–64.

Kite, P., Kerr, K., Taylor, G. & Millar, M. R. (1991). Rapid diagnosis of human listeriosis using the polymerase chain reaction. In: *PCR Topics. Usage of Polymerase Chain Reaction in Genetic and Infectious Diseases*. Eds A. Rolfs, H. C. Schumacher and P. Marx, Berlin, Springer-Verlag.

Kobayashi, T., Enomoto, S., Sakazaki, R. & Kuwahara, S. (1963). A new selective medium for pathogenic vibrios: TCBS agar (modified Nakanishi's agar). *Japanese Journal of Bacteriology*, **18**, 387–91.

Lammerding, A. M. & Doyle, M. P. (1989). Evaluation of enrichment procedures for recovering *Listeria monocytogenes* from dairy products. *International Journal of Food Microbiology*, **9**, 249–68.

Lee, W. H. & McClain, D. (1986). Improved *Listeria monocytogenes* selective agar. *Applied and Environmental Microbiology*, **52**, 1215–7.

Lovett, J. (1988). Isolation and enumeration of *Listeria monocytogenes*. *Food Technology*, **42** (April), 172–5.

Lovett, J., Francis, D. W. & Hunt, J. M. (1987). *Listeria monocytogenes* in raw milk: detection, incidence and pathogenicity. *Journal of Food Protection*, **50**, 188–92.

McClain, D. & Lee, W. H. (1988). Development of USDA-FSIS method for isolation of *Listeria monocytogenes* from raw meat and poultry. *Journal of the Association of Official Analytical Chemists*, **71**, 660–4.

Mead, G. C. (1985). Enumeration of pseudomonads using cephaloridine–fucidin–cetrimide agar (CFC). *International Journal of Food Microbiology*, **2**, 21–6.

Mossel, D. A. A. (1983). Essentials and perspectives of the microbial ecology of foods. In: *Food Microbiology: Advances and Prospects*. Eds T. A. Roberts and F. A. Skinner, London, Academic Press.

Mossel, D. A. A. (1985). Media for Enterobacteriaceae. *International Journal of Food Microbiology*, **2**, 27–32.

Mossel, D. A. A., Visser, M. & Cornelissen, A. M. R. (1963). The examination of foods for Enterobacteriaceae using a test of the type generally adopted for the detection of salmonellae. *Journal of Applied Bacteriology*, **26**, 444–52.

Mossel, D. A. A., Kleynen-Semmeling, A. M. C. & Vincentie, H. M. (1970). Oxytetracycline-glucose-yeast extract agar for selective enumeration of moulds and yeasts in foods and clinical material. *Journal of Applied Bacteriology*, **33**, 454–7.

Nieuwenhof, F. F. J. & Hoolwerf, J. D. (1987). Impedance measurement as an alternative to the plate count method for estimating the total count of bacteria in raw milk. *Journal of Food Protection*, **50**, 665–8.

Notermans, S. (1982). Detection of staphylococcal enterotoxins (SE) with special reference to the enzyme linked immunosorbent assay (ELISA). In: *Isolation and Identification Methods for Food Poisoning Organisms*. Eds J. E. L. Corry, D. Roberts and F. A. Skinner, London, Academic Press.

Oblinger, J. L. & Koburger, J. A. (1975). Understanding and teaching the Most Probable Number technique. *Journal of Milk and Food Technology*, **38**, 540–5.

Pettipher, G. L. (1983). *The Direct Epifluorescent Filter Technique*. Letchworth, Research Studies Press Ltd.

Pettipher, G. L. (1986). Review: the direct epifluorescent filter technique. *Journal of Food Technology*, **21**, 535–46.

Phillips, J. D. & Griffiths, M. W. (1989). An electrical method for detecting *Listeria* spp. *Letters in Applied Microbiology*, **9**, 129–32.

Rayman, M. K. *et al.* (1978). ICMSF methods studies. X. An international comparative study of four media for the enumeration of *Staphylococcus aureus* in foods. *Canadian Journal of Microbiology*, **24**, 274–81.

Reuter, G. (1985). Elective and selective media for lactic acid bacteria. *International Journal of Food Microbiology*, **2**, 55–68.

Rodrigues, U. M. & Kroll, R. G. (1988). Rapid selective enumeration of bacteria in foods using a microcolony epifluorescence microscopy technique. *Journal of Applied Bacteriology*, **64**, 65–78.

Rogosa, M., Mitchell, J. A. & Wiseman, R. F. (1951). A selective medium for the isolation and enumeration of oral and faecal lactobacilli. *Journal of Bacteriology*, **62**, 132–3.

Sakazaki, R. *et al.* (1986). ICMSF methods studies. XVI. Comparison of salt polymyxin broth with glucose salt teepol broth for enumerating *Vibrio parahaemolyticus* in naturally contaminated samples. *Journal of Food Protection*, **49**, 773–80.

Scott, P. M. (1988). Detection of mycotoxins in foods. In: *Developments in Food Microbiology — 4*. Ed. R.K. Robinson, London, Applied Science Publishers.

Sharpe, A. N. & Kilsby, D. C. (1971). A rapid, inexpensive bacterial count technique using agar droplets. *Journal of Applied Bacteriology*, **34**, 435–40.

Shaw, B. G., Harding, C. D., Hudson, W. H. & Farr, L. (1987). Rapid estimation of microbial numbers on meat and poultry by the direct epifluorescent filter technique. *Journal of Food Protection*, **50**, 652–7.

Shinagawa, K. (1990). Analytical methods for *Bacillus cereus* and other *Bacillus* species. *International Journal of Food Microbiology*, **10**, 125–42.

Stannard, C. J. & Wood. J. M. (1983). The rapid estimation of microbial contamination of raw meat by measurement of adenosine triphosphate (ATP). *Journal of Applied Bacteriology*, **55**, 429–38.

Terzaghi, B. E. & Sandine, W. E. (1975). Improved medium for lactic streptococci and their bacteriophages. *Applied Microbiology*, **29**, 807–13.

Tham, W. & Hajdu, L. J. (1987). A comparison of six media for isolating *Staphylococcus aureus* from foods. *Food Microbiology*, **4**, 133–46.

Todd, L. S., Roberts, D., Bartholomew, B. A. & Gilbert, R. J. (1987). Assessment of an enzyme immunoassay for the detection of salmonellas in foods and animal feeding stuffs. *Epidemiology and Infection*, **98**, 301–10.

Tranter, H. S. & Brehm, R. D. (1990). Production, purification and identification of the staphylococcal enterotoxins. In: *Staphylococci*. Eds D. Jones, R.G. Board and M. Sussman. Journal of Applied Bacteriology Symposium Supplement No. 19, 109S–22S.

Van Netten, P., Perales, I. & Mossel, D. A. A. (1988). An improved selective and diagnostic medium for isolation and counting of *Listeria* spp. in heavily contaminated foods. *Letters in Applied Microbiology*, **7**, 17–21.

Walker, J. & Dougan, G. (1989). A review. DNA probes : a new role in diagnostic microbiology. *Journal of Applied Bacteriology*, **67**, 229–38.

Walls, I., Sheridan, J. J., Welch, R. W. & McDowell, D. A. (1990). Separation of micro-organisms from meat and their rapid enumeration using a membrane filtration-epifluorescent microscopy technique. *Letters in Applied Microbiology*, **10**, 23–6.

Webster, J. O. J. *et al.* (1988). Improved sensitivity of the bioluminescent determination of numbers of bacteria in milk samples. *Journal of Food Protection*, **51**, 949–54.

Wernars, K., Heuvelman, C. J., Chakraborty, T. & Notermans, S. H. W. (1991). Use of the polymerase chain reaction for direct detection of *Listeria monocytogenes* in soft cheese. *Journal of Applied Bacteriology*, **70**, 121–6.

White, D. G., Matos, J. S., Harmon, R. J. & Langlois, B. E. (1988). A comparison of six selective media for the enumeration and isolation of staphylococci. *Journal of Food Protection*, **51**, 685–90.

Wieneke, A. A. & Gilbert, R. J. (1987). Comparison of four methods for the detection of staphylococcal enterotoxin in foods from outbreaks of food poisoning. *International Journal of Food Microbiology*, **4**, 135–43.

Wilson, G. S. (1970). Symposium on microbiological standards for foods: concluding remarks. *Chemistry and Industry*, **8**, 273–4.

Chapter 5

FACTORY DESIGN AND CONSTRUCTION

When an entirely new factory is to be built many factors must be taken into account in its planning. Of fundamental importance are the choice of site, the nature of the construction materials, the general layout and the various services required. Obviously the financial implications are of paramount importance too, although it would be inappropriate to discuss this aspect here.

5.1. THE FACTORY SITE

Whilst many new factories will involve redevelopment of existing sites the acquisition of new land for building purposes may be necessary. New industrial developments will require planning permission from both national and local government authorities, will need to comply with any legislation covering factory construction and must accord with any local by-laws. Particular care must be taken to ensure that adequate supplies of water, both potable and non-potable, are available; in addition it is important to discuss problems relating to the discharge of waste materials with the local authorities and in this context it may be necessary to treat all effluents before discharge if municipal facilities cannot be utilized. Other considerations, not amplified here, include the availability and quality of labour, transport facilities and site accessibility.

The area around a potential site should be surveyed carefully and an assessment made of any possible hygiene hazards. It is essential to determine what is produced in neighbouring factories and whether there is any chance of atmospheric pollution. The grounds around those factories should be examined particularly in relation to the problem of the dumping of disused machinery, wooden boxes or crates which can act so easily as a harbourage for rodents or a breeding ground for insects. Where any rubbish, although particularly food, is allowed to collect hygiene problems will inevitably ensue.

The factory site selected should be large enough for expansion and should be attractively landscaped with natural features such as trees retained wherever possible; in fact, green fields represent the ideal site in many ways. In their absence the immediate surrounds should always be laid to grass where possible to minimize dust problems except where concrete, paved or asphalt areas are needed; car parks should be situated at a reasonable walking distance from the factory to reduce fumes and noise.

5.2. GENERAL DESIGN PRINCIPLES AND STRUCTURAL TECHNIQUES

A food factory can be said to be an enclosed area into which food materials are brought and modified in various ways to manufacture an end product. But a factory is much more than this: it must be so designed that the various services including heating, lighting, ventilation, transportation, sanitation and waste disposal can operate efficiently; ancillary functions such as staff catering and office accommodation must also be considered. A factory must be as fire resistant as possible, and sound transmission and vibration must be minimized; it must resist the various stresses and strains to which it is exposed; and, finally, the factory must conform to high standards of hygiene.

The above criteria could easily be satisfied by a building which is visually downright ugly and this perhaps explains why so many factories are either banished to the outskirts of the town or sited on an industrial estate. However, industrial building in Britain has undergone a minor revolution since the early 1970s and much of the best new architecture is now industrial. Factories can and should be visually attractive, versatile and comfortable to work in (Fig. 5.1).

In designing a factory two people, the architect and the civil engineer, have essential roles and should work closely together in any project.

Fig. 5.1. A modern food factory with surrounds laid to grass. (By courtesy of Findus (UK) Ltd.)

Perhaps the first decision taken is whether to build a multi- or single-storey unit. Unless there is a land shortage or land is very expensive it is usually much better to have single-storey buildings. Obvious advantages are that materials' handling is simplified and a continuous flow from the reception of raw materials to final product storage is possible; also heavy load supporting floors are easier to construct at ground level. A further advantage is that expansion of the building is facilitated if required.

Another advantage of the single-storey unit is that greater flexibility is achieved and this enables major changes in the process patterns or techniques to be introduced with the minimum of delay and disruption. In contrast, services in the multi-storey building can be arranged more economically and where relatively light loads are anticipated multi-storied factories should be considered. There are, of course, situations where a particular process dictates factory design to a considerable degree; thus, for example, flour mills are conventionally multi-storied since they use gravity feed conveyors for the transfer of material.

Ideally the building should impose no constraints on any process or plant layout. However, all too frequently production lines are sited in unsuitable buildings as an economic necessity. These latter difficulties are discussed more extensively in the next chapter.

The single-storey framed structure is now the most common type of building. Construction can be based on a reinforced concrete or steel frame. The latter is quicker to construct and has the added advantage of being easier to modify after completion. In the framed structure the external walls are essentially a skin designed to protect personnel and equipment inside and this means that external walls can be constructed of relatively light materials such as aluminium, although bricks and cement blocks are often used. The distances between support stanchions can thus be maximized with light materials giving greater adaptability and space; it should always be possible to rearrange a production line or replace processing equipment without affecting accessibility. Machinery cannot be suspended from or connected to the building framework and separate supports are required for such equipment.

5.2.1. Roofing and Lighting

Roofing is normally flat or slightly pitched and is supported by trusses or beams. The roof height is a critical factor and must always be sufficient to accommodate and allow lateral movement of the tallest plant. The roof can be a source of natural light but the quality obtained from the erstwhile popular but now rather outdated north-facing fenestration is variable. Where lighting is from two directions the illumination is more uniform and shadowing is restricted but direct sunlight can cause both glare problems and temperature increases; the latter may potentially pose serious problems in food factories. Single-glazed windows permit excessive heat loss and for this reason should be restricted to 10% of the roof area. Wall windows can be used as an alternative to roof fenestration with the glazed area at least 30% of that of the floor but with single glazing again restricted to 10% of the wall area. Opening windows are not recommended but where used they must be fitted with screens to prevent the entry of birds and insects; such screens must be regularly inspected and maintained. Window frames must be corrosion-resistant and sills should be sloped at 45° to prevent the accumulation of dust and debris.

Artificial lighting has many advantages provided that shadowing and glare are avoided, and the intensity of light is adequate and spread as evenly as possible over the entire working area; furthermore the characteristics of the light can be adjusted so that foods can be efficiently inspected for colour characteristics and quality. Thus windowless factories have rightly become popular although perhaps a combination of natural and artificial light is to be preferred where possible.

Lighting must comply with any legislation in force. In the UK, for example, the regulations state that all fixtures must be captive, the lights should be so positioned that no part of them can fall or be shattered onto food materials and each light should have a minimum horizontal surface so as to prevent the collection of dirt and dust; such legislation appears essential. From a hygienic viewpoint recessed lighting, flush with the ceiling or walls, is to be recommended; ceilings and walls should, in turn, be pale coloured with finishes having proper light reflectance characteristics (Fig. 5.2).

Fig. 5.2. A ceiling with the minimum of overhead obstructions. N.B. Note angled sills. (By courtesy of Birds Eye Walls Ltd.)

5.3. CONSTRUCTION OF CEILINGS, WALLS AND FLOORS

5.3.1. Ceilings and Overhead Fittings

Often the most important zone in any factory process area is that between one's head and the ceiling. In far too many food factories a myriad of overhead pipes, lights, cables and beams is to be seen which can only vitiate the required hygiene standards of the factory; they may also lead to foreign matter entering the food with subsequent complaints from the consumer.

Overhead pipework sometimes passes directly over process lines and water condensed on the pipes can drip on the underlying food and equipment. In a food factory environment this water probably harbours substantial numbers of microorganisms, many of which may be the main spoilage types of the food being prepared. A further hazard of poorly situated pipes is flaking paintwork. Pipes, exposed beams and lights are ideal roosting perches for birds whose droppings will inevitably contain large numbers of bacteria including, not infrequently, *Salmonella* spp.; furthermore, pipes and exposed beams may become runways for rodents. The cleaning and maintenance of overhead fittings can prove troublesome particularly where access is difficult. Accumulation of dust and debris often necessitates frequent cleaning as in, for example, bakeries.

It is obvious from the above that there is a strong case for reducing overhead equipment to the minimum. This is undoubtedly best accomplished by providing a separate service floor (Fig. 5.3) to house ducts, pipework, electric cables, etc., and one of the advantages of this system is that maintenance can continue during production. The incorporation of a service floor of this nature means that ceilings can be smooth and unencumbered thus making cleaning so much easier (Fig. 5.2). As with the recessed lighting mentioned previously, all fittings should be flush with the ceilings and any void spaces eliminated by using appropriate sealing methods. Vertical service ducts should not be attached to walls but should be sited at a distance of at least 4 cm from them to facilitate cleaning; lateral services from these ducts should be as short and direct as possible.

Ventilation must be efficient and ceilings should be constructed of and finished with materials that obviate condensation, paint flaking and mould growth. There are many types of hygienic and durable ceiling materials available to-day which vary enormously in cost. Typically they are of panel construction, the panels being made of various types of PVC, reinforced resins, PVC-faced plasterboard, PVC-foil-faced foam-

Fig. 5.3.   A service floor housing ducts, pipework, cables, etc. with floor forming ceiling to production area. (By courtesy of Findus (UK) Ltd.)

board or plastic-coated steels. Care must also be taken to ensure that the junctions between ceilings and walls are rounded and sealed against dust and water; in this connection thermoplastic materials are used in forming rounded corners and any required angles generally. Rounded junctions are easily cleaned and therefore cannot act as insect harbourages.

5.3.2.   Walls

As with ceilings the aim is to create internal wall surfaces that are smooth and free from cracks and crevices where food materials may lodge. The microbial breakdown of such trapped foods may result in the production of organic acids, etc., which may attack the surface finish; because of this it is desirable that surfaces should be resistant to chemical and biological agents but they should also be impervious to grease and water.

Walls should be constructed with durable materials and in certain situations may have to be additionally protected against more severe impact damage. Whilst damage due to trucks, etc., can be minimized by the use of suitably sited crash barriers and bumpers, areas susceptible to such damage may have to be additionally protected with stainless steel or plastic (i.e. polypropylene) sheeting. When such protection is installed care must be taken with the bedding and jointing as bacterial growth may otherwise occur behind the sheets.

All wall angles, corners and junctions of walls and floors must be imperviously sealed and be rounded for ease of cleaning.

Coving of the wall/floor junctions to a height of *ca* 15 cm is strongly recommended (Fig. 5.4). As with ceilings, ease of cleaning is a prerequisite and wall finishes must be resistant to repeated washing with the various mechanical devices, detergents and sanitizing agents available.

There is a variety of wall surfacing materials available for food premises. Traditionally tiles have been widely used because many types are relatively cheap. Tiles are prone to damage and lack flexibility increasing the chances of cracking. Porous grouting, cracks in tiles and

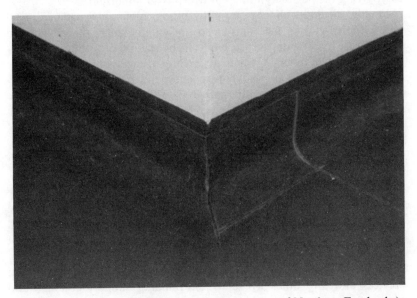

Fig. 5.4. A coved wall/floor junction. (By courtesy of Northern Foods plc.)

the reverse side of loosened tiles are places where food build up can occur with consequent microbial growth and, in the last example, even insect infestation. Since it is also difficult to replace tiles once laid, their widespread use must be viewed with some caution.

Strong, pale-coloured, cement-rendered walls are also popular, especially in abattoirs. It is often preferable to leave such walls unpainted since many paints are liable to flake. However, both painted and unpainted surfaces can be susceptible to the effects of microorganisms and chemicals; they are also readily damaged by impact. These difficulties can be partially overcome by incorporating fungistatic agents in suitable paints (e.g. polymer resin paints) or by the application of chemical coatings to cement-rendered walls.

Modern materials that are used increasingly for walls include polypropylene and different PVCs. Suitable polymers and glass-reinforced PVC can be layered on a thin metallic surface with a cladding of a suitable insulating material. Whatever wall material is used it will be supplied in units which, if damaged, can be easily replaced. Further advantages of such materials are that they are easily cleaned and they have a good resistance to impact. In addition, because these wall units are not load bearing they can be easily moved to alter the layout of the factory, giving the flexibility in design which is so important.

Finally, all doors should be tightly fitting, preferably made of metal and with self-closing mechanisms to prevent the entry of rodents and other pests. Insect entry is also hindered by the use of overlapping plastic strips which function when doors are opened; unfortunately they quickly get dirty and are readily torn so are not recommended—air curtains are the most effective alternative. Door frames should also be protected and stainless steel cladding is recommended.

5.3.3. Floors and Drains

The general properties required for floors are not unlike those for walls; thus floors should be made of materials that are impervious, durable and impact resistant and, in addition, are able to withstand the passage of vehicles and the pressure of standing equipment. Other essential properties are resistance to grease, to cleaning agents including hot water or steam and to biochemical and microbial attack; sealed joints should have similar properties. Floors should also be free from cracks, crevices and any other irregularities. Building support columns and plinths for equipment should be coved where they meet the floor.

Another important property of floors is that the surface should be non-slip and to this end surface additives such as carborundum may be incorporated. Floors should be easy to clean and disinfect and since this is generally achieved by the use of aqueous solutions the floors should be sloped at an angle of between 1 in 40 and 1 in 60 towards suitably sited drains; distances between drains should be restricted to 5 m. Ideally the washing fluids should pass from cleaner to dirtier areas and residual pools of water must be avoided since they are both a safety and a microbiological hazard. Floors in dry areas such as bakeries and warehouses are often constructed of wood and such floors may prove difficult to clean, cleaning being limited to dry vacuuming.

Many types of flooring materials are used with ordinary concrete as a base. In fact concrete may be used as the final surface in certain situations but it has certain limitations. It is prone to be dusty, has little resistance to acids or to biochemical attack, is liable to wear unevenly forming pits and, if not properly laid, may crack due to shrinkage; moisture retention is therefore encouraged which may lead to microbiological hazards.

The quality of the floor surface can be improved by topping the concrete with suitable protective materials. These include epoxy, polyester or acrylic resins, chlorinated and styrene butadiene rubbers and bituminous paints and mastics. The properties of the floor surface required depend on the concentration of resin included in the mix whilst the aggregate used can affect the appearance of the floor; pigments may also be added as colourants. The resins have many advantages over other types of floor surface since they provide considerable resistance to impact, chemicals and microbial attack; they are durable and easy to clean and most provide good non-slip surfaces. These topping materials are, however, generally temperature-sensitive and exposure for other than limited periods to temperatures above 70–80°C must be avoided. Epoxy resins are probably the most durable of the polymers and whilst there is sometimes an odour associated with them, careful selection can obviate this problem; furthermore, excessive smoothness can be avoided by the incorporation of suitable additives. However, under constantly wet conditions they do absorb water and eventually become saturated.

Tiles can provide a relatively long-lasting and impervious surface provided they are carefully laid and properly grouted. Unfortunately, as with wall tiles, there are weaknesses since they crack, lift to be underrun by water and are difficult to replace satisfactorily. Two types of tile are commonly used in the food industry (viz. quarry and ceramic) whilst furans

are perhaps the best grouting materials; however, even with high performance materials the durability of a tiled floor depends primarily on
the quality of grouting. Quarry tiles are cheaper than ceramic tiles and
are characterized by a non-slip surface, relatively good resistance to
impact, generally excellent resistance to chemical and biochemical attack, and ease of cleaning. They are, however, rather noisy to walk on
and this tends to induce fatigue in operatives. Ceramic tiles have all the
characteristics of quarry tiles with the non-slip properties being created
by the inclusion of additives during manufacture; they are, however,

Fig. 5.5.   An open drain of ample size. N.B. Note crash barriers. (By courtesy
of Marks and Spencer plc.)

more expensive. Where very wet floors are encountered floor bricks are often used; in this situation it is necessary to underlay the bricks with a waterproof membrane since water will inevitably penetrate even with the best quality grouting material.

Obviously, efficient drainage is of paramount importance and equipment should not drain directly on to the floor. The drainage system used will vary depending on the food being produced. However, easy cleaning and maintenance of all drainage systems is essential so that totally covered drains should be avoided where possible. In meat processing areas, for example, gullies and open channels are preferred. They should have smooth, vertical sides with a rounded base and be sloped to allow for adequate water velocity. The depth should be sufficient to prevent overflow and at the same time facilitate cleaning. These open drains should preferably be sited adjacent to but not directly against walls (Fig. 5.5) so that flooding of the wall/floor junction and stacking against walls are prevented. Drains should be provided with traps to collect solid materials and water locks to prevent odour feedback; grease traps should also be provided where necessary.

In other processing situations, e.g. fruit and vegetable washing areas, drains with removable perforated metal gratings are often used. These gratings may prove difficult to clean with debris building up in the slots but cleaning difficulties can be reduced by having wider openings at the top than at the bottom. Drain covers are often required to allow transport or personnel to pass over them; such covers should be flush or even slightly above the floor level to protect the edge of the drain from damage.

5.4. VENTILATION AND AIR CONDITIONING

Food processing is a specialized operation and the quality of the final product can be adversely affected by a lack of environmental control in the factory. Adequate ventilation is therefore as essential for hygienic food processing as it is for the comfort of personnel. These requisites may demand the installation of air conditioning, an aspect that is dealt with towards the end of this section.

The importance of efficient ventilation in overcoming condensation problems has been mentioned previously. Steam and moist air condensing on cool surfaces like metal pipes frequently results in severe mould growth and other effects of inadequate ventilation include corroded surfaces, peeling paintwork and disintegrating plaster. Steam and moist air should be controlled by the installation of canopies, trunking and extractor fans around offending equipment which should be grouped together where possible. Obviously all ventilation equipment, including air inlet and extracting vents, should be corrosion-resistant and fitted with louvres and screens to prevent the entry of airborne contaminants as well as birds, insects and rodents. Care should also be taken to ensure that screens are easy to clean and hence air flows are not restricted by clogging.

Another objective of ventilation is to minimize the build up of offensive fumes and odours. This is of vital importance when considering the comfort of personnel but it should also be appreciated that certain foods, e.g. flours and dried milks which contain dispersed fat, pick up odours readily.

Other variables requiring control are temperature, overall humidity, air movement, the proportion of fresh air and the replacement of extracted air; these are all important in the factory environment and requirements may vary from one factory to another.

The optimum environmental temperature varies from 21 or 22°C for personnel in sedentary occupations to 13 or 14°C for those engaged in heavy manual work. Overall humidity should be between 30 and 70% whilst the rate of air change and the amount of fresh air required are dependent on the job being done. High rates (20 changes per hour) are required in cooking areas, although these rates can be reduced if localized extraction systems are used; alternatively, rates as low as 1½ changes per hour can be used in stores and offices. As a general principle, air flow should be in the opposite direction to the process flow; that is, the freshest air is required in the finished product handling area whilst much simpler ventilation, humidity and temperature control will suffice for the department handling raw materials. However, a high proportion of fresh air will always be required where heavy manual work is being undertaken.

From the foregoing it will be seen that there are many inter-related factors affecting the environment within the factory. Natural ventilation may be all that is required in certain situations. This would involve the removal of air through ceiling and wall vents supplemented by extraction fans sited so as to cause the minimum of noise and vibration. Because this simple form of ventilation is often misused with consequential heat losses it is often more economical to obtain the desired environment by the installation of air conditioning. The specification will vary in different factory areas and for greater flexibility individual air conditioning and ventilation plants may have to be installed.

Ideally all air entering the factory should be filtered or treated in such a way as to minimize the number of microorganisms and other contaminants carried. To ensure that air enters only under controlled conditions it is preferable to have a slight excess pressure within the factory relative to the outside. After passing through the factory the air is generally expelled, although in some cases it may be recirculated after further filtering and, if necessary, deodorizing. It will be appreciated that the air systems recommended above are expensive and such arrangements may only be required in limited parts of the factory; however, such systems are increasingly used in modern food factories.

With the extremely high and mounting costs of fuel, energy efficient buildings are essential and so good insulation is vital. Thermally insulated buildings reduce both cooling requirements in hot weather and heating in cold weather. But there are other ways of saving costs such as recirculating the air warmed by lights and machinery so that, all in all, there is much to commend the use of air conditioning.

5.5.   NOISE AND VIBRATION

It may seem at first sight that the relationship between noise and vibration on the one hand and the health and safety of personnel on the other is a tenuous one. However, there is no doubt that fatigue sets in more quickly in a noisy environment than in a quiet one and this can undermine the general wellbeing and safety awareness of workers. Unfortunately in hygienically designed and constructed food factories with flush fittings, unobstructed lines, smooth, hard and easily cleaned surfaces, conditions exist which tend to be noisy.

Noise abatement can be encouraged by designing machines that are quiet in operation, that are sound-insulated or are modified with silencers or attenuators. Many machines create disturbance through vibration and such machines should be placed on resilient mountings. More generally, sound insulation of walls, floors and ceilings by the use of acoustic materials may be necessary. Care should be taken to see that acoustic tiles, panels, etc., conform as closely as possible to the same standards (i.e. ease of cleaning, durability, moisture resistance) demanded of other surfacing materials used in the food processing industry.

BIBLIOGRAPHY

Darrington, P. (1985). Factories. In: *Planning. The Architects Handbook*, 10th edn. Ed. E. D. Mills, London, Butterworths.

Drury, J. (1981). *Factories. Planning, Design and Modernisation*. London, Architectural Press.

Imholte, T. J. (1984). *Engineering for Food Safety and Sanitation*. Crystal, USA, Technical Institute of Food Safety.

Katsuyama, A. M. & Strachan, J. P. (1980). *Principles of Food Processing Sanitation*. Washington, Food Processors Institute.

Lyons, S. L. (1981). *Handbook of Industrial Lighting*. London, Butterworths.

MIDAS Bulletin No. 3. (1978). *Floors and Walls*. Meat Research Institute and the Meat and Livestock Commission.

Robson, J. N. & Barnes, G. (1980). Plant design to minimize non-microbial contamination. In: *Hygienic Design and Operation of Food Plant*. Ed. R. Jowitt, Chichester, Ellis Horwood.

Shapton, D. A. & Shapton, N. F. (1991). *Principles and Practices for the Safe Processing of Foods*. Oxford, Butterworth-Heinemann Ltd.

Society of Food Hygiene Technology. (1987). *Food Premises—Today and Tomorrow*. Conference Proceedings, Potters Bar, SOFHT.

Chapter 6

FACTORY LAYOUT

In the previous chapter criteria for the design and construction of food factories were reviewed. In this chapter the food itself is considered in relation to the factory environment and through the various stages of production. This involves the overall flow of food materials starting with reception and storage, continuing through the processing and concluding with storage of the final product. In addition, an assessment is made of how different departments within the factory are best arranged. Obviously any design plan will depend partly on what is being produced but the demands of good hygiene as well as production efficiency do impose constraints on the juxtapositioning of the various processing departments as well as on other areas of work. The earlier plea for flexibility in factory design should not be forgotten however.

6.1. HANDLING OF FOOD MATERIALS

The handling of materials essentially involves the study of problems related to their flow from reception to final distribution. The smoother the flow of food materials and the less handling they receive, the better the end product is likely to be in microbiological terms. The comparatively recent introduction of fully automated handling, from raw material to final product storage, is therefore welcome and greater use of this method is certain in the future.

239

Handling efficiency is affected by many factors. In the last chapter mention was made of the effects of the limitations imposed on the process and on the plant layout by a poorly designed factory building and clearly this must also adversely affect handling efficiency. Ideally of course, the handling system employed, just as with the arrangement of work areas, should be determined by production needs modified, where necessary, to suit the dictates of hygiene.

Handling efficiency is also affected by factors which are under the domain of work study and stock control managers. Here a knowledge of the quantity, sequence, timing and distance of food flow is required. Starting with the delivery of raw materials the factory should have a storage capacity for incoming materials which is in excess of any delivery load. Transport which is unable to off-load is wasteful in economic terms. In addition, perishable foods will begin to deteriorate if not quickly transferred to appropriate storage or to the processing lines. Raw materials should also arrive in the best form for handling; the weights and volumes of goods and the frequency of their delivery may have to be adjusted for greater efficiency.

Similar principles apply to food materials when taken out of storage for processing. It is important to ensure that raw materials are not withdrawn from storage in excess of processing requirements especially when perishable foods are involved.

Turning to the processing of the food itself, here again the aim is to achieve a smooth flow of operations, both within and between the various process areas. Thus investigational methods used by work study personnel, e.g. flow process charting, are of great value when investigating potential microbial problems on process lines; liaison between the work study personnel and hygiene managers should therefore prove beneficial.

6.1.1. Work Flow Patterns and Plant Layout

Kotschevar & Terrell (1977) listed eight principles of significance when dealing with the efficient flow of work in food service areas. These principles which are almost equally applicable to a food factory process area are:

1. Functions should proceed in proper sequence directly, with a minimum of criss-crossing and backtracking.
2. Smooth, rapid production ... should be sought, with minimum expenditure of worker time and energy.
3. Delay and storage of materials in processing ... should be eliminated as much as possible.

4. Workers and materials should cover minimum distances.
5. Materials and tools should receive minimum handling, and equipment minimum worker attention.
6. Maximum utilization of space and equipment should be achieved.
7. Quality control must be sought at all critical points.
8. Minimum cost of production should be sought.

Of these eight principles the first one listed needs some amplification. In essence the sequence of processing operations should be as direct as possible and the so-called 'straight line' process flow is regarded as the most efficient. This term implies that the process layout is literally

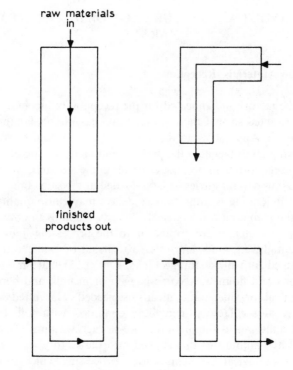

Fig. 6.1. Examples of processing operation layouts.

straight but the term is also used for a variety of other layouts (see Fig. 6.1). These layouts minimize the chances of recontamination of a semi-processed or processed food product by unprocessed or raw materials; that is, cross-contamination is effectively prevented. Raw materials and processed foods are kept as far removed from each other as possible although, as an additional safeguard, the provision of panels separating different parts of the process line is advantageous as it restricts the movement of personnel from one area to another.

It cannot be stressed too strongly that one of the greatest hazards in food production is that of *cross-contamination of cooked or semi-cooked foods by raw foods*. Outbreaks of salmonella food poisoning are frequently caused in this way and it only needs a little common sense and planning to rectify the problem.

6.2. LAYOUT AND INTEGRATION OF DIFFERENT WORK AREAS

6.2.1. Raw Materials: Reception

Areas of the factory site allocated for the reception of incoming materials should be situated so that there is direct and easy access for transport of whatever form. Approaches to the unloading bays should be unrestricted and large enough to facilitate the parking and manoeuvring of articulated and other large lorries. Bays should be of ample size and, where volume demands, allow several lorries to be unloaded at the same time.

Efficient unloading is important as delays may cause quality deterioration in the incoming raw materials. Speedy methods of conveyance of a diversity of materials to storage or to the processing lines are often needed. Liquids, e.g. milk and oils, are carried by tankers and require to be pumped into bulk storage tanks. Dry, particulate materials, e.g. cereal grains and flour, may come bagged or in bulk and here transfer by mechanical or pneumatic means is needed. Harvested vegetables often arrive at the factory in a filthy condition with soil, leaves and stems still adhering to them. Initial pre-storage washing of such foods will reduce the amount of dirt carried into the factory; excess dirt may create microbiological problems during processing which often affects the subsequent shelf-life of the product. In addition, fork-lift trucks and

Fig. 6.2. Raw materials' storage area. (By courtesy of Birds Eye Walls Ltd.)

similar vehicles can carry such dirt on their tyres and later deposit it indiscriminately around the factory; to avoid this the unloading area should always be kept as clean as is practicable.

Storage buildings for incoming raw materials should also be sited so that an efficient flow of materials to the processing areas is maintained. Storage space should be of a size that is sufficient to accommodate all raw materials even at peak periods; all too often inadequate storage space leads to an undesirable overflow of raw materials into production and other areas.

6.2.2. Raw Materials: Storage

The storage rooms used for food materials should provide a clean environment, adequate space for inspection and cleaning (Fig. 6.2), good air circulation, and the correct temperature and humidity. The storage requirements for different categories of foods obviously vary but all storage areas should provide protection from dust, insects, rodents and other pests.

Food materials should not be placed directly on the floor but should be stored on pallets or racks. Stacks should be so arranged that inspection of the upper layers is facilitated. Storage zones should be clearly marked with traffic lanes of an adequate width interspacing them at regular intervals. Goods must not be stacked against walls and in larger storage areas there should be a gap between the wall and the stack that is sufficient for walking down for inspection purposes. Where shelving or racking is used for storage, ground and wall clearances should be at least 30 cm to promote air circulation and assist cleaning; where possible it is better for shelves to be accessible from both sides.

Where chill storage is required, rooms should operate with a temperature deviation restricted to ± 1°C and it is essential that the refrigeration capacity is sufficient to maintain the desired chill temperature even though additional heat is being produced from sources such as motors, operatives and the respiratory activities of the stored foods. Any means of reducing the load on cooling equipment will be advantageous economically. Thus efficient insulation, possibly with automatically operated doors and air locks, is necessary and chill rooms should be sited as far away as is practicable from processing areas where air temperatures are high. To ensure that temperature control is satisfactory at all times, chill rooms must be equipped with adequate monitoring equipment, e.g. temperature/time recording charts.

Cold stores should be cuboid in shape to give the minimum surface area:volume ratio as by this means maximum storage space is obtained for the minimum of insulation; a similar principle can be applied to chill rooms although here it is less critical. Insulating cold stores is expensive but is necessary to obtain a storage temperature which is as constant as possible. Fluctuating temperatures reduce the shelf-life of foods in terms of both quality and weight loss. To control weight losses high relative humidities are required although in this respect it should not be forgotten that air can only absorb low levels of moisture at these low storage temperatures.

Efficient air circulation reduces temperature fluctuations and goods should therefore be stacked in such a way as to encourage rather than impede air circulation. Gaps of at least 30 cm below the ceiling and from walls are essential and on no account should materials be stored directly on floors. Horizontal air circulation is helped by placing strips of wood or inflatable bags every 100 cm or so in depth to separate layers of food packets.

Floors should be durable, smooth, easily cleaned and carefully insulated; in addition, because of the problem of 'frost heave' where the subsoil becomes frozen to a depth of 1–2 m causing weakening of the foundations, a heating system must be incorporated below the level of insulation. Internal cladding for walls and ceiling should also be durable, smooth and easily cleaned; materials such as galvanized steel and aluminium are used but, although more expensive, a PVC-faced sheeting over galvanized steel is particularly recommended.

Design layout of cold stores should aim at keeping the environment as constant as possible. Thus the entry of warm air should be minimized by providing properly insulated double doors with an air lock and, as with chill rooms, siting should be some distance from warm processing areas. Again, cooling capacity must always be in excess of the maximum load and temperature/time recorders are necessary to ensure that the correct conditions are being maintained.

6.2.3. Processing Area

6.2.3.1. General Requirements
The general layout and functioning of the processing line should facilitate the flow of partially processed food material from one operation to the next with the absolute minimum of delay. Delays should also be minimized when different components of a food product are prepared separately and subsequently assembled or mixed together. To assist in this objective it is essential that sufficient space is provided for all processing operations and for the cleaning, maintenance and inspection of all equipment and of the general production area (Fig. 6.3).

The eight principles (Kotschevar & Terrell, 1977) listed in Section 6.1.1 also apply to the processing area. In particular, the 'straight line' flow concept should be strictly adhered to as should the requirement to avoid cross-contamination of cooked or semi-cooked foods by raw foods.

With smaller factories or production lines, equipment and utensils may be used in more than one process. In these situations any equipment so used must be cleaned thoroughly between one operation and the next so that there can be no possibility of the transfer of contaminants from one process to any other.

Foods that may be contaminated with food poisoning bacteria (i.e. raw meats and poultry) should be processed in areas that are kept cool to prevent the growth of such bacteria; for this work a temperature of 10°C should be regarded as a necessity although slightly higher temperatures (11–12°C) may be permitted for limited periods.

All entrances to processing areas should be supplied with adequate washing facilities to be used by *every* employee when entering. These wash-hand basins, actuated by foot or knee taps, should be sited so that it is easy to check that employees are washing properly.

Fig. 6.3. General view of beefburger packing lines. (By courtesy of Findus (UK) Ltd.)

6.2.3.2. Siting of Equipment

Equipment should be sited so that it can be easily operated, cleaned, inspected and maintained. There should be easy access to all internal and external surfaces and parts and the latter should be easily removable. Thus equipment should not be placed close to walls, ceilings or other equipment and a minimum access gap of *ca* 100 cm should be allowed. In addition there should be adequate space (*ca* 20 cm) under equipment for cleaning purposes. Run-offs such as pipes and chutes should also be positioned for easy cleaning, inspection and maintenance whilst adequate space should be allowed for equipment passing through floors or walls for similar practical reasons.

6.2.4. Finished Product Storage

Most of the points raised in Section 6.2.2 on raw materials' storage are equally applicable here and do not need to be reiterated in detail. The principal requirements for final product storage are given in Table 6.1; they are essentially a summary of information given earlier in this chapter dealing with the storage of raw materials. Thus the need for a clean and constant environment is of paramount importance. Sufficient space is also necessary so that inspection and cleaning can be facilitated; again, adequate traffic lanes must be provided. The cold store illustrated in Fig. 6.4 is an example of what can be achieved with finished product storage.

6.2.5. Employee Service and Welfare Areas

A range of amenities must be made available to all employees although the extent of this range will depend on the size of the concern. It has

Fig. 6.4. Frozen foods cold-storage area. (By courtesy of Birds Eye Walls Ltd.)

Table 6.1
Finished Product Storage Requirements

Constant environment
    Correct temperature
    Correct humidity
    Good air circulation
    Adequate space
    Durable construction materials
Clean environment
    Protection from:
        dust and dirt
        insects
        birds
        rodents
        other pests
    Clean surrounds to storage area
    Regular inspection and cleaning
    Easy inspection and cleaning
Careful handling

Efficient stock rotation

Sited at a distance from warm processing areas

Fig. 6.5. A visually attractive dining area. (By courtesy of Birds Eye Walls Ltd.)

long been recognized that morale suffers if employee service and welfare facilities are sub-standard; every effort should be made to see that conditions and facilities are of the highest quality.

The principal amenities should be sited as far as is reasonably possible from work areas so that employees are able to enjoy a totally different and relaxed atmosphere. Views of landscaped gardens and trees are beneficial and the provision of suitable outdoor leisure facilities is a bonus but not a luxury. Indoors, rest and dining areas (Fig. 6.5) should be visually attractive and here again leisure facilities should be contemplated.

Certain facilities will have to be duplicated or multiplied at strategic points in the factory. For example, employees should not have to walk more than 50 m to toilets whilst hand-washing facilities must be readily available in food handling areas (Fig. 6.6). How facilities are best arranged so as to obtain maximum efficiency and personnel satisfaction depends on many factors such as the size of the factory and amount of land available,

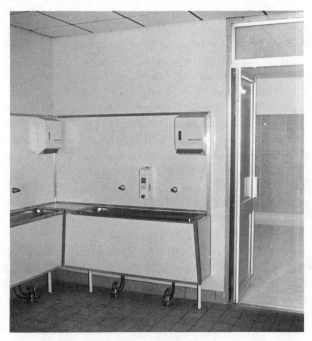

Fig. 6.6. Hand-washing facilities at entrance to processing area. (By courtesy of Birds Eye Walls Ltd.)

the number and quality of personnel employed and how labour is divided amongst the different departments. It may sometimes prove desirable or even necessary to have totally separate facilities (canteens with a common servery, rest areas, cloakrooms and toilets) for groups of employees. This situation arises, for example, in meat and poultry factories handling both the raw and cooked materials, and it is clearly advantageous to separate personnel handling raw materials from those handling cooked products.

At this point it may be advantageous to discuss some of the amenities required in more detail.

6.2.5.1. *Toilets*

Adequate and separate toilet facilities must be provided for male and female employees. The principal toilets should be sited at some distance from the production area. In larger factories additional toilets will supplement them and these should be sited close to but not opening directly onto processing and storage areas. Double self-closing tight fitting doors are required although automatically operated doors are preferable.

Toilets should conform to the same standards of hygiene demanded in food handling areas of the factory, and cleaning and disinfection should be carried out at least once per day. Construction materials for ceilings, walls, floors and cubicle partitions should be durable and surfaces should be hard, impervious to water, smooth and easily cleaned (Fig. 6.7). Junctions between walls and floors should be coved and wall angles rounded. Floors should slope towards drains and efficient ventilation to the outside atmosphere, possibly aided by extraction fans, is vital. All toilet facilities must be well maintained and operating at all times, and in this employees should be encouraged to play a part.

In total, water closets, preferably foot-operated, should be provided at the rate of 1 per 15 female employees whilst 1 urinal and 1 water closet per 25 males is necessary.

Sufficient wash-hand basins, without waste plugs, should be provided in all toilets; in total there should be 1 basin for every 10 to 12 employees handling foods. Hot (*ca* 44°C) and cold water or mixer taps are required which should be knee- or foot-operated; European Community regulations forbid the use of hand-operated taps. Liquid or powdered soap dispensers rather than cakes of soap should be provided together with nail brushes and where more hazardous foods are handled disinfectants should be made available. Hand drying should be by means of disposable paper towels or efficiently maintained warm-air driers and in the former case adequate waste receptacles are required. Employees should be reminded of the

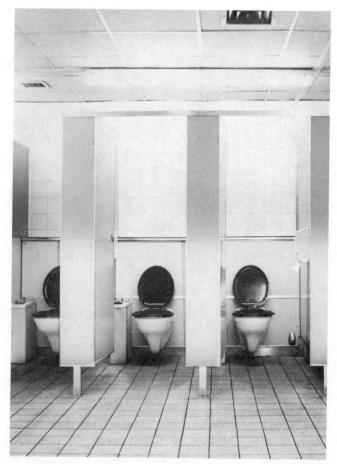

Fig. 6.7. Foot-operated water closets. (By courtesy of Birds Eye Walls Ltd.)

need to wash their hands when leaving the toilet by suitably worded notices but this should be reinforced by routine checks by supervisory staff.

Additional hand-washing facilities should be sited at all entrances to food handling areas. Here foot- or knee-operated rose head mixer taps are preferable as splashing is reduced.

6.2.5.2. Cloakrooms

Cloakrooms are probably best sited near the factory entrance. They should provide adequate locker space and changing/washing facilities

Fig. 6.8. Cloakroom showing lockers with sloped tops. (By courtesy of Birds Eye Walls Ltd.)

for all employees involved in food handling, maintenance and cleaning work. Lockers should house only outdoor clothes and personal items, and should not be used to store food. Locker tops should either be sloped or abutted against the ceiling in order to prevent their use as storage space and as a collecting point for litter (Fig. 6.8). Litter is often thrown behind lockers and so lockers should always be sited centrally in rooms with access routes nearest the walls.

Cloakrooms should contain toilet facilities but on a small scale. Showers should be provided for employees handling raw meats, poultry and fish, for those working in hot or humid conditions and for those working generally with unpleasant or dirty materials such as maintenance engineers and cleaning staff. Showers should be provided at the rate of 1 per 20 such employees.

6.2.5.3. Canteens and Other Food Facilities
Food service facilities of one form or another must be provided for employees. Where catering is based on a canteen or cafeteria type of service, high standards of hygiene both in the construction and running of the facility are required; that is, the general principles of hygiene in the food factory are equally applicable here. Thus dining, service, storage and refuse areas, wash-up rooms and kitchens should be regularly cleaned, a job facilitated

by the use of correct surfacing materials on ceilings, walls and floors, and with angles and junctions appropriately coved and rounded. Whilst good ventilation is important, all windows should be screened and doors should be self-closing. Efficient and regular disposal of rubbish is essential and regular checks to see that standards are being maintained are important.

If full canteen facilities are not provided a snack or rest room should be made available with vending machines installed. Cleanliness is equally important here and waste receptacles for plastic cups, sandwich wrappers, etc., should be provided as should simple hand-washing facilities.

Drinking water should be readily available from suitably sited fountains throughout the factory. Design features should ensure that cross-contamination from one user to another is minimized, and regular cleaning and maintenance of fountains is essential.

6.2.5.4. Health Facilities

As a minimum, a first-aid room doubling as a rest room is required. Medium-sized factories should employ a part- or full-time nurse whilst the bigger concerns are also likely to employ a part- or even full-time medical officer. The facilities required would include offices, a waiting room, rest rooms for male and female employees, and toilets. In all factories first-aid kits must be readily available and clearly displayed. Occupational health centres attached to larger factories are becoming more popular and here a much wider range of facilities is available; it is likely that this welcome trend will become more established in the future.

6.2.6. Offices and General Administration Buildings

The principal office accommodation will either be situated within the factory structure, or outside the factory structure but having adjoining walls, or be separated from the factory but on the same site and probably linked to it by a weather-proofed corridor. Of these alternatives there is much to be said for the last as if office areas are more closely integrated into the factory structure noise, vibration and odour problems can become apparent. Much depends on the amount of land available, however. With office accommodation multi-storey buildings are rightly recommended and the space saved can best be used by separating the administration building from the factory proper which is normally a single-storey building anyway.

A plea was made earlier for flexibility within the factory framework. This same flexibility is desirable in offices and this can be facilitated by using movable wall units. Whether the degree of flexibility should go as far as total interchangeability of all main departments (e.g. administration with production and so on) is open to doubt but this view has been expressed.

The principal offices should be situated near the factory entrance. The main foyer should be spacious and attractively planned and decorated — first impressions are important! In the administration building suitable accommodation must be provided for directors, management, accounting and secretarial staff, etc. Provision must also be made for board and conference rooms, toilets, snack or rest rooms, storage rooms, and so on. But whatever the purpose of the area, the dictates of good hygiene should be observed and the design and construction criteria previously outlined should be employed.

Offices for line management, quality assurance and supervisory staff should be located close to the work area so that efficient control over operations can be maintained. Good visibility is therefore important so that these offices are often situated on mezzanine floors. Wherever they are sited it is important that they are kept clean and tidy so that a good example is set for line operatives.

6.2.7. Laboratories

In most cases it can be assumed that laboratories are going to be used primarily for quality assurance work. This will involve relatively routine microbiological, biochemical, chemical and physical testing of raw materials, foods at various stages of processing and finished products but also tests to monitor the efficiency of sanitation. More sophisticated research and development work is unlikely to be conducted in the factory laboratory and, at best, can only be done on a limited scale. Since research is rarely a function of the factory laboratory the widespread use of biologically or chemically hazardous materials is highly unlikely but, in spite of this, the laboratory should ideally be a self-contained unit with its own ventilation or air conditioning and filtering system. If space permits, the laboratory should be linked to the factory by a corridor in the same way as the administration block.

Laboratory staff should have their own lockers, toilets and even showers close to the laboratory. When they enter the factory their laboratory coats should be replaced by factory overalls or coats and duplicate lockers to house these garments should be provided in the factory area. By such means the admittedly slight chance of cross-contamination via operatives' overalls is effectively prevented.

6.2.8. Machine Maintenance and Storage Areas

Mention has already been made of the problems caused by faulty or obsolete equipment being left around in the processing area or in the factory surrounds. There must be a separate storage area for such machinery which should always be thoroughly cleaned before storage. Periodic inspection of stored machinery to check for possible infestation is important. A machine workshop should be sited close to the storage area. Regular inspection by hygiene staff is essential and every effort should be made to see that engineers and fitters are fully aware of the importance of good standards of hygiene.

6.2.9. Integration of the Principal Areas of Work

In this chapter many suggestions have been made concerning criteria for the siting and layout of different departments. It may not be clear how these departments are best integrated in an overall plan and this aspect is discussed before bringing the chapter to a close.

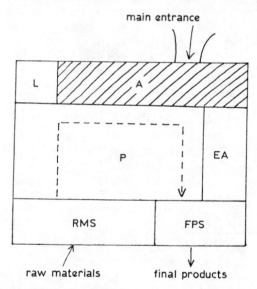

Fig. 6.9. A simple factory layout with all departments housed within a single envelope. In this figure and in Figs 6.10 to 6.13: A = administration, EA = employee amenities, FPS = final products storage, L = laboratory, P = production area, RMS = raw materials storage.

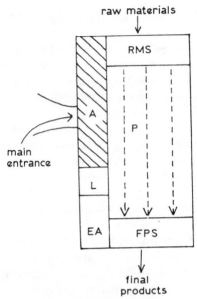

Fig. 6.10. An alternative arrangement of departments under a single factory envelope.

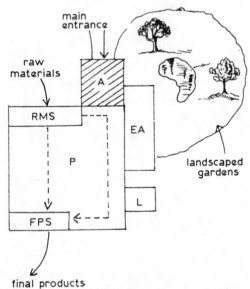

Fig. 6.11. Arrangement with support departments attached to main factory envelope.

There are many acceptable ways of planning the layout of departments in factories but only a few can be mentioned here. Much depends on the amount of land space available and where space is limited all departments may have to be housed within a single factory envelope. This solution is not an ideal one in most situations but can still be accepted provided its limitations are appreciated from the outset. Figures 6.9 and 6.10 illustrate simple layouts of this type. It is assumed with all these figures that the administration/office area is multi-storied whilst the employee amenities area could be within a single- or two-storied structure. All other areas are single storey throughout. In both layouts there is a straightforward flow of materials from raw materials reception to final product storage whilst extension of any work area is not impeded.

As the next stage in design, the materials handling areas are housed within the factory envelope whilst support departments are attached to it. Examples are illustrated in Figs 6.11 and 6.12. Here the quieter departments are further away from the distractions of the materials handling areas and attempts have been made to induce a more restful and relaxed atmosphere by landscaping in the vicinity of the employee amenities area. These advantages are extended in the final layout illustrated

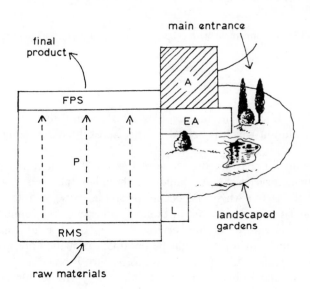

Fig. 6.12. An alternative arrangement to that shown in Fig. 6.11.

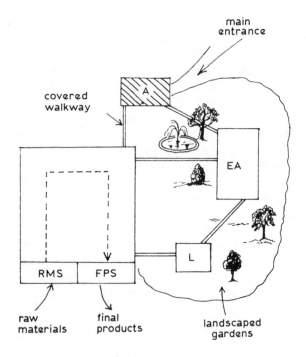

Fig. 6.13. Arrangement with separate buildings for support departments.

in Fig. 6.13. If space and finance allow, this design format usually offers the best solution which is helped by the use of interconnecting covered corridors.

Fire control is beyond the scope of this book but it is an important factor in factory design. All buildings should be planned to minimize the risk of fire but where fire occurs personnel must be able to escape rapidly and safely. The layout of buildings must also be planned so that easy access is gained for fire service vehicles. Further information on this very important topic should be obtained from the appropriate fire protection organization. In the UK this is: Fire Protection Association Information and Publication Centre, Aldermary House, Queen Street, London, EC4N lTJ.

BIBLIOGRAPHY

Cooper, C. E. B. (1973). Cold storage design and administration. *Advances in Food Research*, **20**, 113–52.

International Institute of Refrigeration. (1972). *Recommendations for the Processing and Handling of Frozen Foods*, 2nd edn. Paris, Institut International du Froid.

Pemberton, A. W. (1974). *Plant Layout and Materials Handling*. London, Macmillan.

Woolrich, W. R. & Hallowell, E. R. (1970). *Cold and Freezer Storage Manual*. Westport, Avi Publishing Co.

REFERENCE

Kotschevar, L. H. & Terrell, M. E. (1977). *Food Service Planning*, 2nd edn. New York, John Wiley and Sons.

Chapter 7

DESIGN OF FOOD PROCESSING EQUIPMENT

7.1. INTRODUCTION

Until comparatively recently manufacturers have concerned themselves almost exclusively with the mechanical design and operational requirements of food processing equipment and have tended to neglect hygiene considerations. Equipment should be designed and constructed so that cleaning, maintenance and inspection are facilitated. Parts of the equipment that come into contact with food should be capable of being easily dismantled, (unless clean-in-place systems are being considered), thoroughly cleaned and, if necessary, sterilized. Equipment should protect the food from both external and internal contamination as well as perform the function for which it was originally designed.

Clearly reliability, performance and cost implications must be assessed. Often when the cheapest equipment is purchased it proves difficult to clean, may be unreliable and soon has to be replaced; well-designed, reliable equipment, even if more expensive initially, usually saves money in the long run.

What exactly is meant by hygienically designed equipment? Shore & Jowitt (1971) summarized it in the following way:

> The production of plant which will remain clean during operation or which can be restored to the desired degree of cleanliness with the minimum cleaning effort.

Although this statement emphasizes the gist of the question there is a whole range of factors requiring assessment when considering design and operation parameters for food processing equipment. Different authors and advisory bodies list many desirable features for equipment but, whilst there is general agreement on key points, the complexity of such lists varies enormously.

The seven basic principles for hygienic design agreed by the Working Party appointed by the Joint Technical Committee of the Food Manufacturers Federation (FMF) and the Food Machinery Association (FMA) included in their publication *Hygienic Design of Food Plant* (1967) are a useful starting point in any discussion on hygiene. The principles are:

1. All surfaces in contact with food must be inert to the food under the conditions of use and must not migrate to or be absorbed by the food.

2. All surfaces in contact with food must be smooth and non-porous so that tiny particles of food, bacteria, or insect eggs are not caught in microscopic surface crevices and become difficult to dislodge, thus becoming a potential source of contamination.

3. All surfaces in contact with the food must be visible for inspection, or the equipment must be readily disassembled for inspection, or it must be demonstrated that routine cleaning procedures eliminate possibility of contamination from bacteria or insects.

4. All surfaces in contact with food must be readily accessible for manual cleaning, or if not readily accessible, then readily disassembled for manual cleaning, or if clean-in-place techniques are used, it must be demonstrated that the results achieved without disassembly are the equivalent of those obtained with disassembly and manual cleaning.

5. All interior surfaces in contact with food must be so arranged that the equipment is self emptying or self draining.

6. Equipment must be so designed as to protect the contents from external contamination.

7. The exterior or non-product contact surfaces should be arranged to prevent harbouring of soils, bacteria or pests in and on the equipment itself as well as in its contact with other equipment, floors, walls or hanging supports.

Implicit although perhaps not obvious from these principles is the requirement that equipment should be free of recesses, dead ends, corners and similar areas where microbial growth is likely in entrapped food material.

In this chapter these and other requirements will be considered in more detail but at this stage a paragraph in *Hygienic Design of Food Plant* is apposite and well worth quoting:

> There is no substitute for common sense, and no specification can be complete enough to ensure an hygienic design. Good design requires much attention to detail. It is seldom that a single plant item can be considered in isolation; one must consider the process itself, the environment in which the equipment is placed and, of course, the maintenance which will be provided throughout the life of the equipment. The compatibility of the equipment with the product, the environment and also the cleaning fluids are of vital importance.

7.2. LEGISLATION

In the UK there are no regulations dealing specifically with the design of food processing equipment although there are British and International Standards for certain items such as pipes and couplings which have become widely accepted; in addition, a European Commission Directive, dealing mainly with safety aspects but also including hygienic design principles for certain types of machinery, has been introduced more recently (Machinery Directive, 1989). The legislation that exists is concerned almost exclusively with cleanliness and methods of cleaning. Thus the Food Hygiene (General) Regulations (1970) state that articles of equipment in contact with food must be kept clean and in good repair. They must be made of non-absorbent materials and should be constructed so that cleaning is facilitated. This theme is repeated in more recent legislation concerned with raw meats. Thus the Slaughterhouses (Hygiene) Regulations, 1977, and the Fresh Meat Export (Hygiene and Inspection) Regulations, 1981, state that equipment and fittings should 'be of a durable and impervious material resistant to corrosion and of such construction as to enable them to be kept clean'. The slaughterhouse regulations also require that items such as chopping blocks and cutting surfaces should not be made of wood.

The almost total absence of legislation dealing with food plant design could be a source of criticism. However, the seven basic principles for hygienic design detailed previously may be usefully regarded as a recommended code of practice. Unfortunately, the working party responsible limited themselves to the design of tanks, pumps and pipework and were thus concerned primarily with the handling of liquids. Furthermore, the Joint Technical Committee of the FMF/FMA subsequently disbanded so that no significant progress was made until a fresh working party was appointed by the FMF, the Process Plant Association (formally the FMA) and the Association of Public Health Inspectors. From their deliberations a new set of six principles was agreed in 1978 for the hygienic design of food processing equipment. However the two sets of principles are largely the same and for brevity the latter set (Cook, 1980) is excluded. Of greater interest to-day are the principles listed in the Machinery Directive (1989) mentioned above. The machinery covered in the Directive is classed as 'agri-foodstuffs' machinery, perhaps an example of translation shortcomings of which there are others in the seven principles ('hygiene rules') listed below which must be observed.

(a) Materials in contact, or intended to come into contact, with the foodstuffs must satisfy the conditions set down in the relevant Directives and must be clean before each use.

(b) All surfaces including their joinings must be smooth, and must have neither ridges nor crevices which could harbour organic materials.

(c) Assemblies must be designed in such a way as to reduce projections, edges and recesses to a minimum. They should preferably be made by welding or continuous bonding. Screws, screwheads and rivets may not be used except where technically unavoidable.

(d) All surfaces in contact with the foodstuffs must be easily cleaned and disinfected, where possible after removing easily dismantled parts. The inside surfaces must have curves of a radius sufficient to allow thorough cleaning.

(e) Liquid deriving from foodstuffs as well as cleaning, disinfecting and rinsing fluids should be able to be discharged from the machine without impediment (possibly in a 'clean' position).

(f) Precautions must be taken to prevent any insects or liquids entering inaccessible parts of machinery.

(g) Machinery must be so designed and constructed that no lubricants other than edible lubricants can come into contact with foodstuffs. Where necessary, continuing compliance with this requirement must be checked regularly.

A comparison of the above Directive hygiene rules with the 1967 principles shows a substantial degree of similarity. The Directive is more specific in its requirement (c) concerning the means by which smooth surfaces are obtained and it also specifies that screws, screwheads and rivets may only be used in exceptional circumstances. A further stipulation in the Directive is that non-edible lubricants must not come into contact with foodstuffs. Omitted from the Directive, however, is the recommendation that surfaces in contact with food should be visible for inspection. It is worth noting that the Directive requires Member States to introduce appropriate legislation by the end of 1992.

The position in the UK contrasts sharply with that in the USA. In the USA, standards for the construction and use of a variety of food processing equipment have been established by organizations such as the 3-A Sanitary Standards Committee, the National Canners Association, the American Society of Mechanical Engineers and the Baking Industry Sanitary Standards Committee. The first named organization represents three associations, viz. the US Public Health Service, the International Association of Milk and Food Sanitarians and the Dairy Industry Committee. The 3-A standards were originally introduced for dairy equipment but have become adopted by food processing equipment manufacturers generally in the USA (N.B. copies of published standards are obtainable from the *Journal of Food Protection*, PO Box 701, Ames, Iowa 50010). In fact, the FMF/FMA Working Party's original seven principles were based on those of the 3-A Sanitary Standards Committee although the latter's standards are far more detailed and cover a greater range of components.

As a further example of the detailed legislation extant in the USA, the American Society of Mechanical Engineers could be cited. This society formed a committee concerned with design, construction and manufacturing standards for food equipment; their recommended standards can be accepted as American National Standards. Of particular interest is their publication *Food, Drug and Beverage Equipment* (ASME, 1987) which includes a Standard (ASME/ANSI F2.1—1986) for food equipment of help to manufacturers, users and health and safety authorities; in addition advice is included for users in the selection and installation of equipment, and in modifications to it. The general objectives of this Standard are to provide design criteria which will result in equipment that is:

(a) able to be quickly exposed, using simple tools used by operating or cleaning personnel, for cleaning or inspection when needed.

However, disassembly should not be assumed to be necessary for cleaning all equipment and components where specifically designed to be cleaned and/or inspected by other means;

(b) cleanable and able to be easily sanitized;

(c) made of materials which do not deteriorate in their use environment or cause degradation or contamination of the product;

(d) made to protect the product, while in the equipment, from entrance of and contact with contaminants from any external source. If good design practices indicate that such protection need not necessarily be an integral part of the equipment, then such protection may be external to the equipment;

(e) safe for use by operating, cleaning, and maintenance personnel;

(f) free of areas, inside and out, where product or contaminants can be trapped and stagnate;

(g) free of areas that may harbour vermin.

The Standard goes on to deal with many aspects of hygienic and safety design of equipment including construction materials to be used and concludes with sections on more specific criteria for different categories of food processing equipment.

The above is an example of an American National Standard but other standards may be used only as voluntary guidelines; many guidelines are, however, incorporated as mandatory standards by federal, state or local regulatory agencies.

7.3. CONSTRUCTION MATERIALS

7.3.1. General Requirements

It is necessary to ensure that all surfaces in contact with food should be inert to the food and to cleaning and sterilizing agents under normal conditions of use. Surface material constituents must be non-toxic and must not migrate to or be absorbed by foods. Surfaces in contact with foods should also be smooth, hard, continuous and free from pitting, cracks and crevices. It should be borne in mind that the smoother the surface the easier the cleaning, although the relationship is a complex one depending on the cleaning process employed and the method of obtaining the surface; in this respect electropolishing is preferable to mechanical polishing (Milledge & Jowitt, 1980). One of the principal aims must be to facilitate the removal of food residues during cleaning so that microbial growth is rendered impossible. Construction materials that are used for surfaces in contact with foods should allow the original finish to be maintained and no porosity should develop; in addition, materials should be resistant to deformation, denting, chipping, flaking and delamination (Joint Technical Committee FMF/FMA, 1967).

Surfaces not normally in contact with foods should also be smoothly finished, easily cleanable and made of corrosion-resistant material or rendered corrosion-resistant. Painting of machinery must be limited to non-contact surfaces and such surfaces should not be located above exposed food materials.

With the above requirements in mind it is not surprising that the range of construction materials available is rather limited. Materials commonly used are now considered individually.

7.3.2. Stainless Steel

Austenitic stainless steels are the preferred and most widely used of all surface materials that come into contact with foods. These steels have high percentages of certain alloying elements such as chromium and nickel but have a very low content of carbon. A wide range of stainless steels is available although the so-called 18–8 group (*ca* 18% chromium and 8% nickel) is used extensively. Of this group, alloy grades in the 300 range (e.g. 304 and 316) meet most requirements. Grade 304 is not corroded by most foods or cleaning agents, it produces no product discolouration, cleaning is easy and it is relatively cheap. Where greater corrosion problems are likely to be encountered, such as with brine and markedly acidic foods like vinegar, Grade 316 should be used; this steel has an increased nickel content (*ca* 10%) and also contains molybdenum (2–3%). Virtually complete corrosion resistance is claimed for Hastelloy which contains nickel (56%), chromium and molybdenum (both 16%), iron (5%) and tungsten (4%), although its high cost limits its use.

Stainless steels can also have different surface finishes from descaled to mirror depending on the degree to which they have been ground and polished. Smooth surfaces are required where clean-in-place techniques are used for pipework (e.g. the dairy industry) but it has been found that the surface finish is less important where pressure jet cleaning (see Section 9.10.1) is employed (Timperley, 1984).

7.3.3. Corrosion of Stainless Steel

The comparative resistance of stainless steel to corrosion is due to a protective film of chromium oxide which is formed in the presence of air (i.e. oxygen); however, even the most resistant grades will corrode if not properly maintained. The following types of corrosion are most commonly encountered in the food industry.

7.3.3.1. Pitting
Any damage to the film of chromium oxide may induce corrosion although a self-repair mechanism operates provided sufficient oxygen is available. Food debris or even dust left on the surface can induce corrosion due to the exclusion of oxygen; with food the problem is greater as bacteria growing on the organic matter may well produce acids which increase the extent of the pitting. Pitting can also be induced by physical damage and any rust spot or rough area can easily lead to more serious damage if not treated.

One of the main causes of corrosion is the misuse of cleaning or sterilizing solutions especially sodium hypochlorite. These solutions are sometimes left in contact with surfaces for excessive periods, solutions may be applied at incorrect concentrations or unsuitable agents may be used.

7.3.3.2. Electrolytic Corrosion
Electrolytic or galvanic corrosion occurs when two different metals such as aluminium and iron or even two different grades of stainless steel are moistened by the same solution. Thus if dissimilar metals, comprising part of the same piece of equipment, are treated with a cleaning or sterilizing fluid the fluid may act as an electrolyte and induce this form of corrosion. Electrons flow from the less active (e.g. iron) to the more active metal (e.g. aluminium) and cause corrosion of the latter.

7.3.3.3. Intergranular Corrosion
This form of corrosion is due to the use of a steel with too high a carbon content and can be seen in areas close to welds. It results from carbon being precipitated along grain boundaries as chromium carbide thus reducing the chromium content in the adjacent area and leaving it more susceptible to corrosion; this problem is readily overcome by using low carbon versions of stainless steel, e.g. 304L, in equipment.

7.3.3.4. General Corrosion
This is due to the use of a grade of stainless steel that is insufficiently resistant to the corrosive properties of the food being processed; it can only be corrected by replacing the equipment with equipment constructed with a more resistant grade of steel.

7.3.4. Iron and Mild Steel

Black and cast iron and mild steel have been widely used in the construction of machinery, especially for general framework and for equipment (e.g. retorts) that does not come into direct contact with foods. These materials are very susceptible to corrosion although this can be partially controlled by painting exposed surfaces. Bearings should be nickel plated if there is any chance of contact with food or cleaning agents.

Galvanized iron, which is iron coated with zinc, should not be used for equipment since zinc soon wears off exposing the iron which then corrodes. Furthermore, zinc is toxic (see Chapter 2) and, since it is soluble in fruit acids and in both acid and alkali detergents, its use should be severely restricted in the food processing area. However, galvanized iron is still used in the construction of supporting frames for trays and for trays themselves; this is permissible provided foods do not come into contact with the metal and its shortcomings are fully understood.

7.3.5. Copper and its Alloys

With certain exceptions, such as in brewing, copper is generally unsatisfactory as a construction material for food equipment. With its alloys, brass and bronze, copper is fairly resistant to corrosion and is a particularly good heat conductor. However, vegetables such as peas are discoloured when in contact with copper particularly in the presence of brine. This element also readily forms oxides which not only destroy vitamin C but also oxidize fats and edible oils causing rancidity. If used in food processing equipment copper or brass vessels must be coated with tin.

Monel metal, which is an alloy of nickel and copper (ratio 2:1) has surprising resistance to corrosion in the presence of brine and is used instead of stainless steel for this fluid. However, the presence of copper can still cause the problems mentioned above.

7.3.6. Miscellaneous Metals

As well as zinc, other toxic metals that should not be used are cadmium, antimony, mercury and lead; the last named could be used in solder at no more than 1 part in 20, although welded cans have now replaced the more traditional soldered-seam cans. The use of solder more generally should be strictly limited to joining metal; the solder must be so bonded to the metal that it cannot crack or chip.

Aluminium is popularly used for pots, pans and similar utensils, its low density enabling it to be easily fabricated and its good heat conductance is also an advantage; the use of aluminium is restricted due to corrosion by acids and alkalis. It is also a very active metal and, if present, is often involved in electrolytic corrosion.

Titanium has many advantages but its high cost prohibits its use on a large scale. It has a greater corrosion resistance than high grade stainless steel, is readily cleaned and is much lighter than steel. Tin, like copper, is a durable metal but again should not be used in equipment as a food contact surface; tin plating is, of course, used for the conventional can but this cannot be regarded as equipment.

7.3.7. Plastics

Plastics have become more widely used in the food industry in recent years and there is no doubt they are going to be of increasing importance in the future. They have many advantages being relatively cheap, light, transparent if required, non-toxic and non-tainting, relatively resistant to corrosion, and many are resistant to acids, alkalis and detergents; in addition it is possible to select plastics that are usable over wide temperature ranges. However, plastics are more easily abraded than metals so that cleanability can be adversely affected.

The properties of plastics obviously vary tremendously depending on the raw material used, types of additive incorporated and method of fabrication. Basically, plastics used in the food industry fall into two categories, thermoplastics and thermosets. The former soften when heated and harden when cooled and this process can be repeated any number of times without any appreciable chemical change. Many thermoplastics are based on ethylene, e.g. polyethylenes, polypropylenes, polyvinyl chloride, fluorcarbon polymers and the acrylics, but some are based on other chemicals, e.g. nylon. They are generally highly resistant to acids, alkalis and cleaning agents, they tolerate wide temperature ranges although heat stabilizers may have to be incorporated, and many resist water absorption; polyvinyl chloride and certain nylons are an exception to the last point and polyvinyl chloride may also suffer from microbial attack. Many of these thermoplastics have been used in the construction of tanks, pipes and fittings, and conveyor belts; wooden cutting boards have been partly replaced by hardened but durable thermoplastics, particularly high density polyethylene and this trend can only be welcomed.

Thermosets differ from the above in that they harden when first heated but if heat is re-applied they may be chemically degraded. Thermosets used as construction materials in food equipment include polyesters, epoxy resins and polyurethanes; these materials are generally usable over wider temperature ranges than the thermoplastics but they tend to be more susceptible to attack by acids and alkalis.

In spite of the increasing use of plastics problems remain so that approval for their use must be obtained (see p. 289). Some plastics contain plasticizers (i.e. organic compounds incorporated into plastics to increase flexibility) which can migrate to foods, particularly fats (see Section 8.5.2). Plastic materials remain less rigid than steel so their physical properties may alter with changes in temperature.

7.3.8. Rubber, Glass and Wood

Food-quality natural or synthetic rubber is still used fairly extensively by the food industry. Rubber is an acceptable material for belting provided it is in good condition; when worn or damaged it should be replaced immediately as it can prove difficult to clean in this state. Rubber is also used for seals, gaskets and piping, and here again selection of the right quality is of paramount importance as is the need for frequent inspection to ensure the rubber remains in good condition. Expanded rubber, with suitable filler, has also become popular in recent years for cutting boards.

Glass only has very limited applications as on no account should it be allowed in food production areas unless incorporated in specially designed equipment; any glass so used must be non-breakable and heat resistant. Its main use is for piping to convey liquids such as high salt concentration brines which might otherwise corrode steel. Where glass piping is installed, clean-in-place (CIP) systems should preferably be used. Glass is also used as a lining material in certain steel tanks and vats.

Wood and absorbent materials in general should not be used in food processing. Wood has been used widely for cutting boards but juices from the foods penetrate its surface and are almost impossible to remove by cleaning. Souring and off-odours develop due to the breakdown of the food residues by entrapped bacteria which are equally difficult to remove. The use of wood should be restricted but its incorporation in fermentation vats is acceptable. Many butchers still insist on using wooden cutting blocks and boards; in this situation the hardest woods like maple must be employed although the alternatives suggested earlier are to be preferred on hygienic grounds.

7.4. GROWTH 'POCKETS'

One of the basic requirements of hygienically designed food processing equipment is that it should not harbour food in 'pockets' or other 'dead' areas where bacterial growth may subsequently arise. Food may be retained in these pockets for many hours, or even days if the pockets are so small that cleaning agents cannot reach the debris. During this time food spoilage will occur to a greater or lesser extent. In consequence, when such food material is discharged intermittently into uncontaminated food it could cause spoilage problems. It is essential, therefore, to ensure that all food material is held in processing equipment for a uniform time. This aim is implicit in principle 2 (see Section 7.1) and in the ASME design criterion (f) (Section 7.2) but perhaps some brief explanation on this issue is desirable in order to consider design criteria that should be adopted.

In designing equipment it is essential that all surfaces that come into contact with food are as smooth and continuous as possible. Corners, crevices and recesses should be avoided therefore by suitable curvatures where food material cannot accumulate and cleaning is facilitated; all internal curvatures at corners and at junctions should have a minimum radius of at least 1 cm, although 2 cm is regarded as the optimum by the American 3-A Sanitary Standards Committee; smaller radii are permitted where the function of small parts would otherwise prove impossible. Tanks and similar equipment should also be constructed so that complete self-draining is achieved.

All permanent joints between two surfaces in contact with food should be butt-welded, ground and finished flush with the surrounding surface; all welds should be continuous and smooth, a state that should be checked regularly and so maintained, to prevent any accumulation of food particles in cracks or seams which may otherwise develop.

Dead ends such as thermometer pockets or unused pipework T-pieces must be avoided. The construction materials employed must be robust enough to prevent flexing or bending under process conditions as this may create hollows where foods or cleaning fluids can accumulate. Debris tends to adhere to protruding bolt threads, wing nuts, pot rivets and screws, and these should be avoided where possible in food contact areas. Where construction necessitates some form of bolting the bolt head should preferably be sited on the product side (CFPRA, 1983). Where bolts are used there is always the possibility of a crevice being formed at the junction between the bolt head and the metallic surface of the vessel and this must be avoided by using durable plastic or similar washers. Nuts should always be of the domed type if situated on the product side.

7.5. EASE OF DISMANTLING AND REASSEMBLY OF EQUIPMENT

Whilst both the earlier FMF/FMA principles (Section 7.1) and the 1989 Machinery Directive (Section 7.2) allude to the need for easy dismantling of equipment for cleaning purposes, far too little attention has been paid to this crucial factor. Even with well-designed equipment it is possible for bacteria to build up during a process run and if this build up is accompanied by variations in the residence time of food material a deterioration in the bacteriological quality of the food is inevitable. It is obviously important to minimize increases in bacterial numbers during processing and every opportunity should be taken to use the various break periods for essential cleaning work. Even the short 10–15 min break should be capable of being utilized to clean those equipment surfaces and parts regarded as hygienically hazardous; the cleaning of all surfaces in contact with food should be feasible over the longer break periods.

Unfortunately it takes hours to dismantle many machines for cleaning purposes, complex tools are frequently required and, as a result, cleaning is often delayed. Stripping, cleaning and re-assembly must be made as simple as possible so that those surfaces and parts exposed to hazardous bacterial accumulation can be cleaned within 15 min. To facilitate this aim the number of working parts in food processing equipment should be kept to a reasonable minimum. However, rapid cleaning also entails convenient handling and hence constraints such as the weight and dimensions of individual parts must be borne in mind. Ideally components that require frequent cleaning should be easily managed by one person and suitable racks should be provided to hold dismantled components off the floor. Quick release devices, e.g. captive bolts with coarse threads, should be used to facilitate cleaning; alternatively, clamped joints (Fig. 7.1) may be used and one of their advantages is that they overcome the bacteriological hazards associated with screw threads mentioned earlier. Whatever quick release devices are used it is important to ensure that only the simplest of tools are required, if any, in the dismantling and re-assembly of the equipment.

It may be pointed out here that faulty equipment design is, in part, responsible for large numbers of metal-in-food complaints. As there is often a plethora of nuts, bolts and screws to unfasten when dismantling

Fig. 7.1. A clamped joint used as a quick release device on a screw conveyor. (By courtesy of Birds Eye Walls Ltd.)

certain items of machinery the cause of these complaints can be easily understood and traced; steps should be taken to rectify the faults in the interests of hygiene and common sense. It must also be emphasized that equipment designed for simple and speedy dismantling and cleaning is going to be cleaned more enthusiastically and efficiently than equipment that proves tedious and difficult to handle.

Larger parts of equipment must also be designed so that cleaning, inspection and maintenance are facilitated. All parts requiring cleaning should be cleanable within an hour. Heavier components may be more suitably moved by some form of hoist, on safety grounds, as workers should not have to struggle with parts that are difficult to handle.

In general, equipment used in batch processing is simpler than that used in continuous processing and the former should therefore be easier to design for ready access and cleaning. However, there are many items of equipment (e.g. meat slicing machines, mincers, cream dispensers) where bacteriological build up is rapid and where cleaning difficulties arise because of design failures; it is axiomatic that preferred models must be easily cleanable.

There is no doubt that clean-in-place (CIP) techniques will become more widespread in the future and where prolonged or continuous processing is performed CIP is essential. Until recently these techniques have been used almost exclusively for liquids and pipework but continuous cleaning devices for solid food processing equipment are being introduced. There is no doubt this welcome innovation will, in time, become the norm. Some increase in the price of equipment must be anticipated but if labour costs are reduced an overall saving should be made (Holm, 1980).

7.6. ACCESSIBILITY AND THE SUPPORTING FRAMEWORK

Much of what has been said in the previous section is pertinent to the problems of accessibility of surfaces and components of food processing machinery. This issue has been discussed before but, for convenience, the main points are reiterated: normally equipment should be sited at a distance of *ca* 100 cm from the nearest ceiling, wall and adjacent equipment although greater gaps may be required; in addition, equipment should be raised at least 20 cm off the floor to facilitate cleaning, inspection and maintenance.

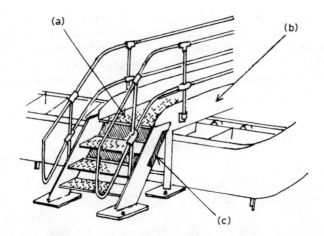

Fig. 7.2. A catwalk illustrating desirable features: (a) non-slip solid plate decking; (b) angled kick-stop; (c) protective backplate. (After CFPRA, 1983.)

Easy access is also helped by reducing floor and wall mountings and supporting framework to a minimum. Framework should be made of circular cross-section tubular steel rather than angle or channel iron as cleaning of the former is easier, it has no horizontal surfaces to collect dust and debris, corrosion is reduced and, because it is stronger and lighter, less supporting framework may be needed. All framework should have the ends sealed or capped to prevent food debris gaining access and it should be welded or otherwise sealed throughout its length to facilitate cleaning.

Supports for heavier floor-mounted equipment should be sealed to the floor so that food debris and pests are excluded and the edging to the seal should be coved to prevent debris accumulating. Where legs are used to support equipment they should be made of tubular steel which should be sealed or have ball feet fitted. The minimum number of such legs should be used and cross-bracing should likewise be reduced to a minimum; a single pedestal floor mounting is to be preferred.

Larger food processing plant such as air driers and cooling tunnels must include entry portals permitting easy access for maintenance and cleaning personnel; such portals should be at least 60 cm in diameter. Taller equipment such as spray driers often presents difficulties but easy and safe entry must be ensured through the provision of ladders or catwalks.

It is important to ensure that catwalks are constructed so that debris cannot be transferred from footwear to underlying equipment or food; thus decking should be constructed from suitable non-slip solid plate rather than mesh and the plate should be angled to incorporate a kick-stop (Fig. 7.2). Stairs, too, should conform to the same standards and protective backplates may be required near food lines (Fig. 7.2).

7.7. EXTERNAL SURFACES

External surfaces which do not come into contact with foods should be corrosion resistant, smooth, easily cleanable and free of protruding parts and crevices where debris might accumulate. Horizontal surfaces should be avoided where possible but if necessary such surfaces should be kept to a minimum and be readily accessible for cleaning purposes. Cheaper grades of stainless steel are often preferable as external surfaces but painted surfaces may be acceptable if durable paints are used. Any surfaces so treated should not be in close proximity to or directly over foods being processed.

Drainage systems from equipment should be designed to prevent splashing and be readily cleanable and drains should be sloped so that efficient discharge is maintained. Drip pans and catch trays for spilled food material may also be required which should be easily removable and cleanable.

So far the general principles of food equipment design have been discussed and these principles apply, where relevant, to all the following items. Therefore, only additional design features for specific equipment are now described.

7.8. DESIGN FEATURES FOR INDIVIDUAL ITEMS OF EQUIPMENT

7.8.1. Tanks, Vats, etc.

Tanks, vats and similar equipment must be constructed so that drains are placed at the lowest point. Thus if there is central drainage the base must be dished to drain to that point whilst if there is side drainage the base must be pitched to the outlet (Fig. 7.3). On no account should pipes be joined to tanks so that pipe ends protrude into the tank itself

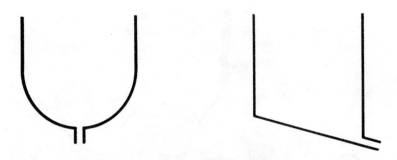

rig. 7.3. Hygienically designed tanks. (By courtesy of Campden Food and Drink Research Association.)

(Fig. 7.4). Such arrangements are good examples of potential growth pockets described earlier whilst corrosion could result from residual cleaning solution or water; in addition, the angles between the pipes and body of the vessel make cleaning extremely difficult. At the risk of unnecessary repetition it might be worth emphasizing that junctions between a vessel and a pipe must be smooth, flush and without crevices. These same remarks apply to temperature probes often used in this type of equipment; the standard fittings for these probes frequently have a small gap between the probe and the body of the coupling where food can accumulate. Improved drainage and cleanability is achieved by mounting probes in a

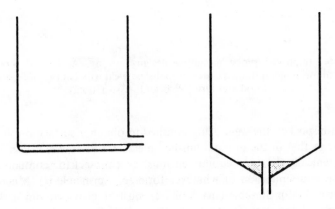

Fig. 7.4. Poorly designed tanks. (By courtesy of Campden Food and Drink Research Association.)

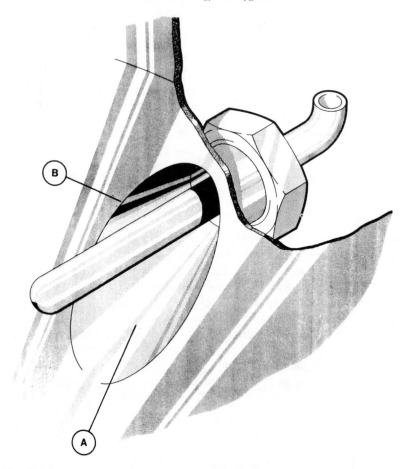

Fig. 7.5. Improved probe mounting design showing, A, conical mounting pocket allowing good drainage and, B, polished weld. (By courtesy of Campden Food and Drink Research Association.)

welded-in pocket, the weld being polished to obtain a surface finish comparable to that of the original finished metal (CFPRA, 1987; Fig. 7.5).

Hygienically designed equipment must also protect the contents from external contamination of whatever form (e.g. principle (f), Machinery Directive, 1989). Suitable covers or lids must be provided which should be made of the same material as the vessel. Covers should be close fitting and easily removable for cleaning purposes on smaller vessels.

Covers on such vessels should preferably be completely removable and unhinged since the hinge can collect debris which might fall into the product when the cover is opened. Where hinges are used they should pivot outwards to avoid product contamination; hinges should be removable and, wherever possible, consist of a simple hook-on type without bolts. Larger vessels are usually fitted with manholes, either in the top or side, which should preferably be *ca* 90 cm in diameter for easy access.

Protection must also be provided against contamination from ancillary equipment used with these vessels such as various types of agitator. All such ancillary items and services linked to the vessel should be located and protected in such a way that contamination of the food by oil, dirt and other foreign matter is prevented. Agitators must be designed so that lubricant contamination is prevented and to further this aim the shaft entry into the product area should be sealed; agitators should also be easily removable for shaft cleaning if of the hook-on type. Bearings should be located outside the product area whenever possible and must be sealed and self-lubricating; where bearings are of necessity situated within the product zone they must be self- or product-lubricated. All bearings should be designed hygienically for easy cleaning, inspection and maintenance; no lubricant should leak from them into the product. Care should also be taken with the design of agitator paddle blades which, where possible, should be attached to the shaft by welding, the welds being ground and polished (CFPRA, 1987). Where blades of necessity are bolted onto the shaft the junction between the blade and shaft assembly and also the bolts themselves can create growth pockets (Fig. 7.6 and see Section 7.4); again exposed threads should be avoided and, where necessary, replaced by hygienically designed domed nuts (CFPRA, 1987).

7.8.2. Pumps

Pumps are used primarily in the food industry for transporting liquids along pipes and will therefore often be an integral part of a CIP system. If such a system is installed the design must ensure contact between cleaning/sterilizing fluids and all surfaces exposed to the product. Where manual cleaning is used pumps should be designed for quick dismantling with the minimum of tools and skills and there should be good access to all product contact surfaces to facilitate cleaning. All such surfaces should be of stainless steel or other corrosion-resistant and inert materials.

Other desirable design features have been listed by the Joint Technical Committee FMF/FMA (1967) and include:

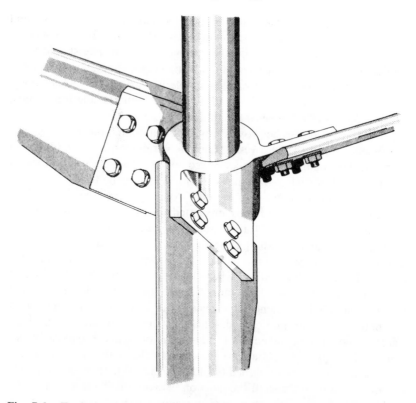

Fig. 7.6. Typical agitator paddle blade assembly. (By courtesy of Campden Food and Drink Research Association.)

1. Passage shapes should be smooth, sharp changes in cross-section should be avoided and dead spaces or turbulence-promoting features minimised....
2. Parts requiring periodic replacement should be easily replaceable and designed so that they cannot be wrongly assembled. Joints and seals should be so designed that leak-proof assembly is most easily achieved with the correct components without recourse to sealing compounds.
3. Screw-threads in contact with food material should be avoided....
4. Location of moving components should be by flats rather than by keyways or splines.
5. Clamp rings or bayonet couplings of smooth, 'clean' shape are preferable to bolts as fastenings for pump bodies.

6. Pumps should be so designed as to be readily emptied of products ... (or, in the case of positive pumps) ... should be either similarly self-draining or capable of being operated so as to be self-emptying....

7. Bearings should be located outside the product zone and be of sealed or self-lubricating type..

8. Wherever possible shaft seals should be of the mechanical type and accessible for inspection, adjustment and maintenance....

9. All external parts should be easily cleanable and capable of with-standing frequent hosing or such other cleaning down procedures as are commonly used in the intended location.

An example of a simple modification in design to improve the hygienic characteristics of a pump is shown in Fig. 7.7. It is worth men-

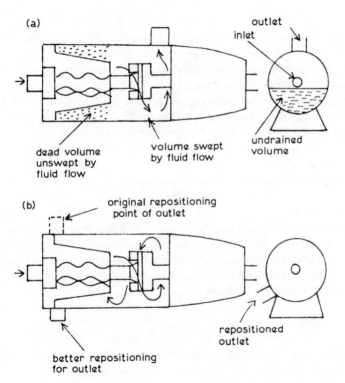

Fig. 7.7. Improvement in pump design by repositioning of outlet: (a) original design; (b) improved design, with full volume swept by fluid flow and self-draining.

Fig. 7.8. Working principle of tri-lobed rotary displacement pump. (By courtesy of Campden Food and Drink Research Association.)

tioning, however, that even hygienically designed pumps can contain growth pockets if incorrectly reassembled after cleaning.

The most common types of pump used in the food industry are rotary displacement pumps and centrifugal pumps. A wide variety of both types of pump is available and well-designed units of both categories exist. In the case of rotary displacement pumps the tri-lobe rotor (Fig. 7.8) is most frequently used (CFPRA, 1982). A popular multi-lobe design is illustrated in Fig. 7.9. and a centrifugal pump in Fig. 7.10. Both types of pump have the following advantages: (1) there are no valves although in the case of centrifugal pumps flow rates are often controlled by valves on the discharge line; (2) they are relatively cheap although rotary pumps are more expensive; and (3) they are capable of handling mixtures of liquids and solids; some models handle dry powders (rotary) and others peas or potatoes in water (centrifugal). Perhaps the most hygienic pumps are peristaltic pumps (Fig. 7.11) since the product only comes into contact with the surface of the flexible tubing; whilst they have been used only on a limited scale by the food industry they are justifiably gaining in popularity.

7.8.3. Valves

Hygienic design features for valves incorporated in pipelines include the need for internal flow passages to be as smooth as possible with minimal frictional resistance. There should therefore be no marked changes

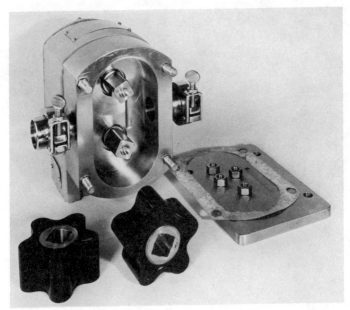

Fig. 7.9. Rotary displacement pump. (By courtesy of APV.)

in cross-section and the bore of the valve should be similar to that of the adjacent pipes; thus with plug and stem or diversion type valves there must be sufficient turbulence to ensure adequate cleaning. Valves should be easily dismantled for cleaning purposes unless a CIP system is installed; they must also be free of growth pockets and crevices, and be self-draining. The valve mechanism must be isolated from the pipeline contents and there should be no threads exposed to these contents; there must be no possibility of external contamination of the product through the valve mechanism. The body should normally be of stainless steel polished on both the inside and the outside.

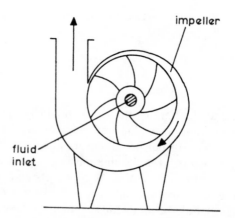

Fig. 7.10. Two-stage centrifugal pump. The liquid enters the impeller axially, is brought into a circular motion by the impeller vanes, and leaves at a higher velocity and pressure than it entered. (By courtesy of Fristam (Pumps) Ltd.)

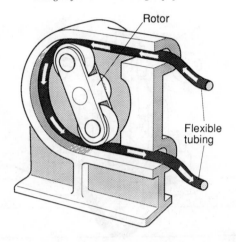

Fig. 7.11. Peristaltic pump.

Other desirable design criteria for valves have been described by the Joint Technical Committee FMF/FMA (1967) and include:

1. Gaskets and joints should be installed to give an effective seal and be flush with adjacent surfaces. They should be positively retained so that they do not distort or move from their position when under pressure or vacuum....

2. Seatings should be designed so as to make an effective seal where shut off is required and should not hinder self-draining. Seatings should give a smooth contour to the bore... In particular, seatings should be designed to resist wear and to allow regular maintenance since wear or distortion will affect hygienic character....

3. Where lubricants are required the design must be such that edible or other approved grades will be suitable.

4. Where part of the valve mechanism is unavoidably within the product area ... it should be capable of being easily dismantled for frequent cleaning.

Many types of valve are used in the food industry but from an hygienic viewpoint those that are glandless are to be preferred. Examples of these are the diaphragm (or membrane) valve (Fig. 7.12) and the air-operated plug and stem type valve (Fig. 7.13); the latter is widely used in liquid handling plants and is suitable for fully automatic lines. Both types of valves have their actuating mechanisms completely separated

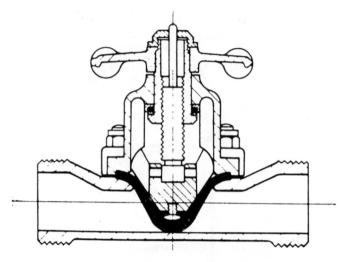

Fig. 7.12. Manually operated diaphragm valve. (By courtesy of Foo
Manufacturers Federation Inc.)

from the pipe contents, they are readily cleanable and are excellent fc
CIP installations. The flexible membranes must be inspected regularly ;
they may get distorted or even ruptured and they should be made of h'
gienic material capable of withstanding temperatures and pressure
likely to be met during processing and cleaning.

Manually operated self-draining, plug cock valves (Fig. 7.14) ai
characterized by good internal flow design, simple construction an
consequent ease of dismantling and re-assembly but their relative e:
pense has restricted their use to the manual type . They are also unsui
able for CIP as the product can get caught between the plug and th
body of the valve when the valve is turned off; bacteria can also gain ac
cess by this route. It is essential to dismantle the valve after use fc
cleaning purposes.

Butterfly valves (Fig. 7.15) are cheap and hygienic but they suff
from the slight weakness that they must be fitted with a rotating sha
which passes into the valve body so that there is a chance of food mate
rial seeping back into the space between the shaft and the body.

Ball cock valves, similar in principle to but less satisfactory than plu
cocks, should not be used due to poor drainage and difficulties in dismar
tling. Gate, globe and similar valves should not be used unless particula
attention has been paid to hygienic design. These valves incorpora

packed glands making cleaning extremely difficult and, furthermore, flow characteristics are often unsatisfactory.

7.8.4. Pipes

Pipes should be designed hygienically and to any standards that are required nationally or internationally. The resistance to flow should be kept to a minimum and there should therefore be no sudden changes in cross-sectional area, or obstructions or abrasions that are likely to hinder the

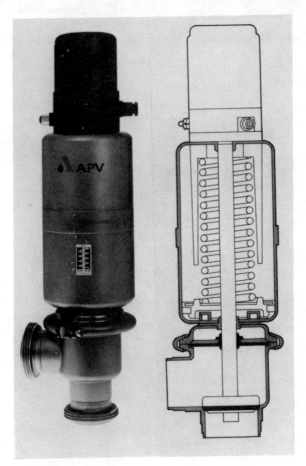

Fig. 7.13. Plug and stem type valve. (By courtesy of APV.)

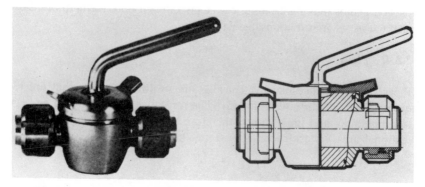

Fig. 7.14. Manually operated plug cock valve. (By courtesy of APV.)

Fig. 7.15. Manually operated butterfly valve. (By courtesy of Alfa-Laval Co. Ltd.)

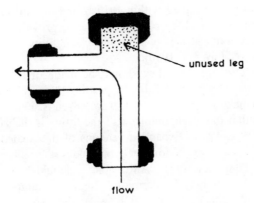

Fig. 7.16. Correct position of pipe T-piece in relation to fluid flow. (After Timperley & Lawson, 1980.)

flow. Cleaning of pipework should ideally be by means of a CIP system and pipes should be pitched so as to be completely self-draining; a fall of *ca* 1 in 100 as a minimum is suitable for this purpose. Pipework should be supported so that sagging does not occur and where valves are fitted additional support must be given. Where plastic pipework is installed especial care should be taken to avoid sagging by increasing the frequency of support. It should be mentioned that where any plastic is used as a food contact surface it must be approved in that country. Where no national standards exist the plastic must comply with internationally recognized standards such as the American Food and Drug Administration (CFPRA, 1987). In general pipes of whatever type should be mounted 4–10 cm from walls to facilitate cleaning, the distance depending on the diameter of the pipe.

Mention has already been made of dead-end T-pieces in pipework which should be avoided wherever possible. Timperley & Lawson (1980) have shown that the shorter the unused leg the easier it is to clean, and that the directional flow of cleaning fluids in relation to the unused leg is important; such fluids should flow into the leg as shown in Fig. 7.16. These authors stress the need to eliminate dead-ends completely but recognize that this is not always possible; for example, if a probe or pressure gauge needs to be fitted then the depth of the dead-leg must be reduced to a minimum.

Pipes are generally made of stainless steel and lengths may be joined together by welding or by couplings which should again meet high standards such as those specified within British or International Standards. Orbital welding has much to commend it as it provides a smooth and crevice-free junction which is of a similar cleanability to the best coupling (Timperley & Lawson, 1980).

Where couplings are used they should be made of stainless steel and fitted securely to the pipe, preferably by welding (Fig. 7.17). It is important to ensure that the grade of stainless steel employed is similar to that of the pipework and that the inside diameter of the fitting is the same as that of the tube (CFPRA, 1987). Expanded versions of the welded coupling are rather less satisfactory as the joint ring protrudes internally making cleaning less easy (Timperley, 1981). All couplings should be crevice-free and only external threads may be permitted. In most examples a joint ring, usually made of rubber, is clamped between the two ferrules which are drawn together by a nut or a split clamp. Careful tightening is necessary since, if over-stressed, the ring may become extruded whilst if too little

Fig. 7.17. Welded IDF (International Dairy Federation) coupling; (a) assembled, (b) separated. (By courtesy of APV.)

torque is applied bacteria can gain access from the outside into the product area even though no leakage of product occurs (Timperley & Lawson, 1980). It is essential that seals are flush with the internal bore of the pipe and are retained in the correct position at all times.

Properly located gaskets of suitable material are to be preferred to 'O' rings because product and bacteria can accumulate in the crevice, up to the 'O' ring on the product side; if 'O' rings are to be used they must be fitted as close to the product area as possible to minimize the depth of crevice formed. Flanges must be carefully located to each other by securing bolts, dowels or spigots to ensure a smooth continuous bore. On no account should caulking be used internally for pipework as it can be easily sucked into the pipe if the coupling is inadequately tightened.

The ring joint type (RJT) coupling (Fig. 7.18) is designed specifically for manual cleaning where it may be adequate being very robust and difficult to assemble incorrectly; unfortunately it does have a crevice which acts as a growth pocket, the 'O' rings are sometimes difficult to remove and it should definitely not be employed in CIP systems.

Where glass or plastic pipelines are installed the same general design principles apply. Glass pipelines and fittings are generally fabricated in borosilicate glass and are normally used in conjunction with a CIP system; to-day, however, because of the concern about glass being used in food processing areas its use is restricted to special commodities (e.g. flavourings in the drinks industry) and as sight glasses mounted directly into pipework. Understandably plastic pipework has become far more popular since it is fairly cheap and light; plastics are, however, unsuitable for use above a 70–90°C temperature range, the maximum being dependent on the plastic material used (CFPRA, 1987).

Examples of further types of pipe connection for both steel and plastic pipework are illustrated in the CFPRA Tech. Memo. (1987).

Where rubber hosing is employed the rubber must again be approved for use. It is important to inspect and maintain such hosing on a regular basis since rubber deteriorates fairly rapidly possibly causing product contamination.

7.8.5. Steam and Air Lines

Although live (i.e. 'culinary') steam is sometimes used directly as a means of heating foods, e.g. in the injection of steam into milk in certain UHT processes, it is more usual to protect the food by an impenetrable barrier as with canned foods. If steam is used directly, traps and filters

Fig. 7.18. RJT coupling: (a) assembled, (b) separated. (By courtesy of APV.)

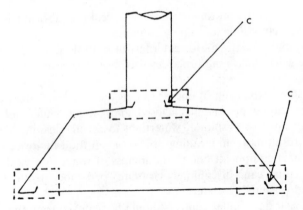

Fig. 7.19. Diagram of hood showing condensate traps (c).

must be incorporated to prevent the condensate and any pipe debris from reaching the product; in addition boiler feed water additives must be suitable for food contact. Where steam is allowed to escape from equipment, hoods should be provided to vent the moistened air direct to the exterior of the building. Hoods should be so designed that condensate, running down the inside, is collected in catch drains which should be pitched to facilitate rapid drainage (Fig. 7.19).

Mechanically introduced air must also be filtered before entering the product area; the filters should be easily replaced or cleaned. Particular care should be taken to see that the air is dry, free from lubricants and odourless.

7.8.6. Motors

The importance of avoiding contamination of food by lubricants has already been stressed in relation to equipment driven by motors, but the actual design of the motor itself and drive mechanisms has yet to be discussed.

When dealing with motors the Joint Technical Committee FMF/FMA (1967) states that hygienic design requirements are few and are confined to:

(a) a smooth, inert crevice-free and easily-cleaned exterior;
(b) adequate clearance for cleaning between the motor body and plinth or floor *or* adequate sealing to the floor to exclude product from this region;
(c) lubrication arrangements which either prevent leaking of lubri-

cant from the motor or ensure that leaking lubricant is both immediately visible and safely contained;

(d) suitability of the motor externals in both design and material for the general cleaning processes used in the area (e.g. hosing down).

Perhaps some expansion of one or two of these points is merited.

Motors should be totally enclosed and, in particular, all electrical equipment must be absolutely watertight to permit cleaning. A common difficulty resulting from cleaning operations in food factories is the failure to restart equipment due to the ingress of water into electrical components; this can discourage cleaning personnel from using the prescribed washing procedures.

Switch gear and similar controls should be removed from the immediate product area. To facilitate cleaning, electrical switch, fuse, relay and terminal boxes should either be sealed to their supports or be so installed that there is a gap of at least 3 cm from the supports; similarly, conduit pipes and covers should be installed so that no crevices are formed.

On no account should lubricants contaminate food or food contact surfaces and it follows, therefore, that motors with gear boxes should not be mounted directly over such vulnerable areas unless provided with drip pans; these should be sloped to drain at the lowest point and should have good drainage so that blocking cannot occur. As with other types of equipment, bearings should be outside the product area and should be sealed or self-lubricating; if bearings must be placed in the product area they should be in an enclosed lubricating system. Similarly drive shafts should be sealed so that lubricants cannot pass into the food and, conversely, shafts must be so protected that food material cannot leak into the bearings or motor; protection can be afforded by thrower rings and housings on the shaft between the motor and the driven casings but if the shaft is enclosed at this point inspection must be facilitated (Joint Technical Committee, FMF/FMA, 1967).

Recommended Codes of Practice have been laid down for the equipment covered in Sections 7.8.1–7.8.4 and 7.8.6; these codes are supported by a number of British and International Standards on sanitary design. Unfortunately similar codes of practice have, as yet, not been promulgated for the remaining items of equipment. Many of these items must be regarded, however, as being more hazardous hygienically than the equipment handling liquids discussed so far.

7.8.7. Size Reduction Equipment

Size reduction is a food processing operation in which food is converted into a relatively constant and smaller form than that of the basic raw materials, the operation being of convenience to the processor or to the consumer. With certain size reduction operations the product marketed is not reduced to an ultimate granular form and operations in this category include chopping, cutting, dicing, mincing and slicing. Everyday examples include potatoes (chipped), green beans (sliced), carrots (diced), meat (minced or comminuted) and fish ('fingered'). Whenever food is so treated the cutting or size reducing surfaces are a serious source of contamination as build up of bacteria on these surfaces is likely to be rapid due to their growth on the juices generally exuded from the cut surfaces of the treated food materials. Build up can be enhanced if the raw material is itself carrying large numbers of bacteria as in the case of fish and meats. The proliferating bacteria are subsequently transferred from the cutting surfaces to the foods so that there is a steady deterioration in the bacteriological quality of the food during a process run. With raw foods that are to be given some form of heat treatment later in the process the position is often less serious but it is inadvisable to place too much reliance on heat treatments unless they are relatively severe; conversely, pre-cooked foods that are reduced in size (e.g. sliced meats and bread) are particularly at risk and often spoil quickly as a direct result of unclean cutting blades.

It is therefore of paramount importance to ensure that size reducing surfaces are kept as clean as possible. Thus cutting blades, dicing and mincing surfaces, etc. should be readily strippable and easily cleaned. It would be preferable if such machines could be designed so that continuous cleaning was a built-in feature. Alternatively, machines could incorporate a duplicate set of size reducing surfaces which would operate alternately, the set not in use being cleaned either manually or, better, automatically; unfortunately such an arrangement would probably be ruled out by excessive cost.

A more practical solution to this problem is to keep the food as cool as possible during size reduction so that bacterial growth is retarded. Many of these size reduction processes cause a marked temperature rise in the food. However, water is often added in these processes, e.g. in the comminution of meat, so that if water is added as ice some form of temperature control is possible. Greater care is always necessary with the higher moisture content foods and equipment such as mincers, comminuting mills and disintegrators, which are so widely used to break down fibrous tissues, require particularly thoughtful design; this is because of the need to design food contact surfaces with small perforations or corrugations to achieve the desired objectives — however, easy access to and cleanability of these surfaces must be a paramount design feature.

Size reduction can also involve a change in the food to granular particles, generally regarded as having a maximum dimension of *ca* 0·3 cm. Common examples include ground coffee, breading crumbs and various crushed grains. To achieve this state disintegrators or crushers such as disc and roller mills are commonly used. Many foods converted by these methods are relatively dry and with the low water activity microbial growth is inhibited. Machines treating these low moisture foods can therefore be run for longer periods without cleaning than those processing moist foods although the general principles of hygienic design must still apply.

7.8.8. Mixers

In the food industry the most common forms of mixing are solids with solids (e.g. dried soups, cake mixes, tea blends), solids with liquids (e.g. canned soups, canned peas, dough), liquids with liquids (e.g. butter, margarine, salad dressings) and liquids with gas (e.g. ice-cream). With so many different materials involved it is not surprising that a multiplicity of mixing devices is available. As with size reduction, mixing may cause a temperature rise so that the pre-cooling of ingredients, or the use of ice where possible, is often beneficial in order to restrict microbial growth.

The mixing together of solid foods is normally performed in drum or rotating mixers which, because of the inherently dry nature of the materials, do not constitute a microbiological problem. The mixing of solids with liquids and liquids with liquids is very commonly performed in bowls or vats with one or more stirring arms or paddles. As mentioned earlier any agitator should preferably either be manufactured as one piece or have the mixing blades welded to the drive shaft. Certain mixers have adjustable blades where the mix can become trapped between the blade boss and the shaft; worse still, with such mixers the blades are retained on the shaft by means of nuts and bolts which should never be used in this assembly. Again, many horizontal mixers have a narrow gap between the agitator and the side of the vessel where debris can accumulate; as it is sometimes impossible to remove the agitator, cleaning is severely restricted in this zone. Clearly any agitators should be easily removable for cleaning and designed with smooth, crevice-free and easily cleaned surfaces. Mixers should also be provided with adequate covers that can be fitted before mixing starts to protect the food from the ingress of foreign matter (CFPRA, 1987); alternatively mixers should be sited in an area where total protection from external contamination is ensured.

7.8.9. Forming and Assembly Equipment

The extrusion of food mixes through openings of the required dimension is a widely used operation in the food industry. Sausages are extruded into their casings whilst spaghetti and macaroni owe their shape to this operation. Mixes of dough and supplementary ingredients used in the manufacture of bread, cakes and biscuits are extruded and deposited in moulds before baking. The accumulation of food debris in depositors and extruders is rapid and hoppers (see p. 304), pistons, piping to the manifold and the manifold itself should be hygienically designed and readily strippable for cleaning. The moulds themselves often build up debris and can also prove difficult to clean if more exotic mould designs are used; it is therefore important that moulds are readily detachable to facilitate cleaning—continuous cleaning by, for example, spray bars is another possibility.

Enrobers and battering machines can be potentially hazardous since some form of food recycling is normally involved. Chocolate enrobing of many types of sugar and flour confectionery is a common practice but, because the low water activity of the chocolate mix inhibits microbial growth, recycling of excess chocolate is safe. In contrast, machines battering fish fingers, fish cakes, chicken portions, etc., are much less safe due to the nature of the batter which allows rapid growth of bacteria. Excess batter normally drains into a receiving well, is filtered to get rid of food debris and then returned into the feeder tank. This recycling procedure, coupled with frequent additions of fresh batter, may continue for several hours during which the build up of bacteria can become very rapid. This build up results not only from the multiplication of the organisms in the batter itself but also from the constant addition of bacteria from the food surfaces due to the 'washing' effect of the batter. Hygienic design should therefore feature the incorporation of a cooling unit to keep the batter temperature at 5°C. As an alternative, batch processing should only be allowed for a limited time and 'topping-up' with fresh batter should be severely curtailed. An important additional safeguard is to reduce the pH of the batter by the addition of, for example, citric acid so that bacterial growth will be restricted.

7.8.10. Mechanical Separators

The initial cleaning of raw materials in washers can be regarded in simple terms as a solid:solid separation, that is food from its contaminants (dirt, microorganisms or even spray residues). The nature of the wash can be important for hygienic reasons as its efficiency may have an influence right down the line.

There are many types of washing machine available designed to deal with different raw materials (e.g. rotary, spray and brush washers). It may perhaps seem unrealistic to stress the importance of their hygienic design when the machines are often involved in the removal of gross dirt and are using recirculated wash water which must perforce have a high microbial count but, in spite of this, easy access and cleanability should be built-in features. Although efficient, the rotating brushes used in the brush washer constitute a weak point in design since they accumulate food debris, wear quickly and must be frequently replaced.

The use of pitched vibrating or reciprocating screens or sieves in the separation or sorting of foods such as peas and many fruits by size and the use of air separators in the removal of damaged food parts from sound food (e.g. pea skins and split peas, grain chaff) are other widely used techniques which should not pose serious hygiene problems; care must be taken to see that the flow of food through the machines is constant, that there are no traps or ledges where food might accumulate, that catch trays are easily removable and that general accessibility for cleaning is an inbuilt feature of the equipment.

The separation of solids from liquids is most commonly performed by filtration which may be batch or continuous. The clarification of fruit juices, sugar syrups and beverages normally involves filtration. Materials used for filters are of many different types and include stainless steel, monel, aluminium, synthetic fibres and fibre-glass; cellulose and cotton cloth should not be used as a filter material for syrups. The characteristics of the product to be filtered have a profound effect on the choice of filter but perforated plates made of metal are most satisfactory hygienically. Suspensions which can be readily filtered usually give rise to a filter cake with a high solids content; in some cases this cake can prove difficult to disperse and clogging of the filter may result. Filter changing is often difficult and continuous filtration linked to a CIP system has many advantages where large-scale operations are involved; in many instances semi-continuous filtration can be installed where the filters are cleaned on a semi-automatic basis. Whatever method is employed, ready replacement or cleanability of filters is crucial for efficient operation and to restrict possible growth of microorganisms on deposited material.

Centrifuges are used for the separation of solids from liquids in mixtures which tend to block filters (e.g. cream separation from skimmed milk and the clarification of fruit purees). Many centrifuges are linked to a continuous process and cleaning difficulties often arise unless CIP or some form of self-cleaning is incorporated. Smooth and polished surfaces (generally stainless steel), correct contouring and design that minimizes turbulence and facilitates cleaning, and ease of dismantling are features of good design which should be incorporated; a sealed impeller drive shaft with self-lubricating bearings is also available.

7.8.11. Equipment Transferring Solid Foods

The methods used for the transfer of solid foods partly depend on the size of the material being handled. For relatively dry and light foods like grain and peas some form of pneumatic transfer is recommended since it is hygienic, dust-free and few moving parts are involved. Transfer can be based on the fluidized bed principle which should be used where possible to convey such foods. As an alternative these foods can be driven through pipes using air blowers or exhausters.

Other handling and transfer devices include conveyors, chutes and elevators of various types. The simplest conveyors (e.g. gravity chutes) rely solely on gravity to transfer materials, often from one floor to another, and their simple design should mean that cleaning is easy provided all food contact surfaces are readily accessible. Screw conveyors are again relatively simple in design and should be free from crevices, hygienically contoured and easy to dismantle (Fig. 7.20). It is important to be able to separate easily the screw from the trough as the gap between the screw flights and trough walls is very small and inaccessible for cleaning purposes; these conveyors should preferably be made of stainless steel although aluminium or mild steel may be acceptable alternatives with dry foods.

Belt conveyors should be made of stainless steel, non-absorbent plastics or food-quality rubber; it is always preferable if the underside of the

Fig. 7.20. Screw conveyor. (By courtesy of Birds Eye Walls Ltd.)

belt is of the same or similar quality to the upper side provided traction is unaffected. Mesh belts are less satisfactory as they can prove difficult to clean but where employed they should be made of stainless steel. Cloth or canvas belts should not be used even if impregnated with waterproofing material.

The whole surface of a conveyor belt needs to be accessible for cleaning and to facilitate this belt guides should be easily removable or hinged to allow the edge of the belt to be cleaned (Fig. 7.21); as a preferred alternative to belt guides the conveyor can be designed with angled idlers or with the edge of the belt turned up (Fig. 7.21). Spreaders, deflectors and scrapers should be made of durable plastic or food-

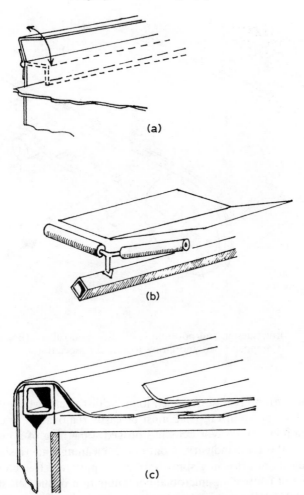

Fig. 7.21. Conveyor belts showing: (a) hinged belt guides, (b) angled rollers, (c) turned up belt edge with angled belt support. (Figures 7.21 (a) and 7.21 (c) after CFPRA, 1983.)

quality rubber; wood is totally unacceptable. These devices should be carefully located so that product build up is avoided (CFPRA, 1983) and they must also be readily detachable and cleanable. Belt rollers and idlers should be made of metal, plastic or food-quality rubber; on no account should wood be used. Rollers and idlers should be designed as a

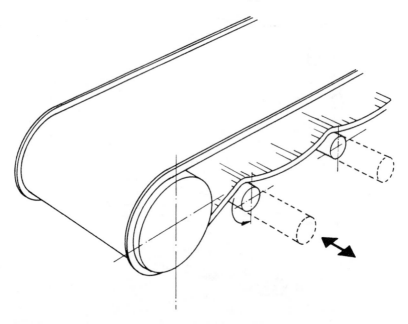

Fig. 7.22. Correct design and positioning of idlers and rollers. (By courtesy of Campden Food and Drink Research Association.)

sealed unit with no recesses or open joints whilst idlers, which should be positioned as shown in Fig. 7.22 must be easily removable for cleaning.

The cleaning of conveyor belts has proved something of a problem in the past to the food industry. Continuous cleaning and rinsing of the belt during its return using spray washers is particularly recommended as a means of controlling microbial contamination of the belt surface; it is important, however, to direct the spray at the correct angle towards the oncoming belt to achieve maximum efficiency (Fig. 7.23). The belt assembly should be constructed so that the underside (i.e. non-food contact surface) can be cleaned after a production run since food debris will inevitably be found here. To achieve this belt tension must be slackened by arranging for the end rollers to be moved into a second ('cleaning') position; cleaning is then facilitated by the side cladding being easily detached. As alternatives open mesh side cladding or strategically placed inspection hatches of an adequate size are equally acceptable. A different cleaning system for belts incorporating scrapers and immersion

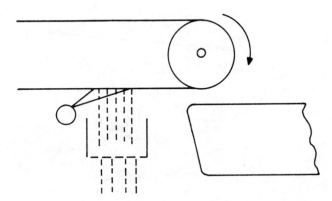

Fig. 7.23. Correct positioning of spray washer on return side of belt. (After CFPRA, 1983.)

tanks holding cleaning and rinse fluids may also be used but it is generally less satisfactory than the spray wash system.

The bucket elevator is another example of a widely used machine where hygienic design has been lacking. Many of the inherent design faults of elevators have been highlighted by the Working Party on Hygienic Design (CFPRA, 1983) who have detailed design improvements which should be made. The traditional bucket elevator has proved difficult to clean due to poor bucket design, side guarding that has prevented access and framework with insufficient clearance from the floor; in addition, catch trays have been difficult to remove and poorly sited whilst continuous cleaning facilities have been inadequate or non-existent. Buckets should be of hygienic design, easily removable if necessary and not made of wood. Amongst the recommendations of the Working Party (CFPRA, 1983) are that: (1) mesh cladding should be more widely employed to facilitate cleaning; (2) the minimum floor clearance should be 20 cm; (3) the cross-member at the frame end should be raised from the floor to the top of the frame to permit a catch tray to be correctly positioned; (4) catch trays at the rear of the elevator should be easily removable; (5) there should be a system of sprays on both sides of the buckets to ensure adequate cleaning.

The simple nature of chutes does not mean that design faults cannot occur; for example, chutes often incorporate deflectors which should ensure more even distribution of the product but often lead to product

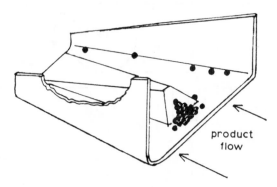

Fig. 7.24. Product accumulating at deflector and at corners of chute. (After CFPRA, 1983.)

accumulation (Fig. 7.24). In general though, chutes should be readily accessible and cleanable provided that the design has allowed for correct contouring, etc. However, chutes are sometimes designed with too narrow a discharge point where clogging of food may occur; furthermore, a narrow discharge can mean that the food may not be distributed evenly over the conveyor which can lead to product accumulation or spillage further down the line (CFPRA, 1983).

Although more loosely connected with the transfer of food, hoppers can be discussed conveniently at this stage. The importance of smooth surfaces, correct contouring, cover design and other hygienic features for this general type of equipment has already been stressed (Section 7.8.1) but there are a few other points that require some comment. For example, the sides of the hopper should be as close to the vertical as possible to ensure efficient product flow; flow may also be hampered by poorly designed attachments, e.g. rubber flaps bolted onto the sides of hoppers, and in these situations the promotion of bacterial growth is almost inevitable. It is equally important that hoppers should not be made too large for the job intended. This is particularly relevant where perishable liquids are involved, e.g. hoppers holding cream, since if the residence time for the material is excessive substantial growth of bacteria is possible. Two small hoppers, used alternately, with one being cleaned when out of operation will largely obviate this difficulty; as an alternative or in addition, hoppers can be cooled by running a refrigerant

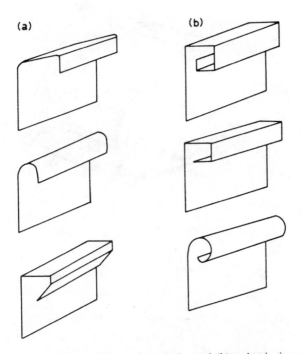

Fig. 7.25. Vessel rims illustrating: (a) hygienic, and (b) unhygienic construction. (After CFPRA, 1983.)

round or through the vessel although cost implications may rule out this often very desirable arrangement. Hoppers should also be fitted with lids to prevent product contamination whilst the upper rims of hoppers should be finished so that cleaning is facilitated; suitable and unsatisfactory designs are illustrated in Fig. 7.25.

When a line is assembled it is necessary for all equipment to be installed and located correctly; this is often not the case where adjoining equipment is concerned and this problem is illustrated repeatedly by the Working Party on Hygienic Design (CFPRA, 1983). One example from this publication will suffice: in Fig. 7.26 it can be seen that, due to the incorrect positioning of the elevator in relation to the hopper, it has been found necessary to install a deflector plate on which product is accumulating and intermittently sliding off; this problem could have been easily remedied by ensuring a discharge into the centre of the hopper.

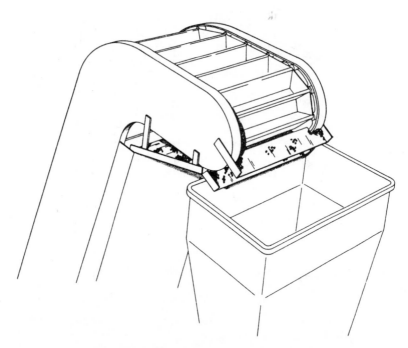

Fig. 7.26. Incorrect positioning of elevator resulting in the use of a deflector plate on which product accumulates. (By courtesy of Campden Food and Drink Research Association.)

7.8.12. Heating Equipment

Heating equipment covers such items as heat exchangers, blanchers, driers, ovens, fryers and boiling pans. These items supply heat to foods using different heating techniques and, from a hygienic standpoint, the inherent dangers vary from the negligible to the considerable. These items also vary enormously in the complexity of their design. However, the general principles of hygienic design still apply so that all equipment should be easily cleaned, and disinfected or sterilized if necessary, should allow easy access to all food contact surfaces and these surfaces should be inert, smoothly finished and free of crevices or other potential growth pockets.

The number of microorganisms surviving, or even developing in, any particular heat treatment will clearly depend on the temperature/time factors involved in that particular process. With foods heated to near

100°C only spores are likely to survive but as the temperature drops an ever increasing variety of viable microorganisms is likely to be present. Even a few minutes at 70°C will restrict the residual flora to spore formers although at this temperature some of the more resistant vegetative cells may also survive. After relatively severe heat treatments of this kind it is important to avoid post-heating contamination by freshly introduced microorganisms; thus ovens, for example, should be loaded at one end and unloaded at the other so that a double set of doors should be provided. It is also equally important to ensure rapid cooling in the 70–20°C range to prevent the growth of any surviving thermophiles or mesophiles. Where temperature-sensitive foods are being processed the lower heating temperatures (< 70°C) and/or shorter times employed reduce the extent of microbial kill-off considerably and thermophiles, and even mesophiles, may survive and become established in the equipment unless cleaning and sterilizing methods are adequate.

Equipment in which foods are heated to temperatures above or around 100°C include batch and continuous retorts and the hydrostatic pressure sterilizer; such equipment, which is used in the processing of canned foods, poses no microbiological problems. Pressure vessels can also be obtained in which foods, liquid, semi-liquid and solid, can be rapidly heated to and cooked at high temperatures (110–120°C) whilst being agitated by variable speed paddles which, in addition, can be adjusted to give a variety of mixing patterns; rapid cooling to below 20°C in a few minutes is facilitated by circulating cooling water through channels in the walls of the vessel and by applying a vacuum to induce evaporative cooling. The whole operation can be performed in the one closed vessel thus eliminating any risk of contamination; a further advantage of this system is that the rapid heating and cooling profiles minimize the possibility of growth of *Clostridium perfringens* which is such a danger where the cooking of meats is concerned (see Chapter 2). Such vessels show what can be achieved in the hygienic design of equipment and an example is given in Fig. 7.27.

Another development of interest is that known as direct resistance heating (DRH; Alwis & Fryer, 1990). In DRH the food is made part of an electric circuit around which current flows. The electrical energy is converted to thermal energy, and solid and liquid food components are heated at the same rate; this eliminates the problem of gross overprocessing of a liquid component in a particulate food where sterility is required. A further benefit of DRH is that there are no heat transfer surfaces so that product deposition and overheating on such surfaces is

Fig. 7.27.   Hygienically designed pressure vessel. (By courtesy of A. Johnson &
Co. (London) Ltd.)

avoided. A commercial process ('Ohmic') has been introduced commercially by APV Baker, under licence from the UK Electricity Council
(Biss *et al.*, 1989).

Heat exchangers are designed to transfer heat from one fluid to another
without mixing the fluids, so that heat is conducted through an intervening material, usually stainless steel. Three types of exchanger are commonly used, viz. tubular, plate and screw. All should be designed to meet
high standards of hygiene as they are often only heating foods to sublethal temperatures. They are used in the preparation of milk, dairy products, ice-cream, liquid egg, soups, chocolate, margarine and many other
commodities. The HTST pasteurization of milk may be cited as an

example of such a treatment, the milk being heated to and held at 72°C for 15–20 s in a tubular heating system linked to a plate heat exchanger. As heat exchangers are often used to re-cool the foods after heating as with pasteurized milk, the need for hygienic design and correct operation of the equipment is of paramount importance if post-heating contamination or microbial proliferation is to be avoided. Unfortunately there can be problems even with heat exchangers. Cleaning can prove troublesome with UHT milk due to product adherence on the heating surfaces whilst many otherwise excellent exchangers are not completely self-draining.

Blanchers supply heat directly to the food by steam or hot water. At the operating temperatures, *ca* 90°C for a few minutes, there will be a substantial reduction in the numbers of microorganisms present; however, the residual flora poses a potential threat. At the discharge end of the blancher temperatures may be conducive to the rapid growth of thermophiles and as the product cools during passage different organisms may become established. Bacteriological problems can be largely overcome by rapid cooling of the product post-blanching and this can be facilitated by ensuring a large excess of cooling water running counter to the flow of product. Further benefits can be obtained by ensuring that both product entry and discharge chutes are sloped sufficiently to prevent product accumulation and by installing continuous cold water sprays at the discharge end to improve cooling and to suppress bacterial build up.

Hot-air driers (e.g. tunnel, drum, fluidized bed) may again be contaminated with thermophilic bacteria so that design must allow for ready access to and cleanability of food contact and other surfaces. Baking ovens are less likely to become contaminated because of the nature of the food material but design should always permit easy and thorough cleaning.

7.8.13. Safety

In this chapter the hygienic design of food processing equipment has been discussed and the importance of easy access to food contact surfaces has been stressed. It is, however, necessary to protect the operator or cleaner from potentially dangerous parts of the equipment by the provision of guards which should not be easily removable. A wealth of legislation exists which aims to ensure the employee's safety and well-being while at work; such legislation in the UK was reviewed by Daniels (1980). As was mentioned previously (Section 7.2) a European Community Directive (Machinery Directive, 1989) deals with safety aspects. Included in the Directive are essential safety requirements covering issues such as safety design and construction criteria, installation and maintenance requirements, guarding, electrical motors and noise control. Of interest is the introduction of an 'EC' mark which will be attached to machinery conforming to the essential health and safety standards; these should include the additional criteria specified for food processing machinery laid down in the Directive (see p. 263) and thus the insignia may well have some real significance in terms of hygienic design of equipment.

The United States lays down similar safety requirements to the above in the F.2.1—1986 Standard (ASME, 1987). Understandably much stress is placed in both documents on the need to guard potentially dangerous items or parts of equipment. In certain circumstances guards may need to be easily removable provided it is safe to do so. This creates an apparent conflict between the dictates of hygiene and safety but, as was stressed by Daniels (1980), 'the best standard of safety is obtained by eliminating the hazard rather than by adding a guard'. Machine designers should aim to reduce guarding but, where necessary, it should not interfere with cleaning. Far too often guards are hygiene hazards being made of close-fitting mesh which acts as a dirt and dust trap; guards must be designed with good hygiene in mind.

BIBLIOGRAPHY

Briston, J. H. & Katan, L. L. (1974). *Plastics in Contact with Food.* London, Food Trade Press Ltd.

Campden Food Preservation Research Association (CFPRA). (1982). *The Principles of Design for Hygienic Food Processing Machinery.* Technical Memorandum No. 289.

Katsuyama, A. M. & Strachan, J. P. (1980). *Principles of Food Processing Sanitation.* Washington, Food Processors Institute.

Milledge, J. J. (1981). The hygienic design of food plant. *Proceedings of the Institute of Food Science and Technology,* **14,** 74–86.

Stinson, W. N. (1978). Sanitary design principles for food processing plants. *Food Processing,* Mid-July, 98–108.

REFERENCES

Alwis, A. A. P. de & Fryer, P. J. (1990). The use of direct resistance heating in the food industry. *Journal of Food Engineering,* **11,** 3–27.

American Society of Mechanical Engineers (ASME). (1987). *Food, Drug and Beverage Equipment,* ASME/ANSI F2.1.—1986, New York, ASME.

Biss, C. H., Coombes, S. A. & Skudder, P. J. (1989). The development and applications of ohmic heating for the continuous processing of particulate foodstuffs. In: *Process Engineering in the Food Industry*. Eds R. W. Field and J. A. Howell, London, Elsevier Applied Science.

Campden Food Preservation Research Association (CFPRA). (1982). *Pumps for the Food Industry*. Technical Memorandum No. 285.

Campden Food Preservation Research Association (CFPRA). (1983). *Hygienic Design of Food Processing Equipment*. Report of the Working Party on Hygienic Design. Technical Manual No.7.

Campden Food Preservation Research Association (CFPRA). (1987). *Hygiene Design of Liquid Handling Equipment for the Food Industry*. Technical Manual No. 17.

Cook, K. (1980). An 'emblem' to denote hygienic standards of food plant design. In: *Hygienic Design and Operation of Food Plant*. Ed. R. Jowitt, Chichester, Ellis Horwood.

Council Directive, 89/392/EEC. (1989). On the approximation of the laws of the Member States relating to machinery (Machinery Directive). *Official Journal of the European Communities*, **L183**, 9–32.

Daniels, G. G. (1980). Health and safety at work and hygiene requirements. In: *Hygienic Design and Operation of Food Plant*. Ed. R. Jowitt, Chichester, Ellis Horwood.

Holm, S. (1980). Hygienic design of food plants and equipment. In: *Hygienic Design and Operation of Food Plant*. Ed. R. Jowitt, Chichester, Ellis Horwood.

Joint Technical Committee, Food Manufacturers Federation (FMF) and Food Machinery Association (FMA). (1967). *Hygienic Design of Food Plant*. London, FMF/FMA.

Milledge, J. J. & Jowitt, R. (1980). The cleanability of stainless steel used as a food contact surface. *Proceedings of the Institute of Food Science and Technology*, **13**, 57–62.

Shore, D. T. & Jowitt, R. (1971). Design and operation of food plant. In: *Hygiene and Food Production*. Ed. A. Fox, Edinburgh, Churchill Livingstone.

Timperley, D. A. (1981). Modern cleaning and recovery systems and techniques. *Journal of the Society of Dairy Technology*, **34**, 6–14.

Timperley, D. A. (1984). Surface finish and spray cleaning of stainless steel. In: *Profitability of Food Processing*. Institute of Chemical Engineers Symposium Series No. 84.

Timperley, D. A. & Lawson, G. B. (1980). Test rigs for evaluation of hygiene in food plant design. In: *Hygienic Design and Operation of Food Plant*. Ed. R. Jowitt, Chichester, Ellis Horwood.

Chapter 8

QUALITY ASSURANCE AND PRODUCTION CONTROL

8.1. ASPECTS OF QUALITY ASSURANCE

8.1.1. Quality Control and Quality Assurance

In the past the term 'quality control' was used to encompass all those controllable factors that determined the quality of the product up to the point when it arrived in the hands of the consumer. Later the term 'quality assurance' became popular, so much so that it virtually superseded 'quality control' in the jargon of most companies. More recently the term 'total quality management' has been introduced. Are these terms synonymous or is it possible to differentiate between them?

Certainly in the case of the first two terms there was a distinction but this has become blurred with time. Quality control is concerned with: (a) the control of the actual processing of the food; (b) raw materials and final products to ensure that they comply with standards laid down; and (c) line sanitation. Quality assurance deals with much wider aspects such as: (a) the evaluation of raw material and final product standards; (b) the design of the factory; (c) process line layout; and (d) the design of machinery. Quality assurance is also concerned with the packaging, storage and distribution of the product and even with retailing. As was stated so succinctly by James (1971):

> Quality Control asks 'Are we doing things right?' Quality Assurance asks 'Are we doing the right things?'

312

The third term 'total quality management' is in many ways similar in emphasis to quality assurance although it is perhaps more dynamic in approach; collective responsibility and the need for continual improvement are stressed. It has been defined as 'a continual activity, led by management, in which everybody recognizes personal responsibility for safety and quality' (Shapton & Shapton, 1991). The aim is 'to get it right first time' and necessarily involves the company as a whole achieving uniformity and quality of a product and thus safety is maintained.

No matter which term is used the importance of food hygiene with all its ramifications is paramount. However, whilst food hygiene should clearly include the retail market, it is at least debatable whether this is a responsibilty of company management. Quality assurance at this final pre-consumer stage is often impracticable because of the large number of retail outlets and the responsibilities would be better placed in the hands of the retailer. However, it is at this point that mishandling of the product, over which manufacturers have virtually no control, so frequently occurs and food companies are rightly concerned that so many customer complaints result from incorrect or excessive storage of their products; these complaints are usually, although not entirely, associated with smaller retail outlets.

8.1.2. The Definition of Quality

The quality of a product may be defined as its measurement against a standard regarded as excellent at a particular price which is satisfactory both to the producer and to the consumer. The aim of quality assurance is to ensure that a product conforms as closely as possible and consistently to that standard at all times.

Quality can be measured in terms of the senses (e.g. taste panels), chemical composition, physical properties and the microbial flora, both quantitative and qualitative. In this book microbiological aspects are being emphasized but the other criteria must be taken into account in any valid total quality assessment.

8.1.3. The Achievement of Quality

An 'excellent' quality at a specified price can only be achieved by answering in the affirmative the questions asked by Quality Assurance and Quality Control—'yes, we are doing the right thing and we are doing things right'. To answer thus means that a successful quality assurance scheme (and likewise a food hygiene scheme with all its ramifications) must be operating with the full and sincere support of top management and all those concerned with the implementation of the scheme. There must, of course, be full control over all aspects of production so that a consistency of product quality is maintained. This necessitates strict control over the initial quality of raw materials, over the process itself and over packaging and storage conditions. In microbiological terms the build up of bacteria during a process run must be monitored at critical points along the process line at suitable time intervals. These checks may include determinations of the total numbers of viable bacteria at appropriate incubation temperatures, the total numbers of coliforms and enterococci present, and more selective testing for spoilage or food poisoning bacteria. In this way specific trouble-spots can be identified but these checks should be linked with those performed on raw materials and final products.

Unfortunately in practice many microbiological testing schemes, which in theory at least effectively monitor the quality of food production, may be deficient in one or more areas; only when unexpected problems arise are these deficiencies highlighted. It has been claimed that the efficiency of microbiological control can be improved significantly by a logical and systematic stepwise analysis of the risks involved in any process. Thus those stages of the process which are critical in terms of quality and where control should be tightest are identified. This is the basis of the Hazard Analysis Critical Control Point system (HACCP) which is reviewed more fully below.

8.1.4. Cost Implications

Costing on any product must incorporate an allowance for prevention and assessment elements. Prevention costs include items such as training programmes for employees and cleaning and maintenance, whilst assessment costs include final product testing, line and sanitation checks and quality audits in general. These costs must be credited against losses resulting from the reprocessing of sub-standard goods, customer complaints, machinery breakdown, etc., more significantly, costs can be credited against loss of brand image resulting from court cases brought by local authorities or even food poisoning outbreaks at a national level! There can be no doubt that the extra expense of maintaining a product at a high and consistent level far outweighs the possible financial consequences of short-term savings in quality assurance or food hygiene expenditure.

8.2. HAZARD ANALYSIS OF FOOD

The Hazard Analysis Critical Control Point (HACCP) concept was introduced in the United States in 1971 at the Conference on Food Protection (FDA, 1972) where it was 'recommended for widespread use'. In general the US food industry showed little interest but microbiological problems with low-acid canned foods, particularly mushrooms, led to the FDA promulgating specific regulations for control embodying HACCP principles. Their successful introduction in the canning industry inevitably led to pressure for their wider acceptance by the food industry (Silliker, 1987); the National Research Council recommended that regulations should be introduced requiring that the HACCP system be used throughout the US food industry (NRC, 1985). At the present time the US is moving towards the implementation of appropriate mandatory regulations in the meat processing industry supported by a detailed training programme for inspection and industrial personnel. Furthermore the European Community is expected to promulgate legislation incorporating the HACCP system well before the turn of the century, whilst in the UK the first report of the Richmond Committee on the Microbiological Safety of Food (Richmond Committee, 1990) recommends that HACCP should be widely used in the UK food industry.

The HACCP concept was originally described by Bauman (1974) as 'a preventative system of control, particularly with regard to microbiological hazards'. The author defined Hazard Analysis as 'the identification of sensitive ingredients, critical process points, and relevant human factors as they affect product safety'. Critical Control Points were described as 'those processing determiners whose loss of control would result in an unacceptable food safety risk'. Thus the HACCP system is concerned with an analysis of hazards and an assessment of their danger, and the identification and subsequent monitoring and strict control of critical control points. The analysis should be performed by a multi-disciplinary team of experts in the process; the team should include a process engineer, a production manager, a hygiene specialist, a microbiologist and perhaps a chemist. The growth, harvesting, processing, packaging, storage and distribution of food components or food products should be included in the hazard analysis which can even embrace customer abuse. Typically the team will concentrate on hazardous raw materials or animal meat entering a factory and how the various processes will affect contamination, growth or survival of pathogenic or spoilage organisms (Simonsen *et al.*, 1987). It was argued realistically by Spencer (1984) that contamination must include microbial toxins and that processing must destroy all those microorganisms and toxins which are deemed hazardous; also stressed was the need to consider the possibility of contamination by pathogenic or spoilage organisms after processing and the growth of such organisms in the final product.

8.2.1. Critical Control Points

Having performed a detailed hazard analysis it is then necessary to identify critical control points (CCPs) where control of the hazard can best be achieved. For example, raw materials may require testing to check against agreed specifications; such materials often constitute a hazard and hence can be identified as a CCP. However, if that material is heated during processing in such a way as to *guarantee* the elimination of the hazard the heating step then becomes the CCP. In fact, two types of CCP have been recognized (ICMSF, 1988). The first (designated CCP1) is used to indicate that control at this point will ensure control of the hazard; an example is the heating step just referred to. The second type of CCP (designated CCP2) merely minimizes the hazard but elimination is not assured. An example is the control of salmonella contamination on raw beef carcasses received at a factory: by monitoring the level of carcass contamination from different sources the best suppliers can be found—clearly sound animal husbandry and salmonella-free animal feedingstuffs would be further means of control.

Heating processes are the most common CCP followed by refrigeration and then by preservation techniques such as salting to reduce a_w or the addition of preservatives. Thus a point is reached where process line monitoring becomes paramount and less and less emphasis is placed on final product testing. General factory sanitation, design and cleaning of food processing equipment, and the personal hygiene of employees should also be included in the appraisal of hazards.

Baird-Parker (1990) has succinctly summarized the measures required in applying the HACCP scheme and these are listed in Table 8.1.

8.2.2. Hazard Categories of Foods

Foods and their constituent raw materials and ingredients can be grouped into hazard categories based on answers to three questions: (1) does the food contain a hazardous ingredient (e.g. raw meat) which may

Table 8.1
Rules in Applying HACCP[a]

Define scope of hazard analysis, e.g. pathogenic microorganisms, pesticide residues, etc.

Set up a multi-disciplinary team to carry out a hazard analysis.

Carry out a detailed analysis of the actual or proposed food processes to establish the precise equipment, production facilities and operation practices. Also obtain relevant information on raw materials, partially processed foods, packaging, distribution, sale and use conditions.

Make a detailed flow diagram of the food operation.

Carry out hazard analysis of the process and rank all hazards according to their severity and probability of occurrence.

Identify a CCP, i.e. where the food process control can be achieved over an identified hazard.

Specify criteria for control at each CCP; the greater the concern the tighter should be the control.

Identify effective means of monitoring so that control is achieved.

Identify actions to be taken if control tolerances are exceeded.

Document all control and monitoring procedures.

Train personnel to carry out their control and monitoring duties.

---

[a]From Baird-Parker (1990).

contain a harmful microorganism or toxin; (2) is there no processing step that will destroy such harmful organisms or their toxins; (3) can the food allow growth of such organisms during distribution or whilst in the hands of the consumer? If all three questions are answered in the affirmative for a product it would have a hazard class '+++'. Thus '+o+' means that the product is heated so as to destroy harmful organisms, 'o++' means there is no harmful ingredient but hazards are possible during processing or later, and 'ooo' means that the food is safe in all aspects (Corlett, 1987).

Table 8.2
Hazard Category Assignment for Typical Foods[a]

Food	Contains hazardous ingredient(s)	Lethal heating stage	Later growth in product possible	Hazard category
Raw frozen chicken	+	+	+	II
Raw oysters	+	+	+	II
Precooked steak pie	+	0	+	III
Frozen peas	0	+	+	III
Milk chocolate	+	+	0	III
Canned Scotch broth	+	0	0	IV
Canned strawberries	0	0	0	V

N.B. Hazard Category I covers foods intended for 'at-risk' groups (i.e. infants, the aged and infirm) containing one or more hazards.
[a]From Corlett (1987).

From these general hazard characteristics five Hazard Categories can be formulated and examples of foods illustrating each category are shown in Table 8.2. It will be appreciated that whilst defining the hazard category of a food product it is necessary to scrutinize all included ingredients; as a result the individual ingredients are allocated to a hazard category so that with multicomponent products the category list can be quite extensive.

8.2.3. The Application of HACCP: Flow Diagrams

Frozen green beans, a simple single-component food, are used to illustrate the practical application of the HACCP concept. Amongst the rules listed in Table 8.1 is that referring to the preparation of a detailed flow diagram and this aspect is now discussed more fully.

Figure 8.1 shows the basic steps in the preparation of frozen green beans and identifies potential hazards and CCPs where those hazards can be either eliminated or at least minimized. The figure is largely self-explanatory but a few points require amplification. The flow diagram concentrates on microbiological hazards and their control but it mentions other aspects which require inclusion such as the presence of chemical residues. Excluded from the figure is storage and distribution. With such frozen foods this should be a safe operation but the final stage in the chain, consumer handling, can be hazardous due to product abuse.

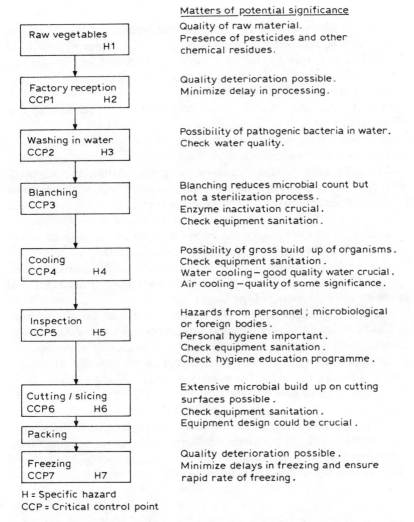

Matters of potential significance

Quality of raw material.
Presence of pesticides and other
chemical residues.

Quality deterioration possible.
Minimize delay in processing.

Possibility of pathogenic bacteria in water.
Check water quality.

Blanching reduces microbial count but
not a sterilization process.
Enzyme inactivation crucial.
Check equipment sanitation.

Possibility of gross build up of organisms.
Check equipment sanitation.
Water cooling – good quality water crucial.
Air cooling – quality of some significance.

Hazards from personnel; microbiological
or foreign bodies.
Personal hygiene important.
Check equipment sanitation.
Check hygiene education programme.

Extensive microbial build up on cutting
surfaces possible.
Check equipment sanitation.
Equipment design could be crucial.

Quality deterioration possible.
Minimize delays in freezing and ensure
rapid rate of freezing.

Raw vegetables — H1

Factory reception — CCP1 — H2

Washing in water — CCP2 — H3

Blanching — CCP3

Cooling — CCP4 — H4

Inspection — CCP5 — H5

Cutting / slicing — CCP6 — H6

Packing

Freezing — CCP7 — H7

H = Specific hazard
CCP = Critical control point

Fig. 8.1.   Flow diagram for frozen green beans identifying hazards and critical
control points.

No attempt has been made to distinguish between the two categories
of CCP or to identify points at which tighter control, and hence a
higher level of monitoring, is necessary. It is likely that the routine microbiological testing of water quality will be performed less often than

tests on slicing/dicing equipment and at the point where beans are cooled; but where water is re-circulated for further use more frequent testing may be necessary. To maintain the desired degree of control at points where microbial build up is rapid the frequency of cleaning machinery, pipework, etc., should be increased; this may in turn allow for reduced levels of microbiological testing. The rationale of equipment sanitation is explained more fully in Chapter 10 and the reader may find it helpful to refer to it for more information.

As a result of the controls now exercised on the total food process the traditional microbiological tests on final products become less important. It has been suggested that more useful information on such products can be derived by physical or chemical tests such as those monitoring water activity, concentration of preservative or pH (Spencer, 1984). There is a lot of truth in this but there will be an understandable reluctance, at least amongst food microbiologists and quality assurance staff, to accept the total elimination of traditional final product testing. Where such testing is deemed necessary it should be included as an integral part of the HACCP scheme, the data collected being used as an aid in the overall monitoring of the food process.

The introduction of the HACCP scheme should enable a strict control to be maintained over the total food process. In the remaining sections of this chapter the main stages in food processing are considered in turn stressing, but not exclusively, potential microbiological hazards and their means of control. No attempt is made to discuss these hazards and critical control points in the light of the HACCP concept; from the foregoing readers will have appreciated its advantages. In fact the HACCP approach emphasizes the need to involve a multidiscipline team starting with no preconceived ideas. However, to the uninitiated some guidance on likely hazards and their control should be beneficial and it is with this in mind that the following sections are written.

8.3. RAW MATERIAL CONTROL

8.3.1. Inspection

The purchase of raw and other materials required for processing is a major expenditure item for food manufacturers and it seems sensible therefore to keep a tight control on incoming materials. For this reason detailed specifications are prepared and agreed with suppliers which act as a safeguard for all parties concerned. The microbiological aspects of this subject have been dealt with in Chapter 4 but a variety of other tests, chemical and physical, may have to be performed before clearance is given for raw materials to be processed. Incoming goods should also be examined for evidence of pest infestation, decomposition, damage and contamination; in the case of the last named, inspection for soil, lubricants, chemicals and other foreign matter should be performed. Whatever test is performed it should have some value in terms of quality assurance and specifications should include all such tests so as to protect the manufacturer.

Inspection of incoming materials should be diligently performed as soon as possible after their receipt; any worthwhile inspection should ensure that there is little chance of defective goods getting into storage. Where sub-standard goods are received they should be handled and disposed of according to a previously agreed policy, aspects of which must involve the supplier. The safest policy is undoubtedly to destroy defective materials as soon as possible but all too often these materials are left around in a compound where they attract pests which cause hygiene problems elsewhere.

Many foods deteriorate during storage and a properly monitored storage with a stock rotation system is essential. This has been facilitated by the introduction of computers and there is no doubt that automation is reducing the extent of the quality assurance facilities required.

8.3.2. Ambient Temperature Storage

There are many goods that will keep satisfactorily in the 10–20°C range provided storage areas are dry, well ventilated and conform to the standards described in Chapter 6. Goods tolerating the upper end of the temperature range include sugar, salt, vegetable oils and canned foods. Storage at below 15°C is preferable for cereals, dried fruits and spices as above 15°C infestation problems may occur due to the growth of insects from eggs which are occasionally present in these foods.

Powdered and granular foods should preferably be stored in metal containers with closely fitting lids. These receptacles are better than wooden cases, fibre cartons or bags which are liable to attack by pests. For all types of container, storage above floor level is recommended and they should be so arranged as to facilitate inspection.

8.3.3. Chill Storage

Storage under chill conditions implies an approximate temperature range of between −3 and 7°C although refrigeration is typically restricted to a narrower range of between 1 and 5°C. These temperatures totally suppress the growth of most food poisoning bacteria but psychrotrophic bacteria and many moulds are able to grow and cause spoilage. Chill storage is thus used as a short-term preservation technique for many perishable and semi-perishable raw foods such as poultry, fish, many dairy products and certain fruits and vegetables. Typical storage lives for various foods are presented in Table 8.3.

Chill rooms are frequently used to cool cooked foods. If large quantities of hot food are treated in this way the time taken to cool to chill temperature could be well in excess of 24 h. This procedure can be extremely hazardous and may be accompanied by the substantial growth of food poisoning bacteria, particularly *Clostridium perfringens* (see Chapter 2). Where pre-heated foods need to be cooled, the cooling must be as rapid as possible. It is well within the scope of modern equipment to effect the cooling of bulk quantities of food from 100°C to 5°C within an hour. Many smaller food processing companies do not possess such equipment but rapid cooling remains imperative. Thus hot foods should be cooled to *ca* 5°C within 3 h or at least to 20°C in 1·5 h and to achieve this efficient precooling units such as heat exchangers or air coolers must be provided.

Table 8.3
Typical Storage Lives for Selected Foods Held at Either 4 or −18°C

Commodity	Storage life at	
	4°C (days)	*−18°C (months)*
Fresh beef carcase	14	8
Eviscerated poultry	10	3
Fish	5	2–6
Pasteurized milk	7	—
Butter	42	4
Oranges	42	12
Raspberries	4	12
Tomatoes (ripe)	7	—
Cucumbers	7	—
Carrots	10	12
Cauliflower	10	10

Another possible effect of using chill rooms for cooling hot foods is that the refrigeration capacity may be inadequate for the purpose resulting in a significant temperature rise and a consequential shortening of the shelf-life of the foods in storage. Temperature increases are also possible in chill rooms when vegetables and fruits are stored before blanching. During this period the foods are producing heat by their respiration. This temperature rise is accompanied by nutrient losses; for example, 25% of vitamin C in green peas is lost over a 2 day period without refrigeration (International Institute of Refrigeration, 1964). Where such temperature variations are possible, temperature control and air circulation must be efficient to restrict these variations to ± 1°C. Minimum deviations can be ensured by suitable storage arrangements for foods facilitating air circulation, as described previously in Chapter 6.

8.3.4. Deep Frozen Storage

The temperatures used in this form of storage normally vary between −18 and −24°C, although for raw fish significant extension of shelf-life can be gained by storing at −28°C. At these temperatures all microbial growth is suppressed although their enzymes may remain partially active. Of greater significance are enzymes naturally present in fruits and vegetables which must be inactivated prior to freezing. With vegetables this is normally done by blanching at *ca* 90°C for a few minutes although there are certain frozen vegetables (e.g. carrots) which do not require blanching; with fruits, sugar or syrup is added to stabilize the material.

As the data in Table 8.3 indicate, deep frozen storage is a relatively long-term method of food preservation. However, on economic grounds such storage is not used widely for raw materials but rather for finished products, part-processed foods and specialized materials. There are exceptions such as frozen boneless boxed meat and frozen liquid whole egg and it is likely that most manufacturers will have deep frozen storage facilities.

8.3.5. Storage Zones

The greatest care taken over the storage of materials can be vitiated if the storage rooms and their surrounds are not kept clean; rubbish, food debris and unwanted packaging materials should not be allowed to accumulate and attract pests. Where spillage of food material has occurred it should be cleared away immediately and the area thoroughly cleaned; frequent inspection of storage rooms and their surrounds is necessary in order to ensure a clean environment for food production.

Non-food materials such as lubricants and cleaning fluids must be stored outside the production area in a room specifically allocated for this purpose; regular checks should be carried out to see that only the correct items are being stored there.

8.4. PROCESS CONTROL

Food processing essentially utilizes a number of basic operations or manipulations which are performed on food materials to give a product. With the multiplicity of operations available together with the large range of food materials requiring processing it would be quite impossible to consider this subject in detail. Instead a number of specific examples are taken which illustrate those points in the different processes where serious microbiological hazards can be expected to arise and where the strictest control is therefore needed. The introduction of computer process control systems, both for batch and continuous processes, has undoubtedly helped to answer that need (Trystram, 1986).

The initial control of raw materials by inspection has already been considered but this is designed primarily to check on materials in bulk rather than to examine in detail each and every unit. At a later stage, immediately prior to or during processing, further inspection and sorting of the food is often beneficial. Where possible this should be performed automatically as in, for example, the photo-electric sorting of peas and beans (baked); personnel employed inspecting foods on continuous belts are notoriously inefficient and can be regarded as little more than a palliative. Obviously any efficient screening which improves the consistency and quality of raw food materials must be advantageous in terms of final product quality.

Reverting to the process itself, one of the greatest dangers is the cross-contamination of cooked or semi-cooked foods by raw foods particularly when the latter may be carrying food poisoning bacteria; it cannot be stressed too strongly that this must be avoided at all costs. Other obvious hazards which should be avoided are the inadequate heat treatment of foods, their inefficient cooling, post-heating contamination and processing or holding food for lengthy periods in the 20–50°C range and, particularly, in the 30–45°C range. The last-named hazard requires some amplification and is discussed more fully in the next section.

8.4.1. Process Temperature/Time Relationships

Foods held or being processed at suitable temperatures encourage the growth of bacteria best able to grow at those temperatures; thus food processed in the 30–45°C range is particularly likely to encourage the growth of mesophilic bacteria amongst which are the food poisoning bacteria. If a process necessitates operating at this temperature, food so processed should be held at such temperatures for the shortest possible time and then only after appropriate microbiological checks have established the inherent safety of the procedure.

In practice this means that if a continuous process lasts for, say, 3½ h at 37°C, there could be a thousand-fold increase in the number of mesophiles present, clearly undesirable; however, the same process at 25°C would result in a ten-fold increase (i.e. at 37 and 25°C the same increase in bacterial numbers would take 1 and 3 h, respectively). Examples of time/temperature relationships are given in Table 8.4.

Many processing operations involve either the continuous or the intermittent removal of food materials from equipment, so the build up of bacteria may not be quite so rapid. However, the residual food in such equipment represents a potential source of actively growing bacteria and numbers may only be effectively controlled either by efficient cleaning at frequent intervals or by reducing the operating temperature. Related to

Table 8.4

Relationship of Relative Increases in Bacterial Numbers *(E. coli)* at Different Temperatures and Times

Time (h)	Temperature (°C)			
	37	30	25	15
0	1	1	1	1
1	8	4	2	1
2	64	16	4	2
4	4 000	250	16	4
6	250 000	4 000	64	8

this is the problem of 'reworked' foods, i.e. foods that are recirculated to save waste. In economic terms this widely used procedure must be regarded as essential but from a microbiological viewpoint it causes some concern. To be acceptable it is essential to determine the microbiological quality of the food material in these situations and confirm that there is no significant increase in the numbers of organisms present in the food.

Process delays (e.g. due to machinery breakdown) pose special problems where computer integrated control is lacking. It is often impossible to predict the time of delays but, for whatever period they may persist, bacterial growth may continue with increases in numbers determined by the temperature of the food and the length of the delays. Decisions must be made, based partly on microbiological criteria, about the possible disposal of the semi-processed food created by such circumstances but it should be remembered that the food, if retained, may cause gross contamination of other food materials or equipment further down the process line. A similar, but controllable, problem is that of equipment containing partly processed food being left idle during coffee breaks or even the lunch hour. The latter practice, in particular, is to be strongly deplored unless it can be shown by appropriate microbiological tests that quality is not adversely affected. A simple example of the unequal treatment of foods which may result in significant microbiological growth is that of the stacking of containers holding part-processed food at points along a line. Containers may build up to 10 or more before being removed and clearly the residence time for food in the bottom container may be 1 h or more longer than the last stacked container.

Although the most hazardous temperature range is between 30 and 45°C, rapid growth of bacteria is possible outside this range. In particular, thermophiles grow rapidly in the 45–70°C range and such bacteria are readily isolated from tanks and dispensing equipment where stews or gravies have been maintained for long periods within this temperature range. By reducing holding times for food in equipment and by raising the temperature of the food above 70°C growth of thermophilic bacteria can be successfully controlled.

8.4.2. Examples of Critical Points in Processes

8.4.2.1. Milk
The principal heat treatments given to milk and the potential spoilage characteristics have already been described in Chapter 3 to which the reader is referred. The bulk raw milk received at processing depots must

be of a relatively consistent quality as the process is calculated on that basis. The critical points in the process are to ensure that the heat treatment given satisfies the pasteurization regulations, that cooling is rapid and that post-pasteurization contamination from the equipment, bottles or caps is avoided. Methods of process control are almost entirely automatic and, with CIP techniques widely utilized, hygiene problems rarely occur (Lloyd, 1984).

8.4.2.2. Beer

Beer making involves a stage where the wort is boiled and this effects sterilization. The wort is then cooled and piped to fermentation vessels where the yeast culture is added. As has been mentioned previously (see Section 3.11) an important aspect microbiologically is that the required characteristics of the yeast culture (typically *Saccharomyces cereviseae*) are extant. Regular verification of its fermentation ability, flocculation characteristics, oxygen requirements, resistance to infection and stability is important (Thompson, 1987).

Traditional ales are given no further heat treatment after the yeast culture has been added and they are thus liable to contamination from improperly cleaned pipework, the fermenting vessels and, later, from storage tanks and the racking machines used in the transfer of beer to the metal casks. At one time the casks were made of wood and proved difficult to clean but the newer metal casks can be readily cleaned and sterilized.

The need to obtain as clear a product as possible has meant that most beers are centrifuged and filtered during processing. This, together with ageing performed at *ca* 0°C for a few days, reduces the possibility of microbial contamination but even with these operations a light 'pasteurization' is performed in an effort to extend the shelf-life of the product. In spite of these measures, if cleaning is inadequate, contamination by acetobacters and lactic acid bacteria (i.e. lactobacilli and pediococci) can occur followed by their growth during low temperature ageing of the beer. Again many of the processes are automatic and CIP techniques are widely used. However, cleaning of fermentation and storage tanks often has to be performed manually and access problems manifest themselves in such techniques. Careful monitoring of sanitation standards is therefore crucial.

8.4.2.3. Bread and Confectionery Items

As was mentioned in Chapter 3, bread making is a process which may be hampered in microbiological terms by mould spoilage almost exclusively due to the low a_w of flour and the finished product. Microorganisms die off in flour containing less than 12% moisture and insect infestation then becomes the chief problem. Infestation may be derived from wheat or from the mill. A process known as entoletion where wheat flour is propelled with great force against a hard surface deals with the former problem whilst regular fumigation of the mill deals with the latter (Kent, 1986). When the moisture content is raised to above 15% moulds can start growing and cause spoilage. A greater hazard is that of the possible presence of mycotoxins and regular monitoring of raw materials using assay procedures mentioned in Section 4.3.9.7. is advisable.

When the fermented dough is baked the internal temperature of the bread reaches about 100°C so that all microorganisms other than bacterial spores should be killed. The critical points therefore are to ensure that cooling after baking is performed in a clean and dry environment and that there is no post-baking contamination from slicing and wrapping machines; these machines should be frequently cleaned.

Turning to confectionery items, here the control of certain raw materials assumes major importance. Eggs, mainly whole or whites, in the raw, frozen or dried form, can be hazardous. Frozen egg products should be carefully defrosted under controlled conditions at temperatures of 5–10°C where possible, although short time forced thawing is permissible in an emergency. Once the eggs have been thawed, or rehydrated in the case of dried products, they should be used as quickly as possible. For example, glazes for buns can be made from diluted egg and are subject to rapid bacterial spoilage; it is absolutely essential that such glazes are discarded at the end of the production run and not transferred to the chill to be topped-up for the next day's shift.

Cream, both real and artificial, can be hazardous and therefore the cream in dispensing units should be checked carefully during a production run to ensure that the product remains stable in bacteriological terms; frequent cleaning, perhaps hourly, may be necessary to guarantee that no increase in bacterial numbers occurs although much will depend on the operating conditions. Cream dispensers, so widely used, are a well-known hazard and regular washing of these dispensers may prove inadequate. A rigorous cleaning and disinfection protocol is essential at short intervals so that duplicate or even triplicate hand dispensers should be supplied for each employee.

8.4.2.4. Canned Foods

Canned food is an excellent example of a product which is inherently safe *provided that* potential hazards are identified and there is strict control at critical points in the process. Measures that must be adopted have been discussed more fully in Chapter 3 and only the salient points need be reiterated here. These are that: (1) commerical sterility must be achieved—this is dependent not only on a suitable heat treatment of properly mixed food in correctly filled and sealed cans but also on clean raw materials and processing lines; (2) cooling, post-retorting must be rapid to minimize deleterious changes in the organoleptic properties of the food brought about by the heat treatment and to prevent possible growth of heat-resistant thermophiles; (3) seam leakage (i.e. post-heating contamination) must be avoided where possible by careful monitoring of the bacteriological quality of the cooling water and of the can seam dimensions; (4) surfaces with which cooling cans come in contact must be kept clean to avoid post-heating contamination; (5) cans must be handled so as to minimize physical damage.

The outbreak of botulism in Birmingham in 1978 caused by canned salmon is a graphic example of what can happen when quality standards fall on the process line; it also acts as a forceful reminder of the need for constant vigilance of processing conditions in the broadest sense.

8.4.2.5. Dehydrated Vegetables

There are many processing operations common to both frozen and dehydrated vegetables which can be conveniently grouped together. Both preservation techniques include washing, blanching, grading and inspection whilst additional operations common to many vegetables include trimming, slicing and dicing. As previously indicated these latter size reducing operations cause a build up of bacteria on the equipment contact surfaces and on the food so treated. Blanching substantially reduces the bacterial load by some 99.9% and it has been suggested that where several operations are involved in the processing of vegetables it may be better to blanch as late as possible in the process; is this a tacit admission of failure to design a process line to an acceptable standard of hygiene? It should be remembered that the blanching operation itself induces a microbial build up on the cooling side and equipment surfaces should be kept clean at this stage, possibly by the installation of a continuous cleaning system.

With dehydrated vegetables the conditions are very different from frozen vegetables in the final stages of processing; with temperatures not exceeding *ca* 70°C in the food there is an opportunity for thermophiles, in particular, to flourish. The advent of fully automated continuous belt and fluidized bed driers has undoubtedly improved process control but, even with continuous cleaning systems installed, thermophilic bacteria and spore formers have proved difficult to eradicate from the product; fortunately their presence is of little consequence to the consumer.

8.4.2.6. *Frozen Fish Fingers and Steaks*

Since fish is an extremely perishable food, processing of this material should be performed at as low a temperature as possible and preferably in the frozen state throughout; a further advantage of processing in this way is that growth of *Vibrio parahaemolyticus*, if present, should be suppressed. It is essential that fish arriving at the factory should be fresh and of good quality; to ensure this it is better if fish are part-processed and frozen in blocks as soon as is practicable after catching. The conventional method of storing newly caught fish, packed in ice in the holds of vessels, merely retards the rate of spoilage so that fish quality can vary enormously depending on the time of storage in this condition (see Chapter 3).

In the preparation of fish fingers and steaks the blocks of frozen fish are cut automatically in different planes by knives until the required dimensions are obtained. Blade friction causes a temperature increase which, together with the accumulation of bacteria on both blade and cut fish surfaces, constitutes a potential hazard; continuous cleaning of the blades is therefore advisable. At the next stage the still-frozen portions are battered and crumbed, and the inherent difficulties of a continuous battering system have already been stressed (see p. 297). As well as having to control the proliferation of psychrotrophic spoilage bacteria in the batter mix, the mix could be contaminated with salmonellas and other food poisoning bacteria derived from the ingredients of the batter; the strictest control at this point is therefore imperative. The battered and crumbed frozen portions are next deep fat fried at *ca* 190°C for 45–90 s depending on the size of the portion. The frying has only a minor killing effect on the bacterial flora but at least it reduces the bacteria where they are most prolific, i.e. in the batter fraction and on the cut surfaces of the fish portion. After frying, cooling and freezing must be performed expeditiously, otherwise total thaw out of the portion may occur with consequent bacterial growth on equipment surfaces holding the fish portions and in the portions themselves. However, where installation of a fully automated process is supplemented by continuous cleaning no hygiene problems should arise provided that a tight control is maintained over all aspects of processing including sanitation.

8.4.2.7. *Fresh Sausages*

With traditional British fresh sausages made from cheaper joints of either pork or beef, contamination of the meat with salmonellas constitutes the major hazard. This is exacerbated when suitable temperatures enable salmonellas to multiply both in the food and on dirty equipment; lack of control over the process no doubt primarily explains why surveys carried out on the incidence of salmonellas in sausages report isolation rates varying between 0 and 65% (Roberts *et al.*, 1975; Banks & Board, 1983). The quality of meat used also affects the incidence and it is noteworthy that the latter authors found 95% of mechanically recovered meat samples to harbour salmonellas. To suppress the growth of these and other bacteria processing should be performed with materials cooled to, and where possible, maintained at, 10 or even 5°C; to facilitate this, processing rooms should be kept at temperatures below 15°C whilst in butchery rooms temperatures should not exceed 10°C.

Meat is comminuted in a chopping bowl, a process that can cause temperature rises of *ca* 7°C unless flake ice or iced water is added (Dyett, 1971). Unfortunately ice can be heavily contaminated with bacteria so that it has been argued that it is better to allow a temperature rise to occur; obviously the answer is to ensure ice of a better quality! The remaining ingredients (cereal, spices and various preservatives) are then mixed in. Of the preservatives, sodium metabisulphite is of special interest; it is a source of SO_2 which is permitted as an additive to a maximum level of 450 ppm. Dyett & Shelley (1966) found that SO_2 had a marked inhibitory effect on the growth of the Gram negative flora in sausages stored at both 22°C and at 3–5°C; when the preservative was added a 24 h shelf-life extension was obtained at 22°C whilst a 3 day extension was obtained with sausages stored under chill conditions. Dealing specifically with *Salmonella* spp, Banks & Board (1983) found that levels of over 20 ppm SO_2 were sufficient to suppress growth; the use of sodium metabisulphite would therefore seem fully justified from a microbiological standpoint.

Spoilage in the form of souring is caused by the growth of various Gram positive organisms (e.g. *Brochothrix thermosphacta* and lactobacilli; Dowdell & Board, 1971). This souring can be noted before any significant growth of salmonellas is possible so that there is a built-in warning system. Thus the addition of sodium metabisulphite to a raw meat product makes it feasible to safely market that product even though low numbers of salmonellas may be present in the fresh meat.

Reverting to the production process, the mixed sausage meat is extruded into the casings after which the formed product should be chilled immediately; since the extruders rapidly accumulate debris, their frequent cleaning is necessary. After overnight chill storage of the sausages they are transported to retail outlets where they are sold either chilled or at ambient temperature.

8.4.3. Design of and Critical Points in Pork Pie Processing

Because pork pies have been so frequently implicated in food poisoning outbreaks they are an ideal product to illustrate critical points in a process in slightly more detail.

As can be seen in Fig. 8.2 the general pattern of production flow is on a U-basis and is therefore similar to that shown in Fig. 6.9. It is assumed in Fig. 8.2 that the principal offices, employee amenities and laboratory are housed elsewhere and only offices for staff directly concerned with processing are housed within the building. Toilets and rest rooms for employees have been shown at strategic points; these would be used only by the personnel working in areas closest to them. Separate amenities, including showers and locker rooms, are required for personnel handling raw meats. Such personnel should be provided with distinctive clothing and their movements must be restricted to the raw meats section within the factory processing area; likewise, personnel from the cooked meats section should not be permitted in the raw meats section. Meat handling personnel should only be permitted into the main canteen after showering and changing into outdoor clothes; such practices prevent the possible spread of food poisoning bacteria. Alternatively separate canteen facilities could be provided as was suggested in Chapter 6.

There are three separate components to be considered in the preparation of pork pies, viz. meat, pastry and gelatin. The meat will be of a cheaper quality and there is always the chance that it may be contaminated with salmonellas. Since the raw meat handling section creates the greatest hazards it should be housed separately at one end of the building. Ideally, as with sausages, the preliminary operations (cutting, mincing and mixing with curing salts, like nitrite, which help to stabilize the product) should be performed on meat held at 10°C. To facilitate this the meat preparation area should itself be maintained at 10°C so that adequate insulation between this area and the meat cooking area is required. However, unlike sausages, the meat is subsequently cooked so that to maintain it at the prescribed temperature may be regarded as an unnecessary refinement but still one which is desirable.

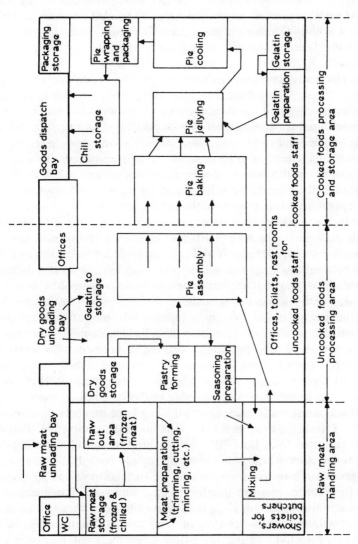

Fig. 8.2. Diagram of suitable production flow in pork pie processing factory.

The second component, pastry, composed of flour, fat and water, poses no serious microbiological problems. However, routine testing of these components is advised since unexpected contamination may occur such as with pre-cooked pastry which can be spoiled by lactic acid bacteria if machinery cleaning is inadequate. It will be noted (Fig. 8.2) that the dry goods unloading bays are separated from the raw meat unloading bays for hygienic reasons. It is essential that the pastry-forming section is physically separated from the raw meats section and that access between the two is restricted. The meat filling is added, after portioning, to the pastry casing in the assembly area after which the pies are baked. The ovens used for this purpose should be loaded on one side and unloaded on the other to prevent cross-contamination; separate sets of employees should clearly also be used for the two operations. This baking will be sufficient to destroy all vegetative cells present in the meat and pastry as well as vegetative cells picked up during assembly; however, many spores are likely to survive the baking process.

The final component, gelatin, can prove hazardous as it is an excellent growth medium for bacteria. Particular care must be taken to see that salmonellas do not gain access and this can be aided by maximizing distances between raw meat and gelatin processing lines. A second difficulty with gelatin is that it is molten at temperatures around 40°C and that these temperatures are highly suitable for the growth of salmonellas. Unfortunately gelatin is sometimes injected into pies at these temperatures with disastrous consequences as the following example illustrates.

An outbreak of food poisoning occurred in Leeds in 1976 when 210 cases were reported, many of whom were admitted to hospital. The vehicle of infection in this case proved to be the gelatin (DHSS, 1977; 1978). It was found that over a number of days the heating cycle for the gelatin was consistently inadequate allowing large numbers of salmonellas (i.e. *S. heidelberg*) to build up. It is imperative, therefore, that the gelatin is heated to at least 70°C before injection to kill any vegetative cells present after which pies should be rapidly cooled; this is also advantageous since re-contamination of pie surfaces post-baking is common. In practice, pies are usually injected whilst hot but injection of cool pies is also practised. High standards of hygiene are again essential and the maintenance and control of these can be facilitated by the installation of semi-automatic processing after the meat has been prepared. Whatever means are employed, transfer of the finished pies to the wrapping machines and thence to chill storage should be as rapid as possible. Chill storage should be of limited duration only after which the product should be dispatched from bays which are sited at some distance from the unloading bays.

Obviously there are alternative layouts to the one illustrated but the general principles of hygiene and production flow must be adhered to.

Before leaving this subject it may be of interest to remind readers that gelatin can be responsible for food poisoning outbreaks by means other than that described above. Thus outbreaks involving over 200 cases in Birmingham in 1976 and over 60 'scattered cases' in southern England in 1977, caused respectively by *S. senftenberg* and *S. eimsbuettel*, were traced to infected gelatin. However, in both these outbreaks the primary cause was found to be a design fault in the automatic jellying machines; small crevices existed in the machines which could not be cleaned properly and the residual gelatin supported the growth of the salmonellas. This is a timely reminder of the need to design food equipment hygienically and to monitor closely all aspects of production.

8.4.4. Waste Materials

Waste food materials should never be allowed to accumulate in the processing areas and regular cleaning is essential. Cleaning materials should not be left in such areas unless proper storage facilities are provided. When waste material is being disposed of it should be by the shortest route possible, out of and not through the processing area; suitable exit points should therefore be provided. To further minimize the spread of dirt no vehicles or personnel should be permitted to use any processing area as a through route.

8.4.5. Maintenance of Processing Lines

Maintenance engineers and fitters are often unwittingly the cause of hygiene problems. The fact that they often have to work under difficult and cramped conditions is no real excuse. They are frequently clothed in dirty overalls, bring greasy tools and rags into the processing area and fail to clean up when finished. Alterations are occasionally made to equipment in a piecemeal fashion and without due regard to the dictates of hygiene. Equipment modification may affect the microbiological quality of the food so that quality assurance management must be informed of any such changes beforehand; in fact a reappraisal of any hazard analysis performed may well be necessary for approval to be sanctioned.

When faulty equipment has to be set aside awaiting more extensive repairs the production area is not a suitable place for storage; to exacerbate matters the equipment is often stored uncleaned and becomes infested. Malfunctioning equipment must be cleaned and then removed to the maintenance section which should be located well away from the production area. Obviously it would also help if engineers and fitters were not forgotten when hygiene courses are run.

8.5. PACKAGING

8.5.1. Required Properties of Packaging Materials

The chief purposes of packaging are to contain and to protect the food product from deterioration in quality resulting from the activities of microorganisms or pests, or from chemically or physically induced changes. Obviously packaging must also serve to identify the product and to make it as attractive as possible to the customer in order to promote sales.

Packaging materials should be hygienic, odourless and inert, not reacting with either the contained food or the surrounding atmosphere. Protection should be afforded against the entry of microorganisms and fumes, and the materials involved may also be required to exclude water and air; materials should not contain toxic components which may migrate into the food. Materials must also be relatively inexpensive, easily assembled into containers, and easily filled and closed. Food containers must be robust enough to withstand rough handling during transport and storage.

8.5.2. Main Types of Packaging Materials and Containers

The type of packaging material used depends partly on the processing method, partly on the final storage conditions required and partly on the properties of the product. Packaging materials and containers must therefore have very special and often different properties and allowance may have to be made for a final heating before serving.

Glass has most of the desirable qualities and is easily cleaned and reused. Plastic films as single components, co-polymers or laminates are widely used and their effects on the microbial flora have been discussed earlier (see Section 3.4.2 *et seq.*). Whilst plastics undoubtedly offer many advantages their use has introduced a wide range of chemical compounds not previously used in the food industry; these compounds may be in close contact with foods for extended storage periods. Particular concern has been expressed about the use of plasticizers which are incorporated into plastics to improve their flexibility, since certain of them have been found to have toxic or carcinogenic properties; furthermore, it is possible for such compounds to be transferred from the packaging material into the food, a process termed migration (Crosby, 1981). Perhaps understandably this phenomenon has been the subject of numerous studies in recent years (e.g. Kozyrod & Ziaziaris, 1989) and legislation has been introduced worldwide prohibiting or regulating the level of suspect compounds in packaging materials.

Cans, conventionally made of steel plate coated with tin, are robust and able to withstand severe heat treatments; however, cross-reactions between the container and the food are again possible and this is prevented by lacquering the internal can surfaces, the types of lacquer used depending on the pH and type of food being processed. Aluminium foil containers and cans are widely used for packaging of prepared frozen foods and beverages, respectively. Aluminium foil can be heated to high temperatures but lacks rigidity and is liable to corrosion; laminates are often incorporated to overcome the corrosion. Paper board packages, often waxed, vegetable parchment, greaseproof and glassine papers are all widely used but, for brevity, cannot be considered here.

8.5.3. Microbiological Aspects of Packaging Materials

Packaging materials are an insignificant source of microorganisms. Only low numbers are usually isolated as the surfaces of most materials are smooth and impenetrable to microorganisms.

Cans and bottles should be rinsed with hot or cold water, or even sterilized before filling; these containers should always be held inverted until the last possible moment. Numbers of bacteria isolated from rinsed cans rarely exceed 100 per can. Plastic films, often cellulose-based, are inherently free from bacterial contamination as they are killed during the preparation of the film. Bacteria also fail to penetrate most films unless the films are physically damaged. Again, aluminium foil is prepared at high temperatures and should be sterile. The manufacture of paperboard involves a drying process at just under 100°C and this only leaves a few spore formers, none of any significance. Cellulolytic bacteria occasionally cause problems during paperboard manufacture but they are either killed during the waxing process or are prevented from reaching the food by the layer of wax.

Of course, the protection afforded by the package depends on the efficiency of the sealing or closing operation. The stable environment required is undermined if sealing machines are malfunctioning. With canned, vacuum-packed and dehydrated foods the whole process is vitiated by failure at this point and careful monitoring is therefore essential.

8.5.4. Storage of Packaging Materials

Packaging materials should be handled and stored in such a way that contamination is avoided. Glass jars and bottles, cans and aluminium foil trays should be stored in their original container until required; particular care must be taken to exclude moisture from the storage area. Unused cans, etc., should always be stored inverted and in a tidy way that facilitates inspection.

Flexible packaging materials should again be stored in their original containers until required. Any partially used bulk packages should be closed and carefully stored until required. In fact as a general rule all unused packaging materials and containers of whatever type should be moved to a definite storage area rather than be left around haphazardly.

8.6. FINISHED PRODUCT STORAGE

The principal requirements for finished product storage have already been summarized in Table 6.1 but the following additional points are worthy of mention.

Final products can be stored in bulk after processing and in this form may require subsequent re-packaging into smaller units. Alternatively, final packaging may take place on the production line directly after processing and those packages may themselves be aggregated to form larger packaging units. Thus the package size handled varies considerably. Particulate materials may be moved by means of gravity chutes, power belts and conveyors, fluidized bed conveyors, vibrators and power elevators. Bulk packaging of particulate materials may be performed before or after passage into storage. Final packaged products may be moved by gravity chutes, gravity roller conveyors, power belt conveyors, elevators or by vehicles such as fork-lift and hand trucks.

Whatever means are employed in the handling of the products, delays should be avoided and efficient dispatch to storage is of paramount importance. Whilst foods are in storage conditions should remain equable, an aim facilitated by computer control (Münch, 1984); efficient stock rotation is also facilitated by fully automatic storage controlled by microprocessors. Frozen foods should be moved to cold stores by conveyors in insulated tunnels and where such foods are to be stored on pallets stacking should only be done in the cold store.

Where damaged or defective products are involved, prompt action is again necessary. Where salvage is possible the products should be immediately re-packaged and dispatched to storage; non-salvageable materials should be quickly disposed of.

Storage rooms should not open directly on to the dispatch area but should be connected by traffic lanes having intervening doors. The dispatch area should provide as constant and as clean an environment as possible for outgoing goods. Foods should not be left around awaiting delivery and debris should not be allowed to accumulate. Pests are a particular problem in this vicinity. Dogs, cats, rodents, birds and insects must be excluded if possible. Dispatch bays should be roofed and have adequately proofed walls and doors. Windows should be screened and other measures that may have to be adopted include the use of toxic baits, insecticides and traps.

8.7. TRANSPORT AND DISTRIBUTION

This aspect is an extension of final product storage so far as the maintenance of a clean and constant environment is concerned; expeditious but careful handling of the foods is also necessary. The problems that arise during transportation are usually similar to those of storage but additional factors play a part.

Transport can be by road, rail, sea or air and where containers are used a combination of two or more transportation methods is often involved. In this context readers are reminded of the use of controlled or modified atmospheres in containers as a means of extending the storage life of raw meats (see Section 3.4.5); such techniques are also used in the transport of many other types of foods including vegetables and fruits. Whichever method is used there is inevitably a pressure on space which can cause losses in efficiency of ventilation and air circulation, and undesirable temperature changes. Less than ideal conditions can be tolerated during distribution but for limited periods only; the longer the period involved in transportation, the more stringent must be the control of the environment during that transportation.

All vehicles and containers used for the transport of frozen foods should be fitted with temperature recorders to monitor the environment. Frozen foods should be transferred, where possible, at a temperature of −18°C from the cold store to the refrigerated vehicle or container. To facilitate this, transfer should be as rapid as possible, the container should be pre-cooled and a port-door should be used (see Fig. 8.3). If possible, the storage space should be fitted to capacity to minimize the load on the refrigeration unit. However, unrestricted air flow is required so that storage units should have air ducts in the floor and walls; in addition, direct stacking on to floors, against walls or up to the ceiling must be avoided.

When unloading frozen foods transfer should again be rapid; this procedure has been helped by the introduction of powered tail-lifts. Distribution to the retail trade often involves unloading small quantities at frequent intervals. Ideally door opening should be kept to a minimum to restrict the entry of warm air but this difficulty can be partially overcome by using curtains across the doorway. Poorly fitting doors

Fig. 8.3. Port-doors used in the transfer of frozen foods to vehicles. These doors help to stabilize product temperature and reduce the amount of warm air entering the cold store. (By courtesy of Birds Eye Walls Ltd.)

and faulty insulation causing severe draughts whilst the vehicle is at speed only add to the load on the refrigeration equipment and this in time can only lead to operating failure and economic losses. It is not surprising that many food manufacturers find it beneficial to employ outside specialist contractors to distribute their products.

Many of the principles described above can also be applied to the transport of chilled foods. Here temperatures should be held at or below 5°C although fluctuations to 6 or 7°C can be permitted for limited periods; here again temperature recorders are helpful in determining the temperature profiles of goods in transit.

Turning to transport in general, infestation of various types in the vehicle storage area should be looked for. Insects are usually the main problem, gaining harbourage in the junction between the wall and floor or between the wall liners, the latter being an area also frequented by rodents. Infestation results from inefficient or infrequent cleaning, which leaves food residues, and from easy access into the storage area. Entry is normally gained through poorly fitting doors or through holes in the exterior surface of the container.

Another problem that arises is when vehicles, possibly due to excess demand, are used to transport different types of food or non-food materials at the same time or in sequence. Vehicles transporting raw carcase meat can be used for no other purpose and this regulation could, with advantage, be extended to certain other foods. Foreign matter complaints arise from the presence of glass, sand, oil and grease in storage compartments and with correct vehicle deployment this can be avoided. Off-odour complaints caused by petrol or diesel fumes or the carriage of non-food materials (e.g. soap) are a further consequence of vehicle misuse.

Microorganisms grow rapidly on surfaces that are poorly ventilated, moist and unclean, and increase the chances of product contamination. Regular cleaning and inspection is therefore essential and checks should also be made on the external condition of the vehicle's food storage area and the general state of maintenance of the vehicle.

BIBLIOGRAPHY

Clark, J. P. & Balsman, W. F.(1990). Computer integrated manufacturing in the food industry. In: *Engineering and Food*, Vol. 1 (Physical Properties and Process Control). Eds W. E. L. Spiess and H. Schubert, London, Elsevier Applied Science.

Clifford, W. H. & Gyeszly, S. W. (1976). Food packaging. In: *Food Engineering Systems*, Vol. 1. Ed. A. W. Farrall, Westport, USA. Avi Publishing Co.

Davis, J. G. (1986). Dairy products. In: *Quality Control in the Food Industry*, 2nd edn, Vol. 2. Ed. S. M. Herschdoerfer, London, Academic Press.

Greensmith, M. (1971). *Practical Dehydration*. London, Food Trade Press Ltd.

Griffin, R. C., Sacharow, S. & Brody, A. L. (1985). *Principles of Package Development*, 2nd edn, Westport, USA. Avi Publishing Co.

International Commission on Microbiological Specifications for Foods. (1980). *Microbial Ecology of Foods*, Vol. 2 (Food Commodities). New York, Academic Press.

Jul, M. (1984). *The Quality of Frozen Foods*. London, Academic Press.

Michels, M. J. M. (1982). Bacteria in frozen vegetables. In: *Bacteria and Plants*. Eds M. E. Rhodes-Roberts and F. A. Skinner, London, Academic Press.

Paine, F. A. (1987). Packaging materials. In: *Quality Control in the Food Industry*, 2nd edn, Vol. 4. Ed. S. M. Herschdoerfer, London, Academic Press.

Rogers, J. L. & Binsted, R. (1972). *Quick-Frozen Foods*. London, Food Trade Press Ltd.

REFERENCES

Baird-Parker, A. C. (1990). HACCP and food control. *Food Control*, 1, 131–3.

Banks, J. G. & Board, R. G. (1983). The incidence and level of contamination of British fresh sausages and ingredients with salmonellas. *Journal of Hygiene, Cambridge*, 90, 213–23.

Bauman, H. E. (1974). The HACCP concept and microbiological hazard categories. *Food Technology*, 28(9), 30–4 and 74.

Corlett, D. A. (1987). Selection of microbiological criteria based on hazard analysis of food. In: *Food Protection Technology*. Ed. C. W. Felix, Chelsea, USA. Lewis Publishers Inc.

Crosby, N. T. (1981). *Food Packaging Materials. Aspects of Analysis and Migration of Contaminants*. London, Applied Science Publishers Ltd.

DHSS. (1977). *On the State of Public Health for the Year 1976*. London, HMSO.

DHSS. (1978). *On the State of Public Health for the Year 1977*. London, HMSO.

Dowdell, M. J. & Board, R. G. (1971). The microbial associations in British fresh sausages. *Journal of Applied Bacteriology*, 31, 317–37.

Dyett, E. J. (1971). Hygiene and meat products. In: *Hygiene and Food Production*. Ed. A. Fox, Edinburgh, Churchill Livingstone.

Dyett, E. J. & Shelley, D. (1966). The effects of sulphite preservative in British fresh sausages. *Journal of Applied Bacteriology*, 29, 439–46.

Food and Drug Administration (FDA). (1972). *Proceedings of the 1971 National Conference on Food Protection*. US Department of Health, Education and Welfare, Washington, DC.

International Commission on Microbiological Specifications for Foods (ICMSF) (1988). *Microorganisms in Foods*, Vol. 4 (Application of the Hazard Analysis Critical Control Point (HACCP) System to Ensure Microbiological Safety and Quality). Oxford, Blackwell Scientific Publications.

International Institute of Refrigeration. (1964). *Recommendations for the Processing and Handling of Frozen Foods*. Paris, IIR.

James, D. E. (1971). Manufacturing raw materials and good manufacturing practices. *Food Technology*, **25**, 1004–5.

Kent, N. L. (1986). Flour and bread. In: *Quality Control in the Food Industry*, 2nd edn, Vol 3. Ed. S. M. Herschdoerfer, London, Academic Press.

Kozyrod, R. P. & Ziaziaris, J. (1989). A survey of plasticizer migration into foods. *Journal of Food Protection*, **52**, 578–80

Lloyd, A. K. (1984). Plant control systems. In: *Engineering and Food*, Vol. 2 (Processing Applications). Ed. B. M. McKenna, London, Elsevier Applied Science Publishers.

Münch, P. (1984). Computer controlled measuring system for the investigation of conditioned storage room. In: *Engineering and Food*, Vol. 2 (Processing Applications). Ed. B. M. McKenna, London, Elsevier Applied Science Publishers.

National Research Council (NRC) (1985). *An Evaluation of the Role of Microbiological Criteria for Foods and Food Ingredients*. Washington, D.C., National Academy Press.

Richmond Committee (1990). *First Report on the Microbiological Safety of Food*. London, HMSO.

Roberts, D., Boag, K., Hall, M. L. M. & Shipp, C. R. (1975). The isolation of salmonellas from British pork sausages and sausage meat. *Journal of Hygiene, Cambridge*, **75**, 173–84.

Shapton, D. A. & Shapton, N. F. (1991). *Principles and Practices for the Safe Processing of Foods*. Oxford, Butterworth-Heinemann Ltd.

Silliker, J. H. (1987). Principles and applications of the HACCP approach for the food processing industry. In: *Food Protection Technology*. Ed. C. W. Felix, Chelsea, USA, Lewis Publishers, Inc.

Simonsen, B. *et al.* (1987). Prevention and control of food-borne salmonellosis through application of Hazard Analysis Critical Control Point (HACCP). *International Journal of Food Microbiology*, **4**, 227–47.

Spencer, R. (1984). Microbiological quality control. In: *Control of Food Quality and Food Analysis*. Eds G. G. Birch and K. J. Parker, London, Elsevier Applied Science Publishers.

Thompson, C. C. (1987). Alcoholic beverages and vinegars. In: *Quality Control in the Food Industry*, 2nd edn, Vol 4. Ed. S. M. Herschdoerfer, London, Academic Press.

Trystram, G. (1986). State of the art of computer control of food processes. In: *Food Engineering and Process Applictions*, Vol. 2 (Unit Operations). Eds M. Le Maguer and P. Jelen, London, Elsevier Applied Science Publishers.

Chapter 9

CLEANING AND DISINFECTION: METHODS

9.1. INTRODUCTION

The operations of cleaning and disinfection are essential parts of food production and the efficiency with which these operations are performed greatly affects final product quality. A prerequisite for an efficient sanitation programme is that the factory and the equipment therein have been designed with high standards of hygiene in mind; the most effective sanitation programme cannot make up for basic deficiencies in equipment and factory design and if design faults exist sanitation can never be totally effective.

The surfaces of the equipment used in the manufacture of foods inevitably become soiled and require cleaning. If not continuous, cleaning must at least be performed at regular and frequent intervals so that a consistently good quality of product is maintained. How this cleaning is done depends principally on: (1) the nature of the soil or contamination to be removed; (2) the type of surface to be cleaned; (3) the materials used for cleaning; (4) the degree of water hardness; and (5) the standard of cleanliness required.

The basic steps in cleaning can be summarized as: (1) the removal of gross soil or dirt; (2) the removal of any residual soil with detergent; and (3) rinsing to remove detergent and soil. The first step is omitted where continuous cleaning or CIP systems are installed. But cleaning often needs to be followed by disinfection (sanitizing) or sterilization which involves two further steps, viz. the disinfection or sterilization of surfaces with suitable agents to destroy microorganisms and the rinsing off of these agents.

Since the reader may well be confused by the terminology used in cleaning and disinfection technology, a short definition of the more important terms is given below.

9.2. DEFINITIONS

Bactericide — A chemical agent which, under defined conditions, is capable of killing vegetative forms of bacteria but not necessarily bacterial spores.

Bacteriostat — A chemical agent which, under defined conditions, is capable of preventing the growth of bacteria (N.B. many bactericidal agents may act as bacteriostats at low dilutions).

Clean surface — One that is free from soil of whatever form and is odourless. Thus it is one from which food debris, detergents and disinfectants have been removed. It will not contaminate foods in contact with it and has residual numbers of microorganisms, if any, that could not undermine product quality during subsequent production. A clean surface is not necessarily sterile.

Cleaning — Covers those processes concerned with the removal of soil from surfaces but not those concerned with sterilization.

Detergent — A substance which assists in cleaning when added to water.

Disinfectant — Originally defined in medical terms as a chemical agent which destroys disease-producing organisms; now more correctly defined as an agent capable of destroying a very wide range of microorganisms but not necessarily bacterial spores.

Disinfection — Covers those processes concerned with the destruction of most microorganisms, but not necessarily bacterial spores, on surfaces and in equipment. Any viable microorganisms remaining are not capable of affecting the microbiological quality of foods coming into contact with the disinfected parts.

Fungicide — A chemical agent which, under defined conditions, is capable of killing fungi including their spores.

Sanitation	— An all-embracing term covering those factors which assist in improving or maintaining man's physical wellbeing including the general cleanliness of his environment and the preservation of his health.
Sanitizer	— A substance that reduces the numbers of microorganisms to an acceptable level (N.B. this term is widely used in the USA and is virtually synonymous with the popular use of the term 'disinfectant').
Sanitizing	— See 'disinfection'.
Soil	— Any unwanted food residue, organic or inorganic matter remaining on equipment and other surfaces.
Sterilization	— The process of destroying all forms of life, including microbial life.
Sterilizer	— A chemical agent capable of destroying all forms of life.

9.3. TYPES OF SOIL

The type of soil to be removed varies according to the composition of the food and the nature of the process to which the food has been subjected. However, the food constituents themselves vary tremendously in terms of their cleanability (Table 9.1) so that a wide choice of cleaning

Table 9.1
Soil Characteristics[a]

Component on surface	Solubility	Ease of removal	Change on heating
Sugar	Water soluble	Easy	Caramelization; more difficult to clean
Fat	Water insoluble, alkali soluble	Difficult	Polymerization; more difficult to clean
Protein	Water insoluble, alkali soluble, slightly acid soluble	Very difficult	Denaturation; much more difficult to clean
Mineral salts	Water solubility varies; most are acid soluble	Easy to difficult	Generally insignificant

[a]From Tamplin (1980).

materials must be available for their removal. Food residues may be dry particulate, dried-on, cooked-on, sticky, fatty or slimy. Such residues may be best removed by physical means or by the use of hot or cold water almost invariably supplemented with detergents of one type or another. The length of time a food residue is left undisturbed also affects the ease of cleaning. For example, fresh raw milk can be readily washed away but if it is allowed to dry greater difficulty will be experienced. This is due to the denaturation of the milk protein and the breakdown of the fat emulsion which results in the fat spreading over other milk particles making them more difficult to remove.

9.4. REMOVAL OF GROSS SOIL

It could be argued that since the main function of a detergent is to facilitate soil removal, any preliminary operation in conventional cleaning is unnecessary and perhaps wasteful in terms of labour utilization. Balanced against this, however, is the fact that if the bulk of the debris can be removed in a preliminary cleaning process a much reduced and more accurately assessable amount of detergent can be used to remove the residual soil; this latter approach is recommended where feasible and it should be initiated as soon as possible after processing has stopped.

The preliminary cleaning of smaller items of equipment may involve pre-soaking in warm or cold water to remove loosely adhering debris. The more tenacious food debris can then be brushed or scraped off by hand in water at *ca* 45°C. Brush bristles should be as hard as possible but should not cause damage to the surfaces to be cleaned. In this respect abrasives such as steel wool and wire brushes should not be used; not only do they damage many surfaces, including stainless steel, but metal particles may pass into foods and be a cause of customer complaints. Scouring powders have a limited use, possibly in conjunction with nylon cleaning pads; where employed, subsequent thorough rinsing is essential. Any cleaning aid causing damage to stainless steel and other food contact surfaces must be avoided since crevices can be formed if unsuitable materials such as steel wool are employed. These crevices and other surface discontinuities are more likely to retain bacteria than unaffected areas (Holah & Thorpe, 1990); bacteria become attached at these sites by means of extracellular fibrils and their removal can prove difficult (Schwach & Zottola, 1984).

High pressure water jets can be used for certain items of equipment although care must be taken to see that the dislodged dirt is contained and that the temperature of the water is suitable; for example, with raw meats too high a temperature will cause protein denaturation and make cleaning more difficult. The sole use of high pressure water jets for cleaning must be regarded as inadequate; only gross dirt is removed and even that can be ineffective when machines are poorly designed resulting in large aggregates of food debris being retained behind cutting blades, inaccessible ledges, etc. With dry materials vacuum cleaning is to be preferred as compressed air jets again tend to spread debris around; such jets are often useful where tenacious debris is involved.

9.5. DETERGENTS

9.5.1. Desirable Properties

Detergents must be capable of removing many different types of soil under a variety of conditions; the list of properties required for a good detergent is therefore an extensive one. Thus, ideally, detergents should be:

1. Readily soluble in water at the desired temperature.
2. Non-corrosive to equipment surfaces.
3. Non-irritating to the skin and eyes and non-toxic.
4. Odourless.
5. Biodegradable; detergents have created foaming problems in effluent disposal systems but such troubles have now been overcome by the use of detergents which are degraded by sewage bacteria.
6. Economical *in use*; the lowest price per unit volume may not necessarily prove to be the most economical in use.
7. Readily rinsable; detergent solutions should be easily washed away so that none remains adhering to the cleaned surface.
8. Stable during prolonged periods of storage.
9. Effective cleaners of all types of soil. Because of the broad spectrum of substances to be removed by detergents the latter must have the ability to:
    (a) Wet the surface of the soil, that is lower the surface tension of the water so that the latter is able to penetrate the soil more readily and remove it from the surface to be cleaned.

(b) Disperse insoluble materials that might otherwise form aggregates and in so doing keep them in suspension so that they are rinsed away before they can be redeposited on the cleaned surface.

(c) Dissolve soluble soils, both organic and inorganic; the more rapid the solution, the better the detergent.

(d) Emulsify fats and oils, that is break up fats and oils into smaller globules and disperse them so that they remain suspended in solution.

(e) Saponify fats, that is convert fats into soluble soaps.

(f) Sequester (i.e. remove or inactivate) calcium and magnesium salts dissolved in hard waters so that their precipitation is prevented and cleaning efficiency is not impaired. This precipitation is exemplified by the scum formed when soap is used for washing with hard water. In essence, therefore, detergents should be able to soften hard water where necessary although it would be anticipated that in hard water areas water softening systems would be installed.

It will be noted that detergents are not expected to possess bactericidal properties although in practice some of them do. However, detergents do physically remove a large number of bacteria during cleaning and this makes subsequent disinfection that much easier.

Since, as yet, no one chemical possesses all the above properties it is necessary to blend different chemicals together to create detergent formulations of the correct balance for a particular cleaning need.

9.5.2. Classification of Detergents

Detergents may be conveniently classified as:

1. Inorganic alkalis—caustic and non-caustic.
2. Inorganic and organic acids.
3. Surface active agents—anionic, non-ionic, cationic and amphoteric.
4. Sequestering agents—inorganic and organic.

9.5.2.1. Inorganic Alkalis

Many detergents incorporate an alkali as one of the principal ingredients. *Sodium hydroxide* (caustic soda) is the strongest of the alkalis and is cheap. It has excellent dissolving properties, is a very strong saponifier and has the added advantage of being strongly bactericidal. It is, however, highly corrosive to metals especially aluminium and extreme care must be taken when handling this detergent as it can cause severe burns to the skin; for this reason protective clothing must be worn together with goggles and heavy-duty rubber gloves when working with this detergent. As with all the alkali detergents sodium hydroxide precipitates insoluble calcium and magnesium salts from hard water so that sequestering agents need to be incorporated with alkali cleaners in any detergent formulation. The principal properties of this and other important detergent ingredients are shown in Table 9.2.

Sodium metasilicate, although a strong alkali, is non-caustic and therefore much less corrosive than sodium hydroxide. In fact sodium metasilicate suppresses the corrosive activity of sodium hydroxide and the two are often combined in detergents for this reason. It is, however, a good cleaning compound in its own right having good dispersive and emulsifying powers and it is readily rinsable; it suffers somewhat by being relatively expensive. *Sodium orthosilicate* and *sodium sesquisilicate* are two substances which have good saponification powers and both are effective cleaners of protein material. Unfortunately both, but especially sodium orthosilicate, suffer by being corrosive to aluminium.

Turning to the non-caustic alkalis, sodium carbonate and trisodium phosphate are the major examples. *Sodium carbonate* (soda ash) is a relatively weak detergent, is somewhat corrosive and precipitates calcium and magnesium salts from hard water. However, it is cheap and has a good buffering capacity (i.e. stabilizes pH) and is frequently included in detergents for this reason. *Trisodium phosphate* (TSP) is a good emulsifier and saponifier, has strong dispersive properties and has the ability to soften water by precipitating the salts as floccules rather than as scale. Although again somewhat corrosive, it is often incorporated in detergents.

9.5.2.2.   *Inorganic and Organic Acids*
Acids are not widely used in the food industry as they are corrosive to a greater or lesser extent and they lack versatility as cleaners; furthermore, many are dangerous and can cause severe skin burns so that protective clothing must be worn. Of the inorganics, *hydrochloric, sulphuric* and *nitric acids* have been used in the past by the dairy industry to remove hard water scale and other mineral deposits (e.g. 'milkstone', a deposit of protein, calcium carbonate and other salts which builds up in pasteurizers when milk films are not completely removed) but, due to the extremely corrosive nature of these acids, they have been largely replaced

Table 9.2

Principal Properties of Important Components of Detergent Formulations[a]

Class	Component	Wetting power	Dispersing power	Dissolving power	Emulsifying power	Saponification power	Sequestration power	Rinsability	Non-corrosion[b]	Bactericidal power
Inorganic alkalis	Sodium hydroxide	1	1	4	1	4	0	1	0	4
	Sodium metasilicate	2	3	3	3	3	1	3	2	2
	Sodium carbonate	1	1	2	1	2	0	1	2	1
	Trisodium phosphate	2	3	2	3	3	2	3	1	2
Acids	Sulphamic acid	1	1	3	1	1	3	1	0	3
	Hydroxyacetic acid	1	1	2	1	1	3	2	2	2
Surfactants	Sodium alkyl benzene sulphonate	4	4	2	4	0	0	4	4	0
	Sodium lauryl sulphate[c]	4	4	2	4	0	0	3	4	0
	Nonyl phenol ethoxylate	4	4	2	4	0	0	2	4	0
	Dodecyl diaminoethyl glycine	4	3	2	4	0	0	3	4	1
Sequestrants	Tetrasodium pyrophosphate	1	2	2	2	2	3	3	4	1
	Sodium tripolyphosphate	1	3	3	2	1	3	2	4	0
	Sodium hexametaphosphate	1	3	1	2	1	3	3	4	0

[a] 4 = excellent; 3 = good; 2 = moderate; 1 = poor; 0 = no activity.
[b] 0 in this column represents the highly undesirable corrosive product, 4 represents non-corrosive.
[c] Unstable in presence of acids.

by milder acids. Amongst these are *phosphoric* and *sulphamic acids* which are less corrosive than those mentioned above and, when coupled with a corrosion inhibitor, can be very effective. However, low levels of stronger acids may be used where deposit build-up is excessive.

Organic acids, which are bacteriostatic in action, are much milder than the inorganic acids and are therefore safer to handle. Amongst the organic acids which have been incorporated in detergent formulations one may find *gluconic, hydroxyacetic, citric* and *tartaric acids*. Acid detergents usually incorporate corrosion inhibitors and wetting agents and as such can be employed in the removal of inorganic deposits and milkstone, and in bottle washing.

9.5.2.3.   Surface Active Agents

Surface active agents reduce the surface tension of water to facilitate wetting. The classical surface active agent is soap which is usually composed of sodium or potassium salts of fatty acids such as stearic, palmitic and oleic acids. Soaps are reasonably effective in soft water but their reduced solubility in cold water constitutes a disadvantage; in addition, soaps form precipitates with calcium in hard water to give insoluble deposits. For these reasons they have been largely replaced by synthetic detergents which are either anionic, cationic, non-ionic or amphoteric depending on their active electrical charge when in solution. When negative charges predominate the surfactant is classed as anionic, when positive cationic, whilst surfactants that do not dissociate in solution are termed non-ionic. Where the predominant charge varies according to whether acid or alkaline conditions prevail the surfactant is then termed amphoteric.

Surface active agents have a molecular structure comprising a hydrophilic (literally 'water loving') portion and a hydrophobic ('water hating') portion. One end of the molecule is thus attracted by water and the other end is repelled but is attracted by fat and oil (i.e. it is lipophilic); thus a bond is established between the water and the oil and this is the basis of the cleaning action of surfactants. The basic reactions are illustrated in Fig. 9.1.

There are many hundreds of surface active agents now available which are incorporated into detergent formulations. Surfactants are usually excellent emulsifying agents, they have good wetting and penetrating powers, they are non-corrosive, non-irritating and readily rinsable. In addition, they are highly soluble in cold water, are largely unaffected by hard water and many are stable in both acid and alkaline conditions; with such diverse properties their widespread use is hardly surprising. Whilst

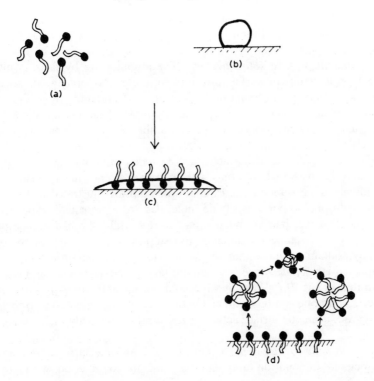

Fig. 9.1. Basic steps in the cleaning action of surface active agents. (a) Detergent molecules with hydrophilic heads and hydrophobic tails. (b) Water drop retains a near spherical shape in the absence of detergent due to an attraction between water molecules creating a high surface tension. (c) The surface tension is broken by the hydrophobic tails of the detergent splitting the bonds between the water molecules. The drop of water collapses and wetting ensues. (d) With greasy materials the hydrophobic tails stick into the fat surface and the hydrophilic heads coat the fat particles. The lightly and similarly charged heads repel each other so that the grease particles remain dispersed and in suspension.

the bactericidal activity of both the anionic and non-ionic detergents is poor, that of some cationics is excellent although the latter are less good as detergents. For this reason the cationic surfactants are used as disinfectants and sterilizers and are discussed later in this chapter.

Many surfactants produce copious amounts of foam especially where considerable turbulence is generated during cleaning. To control the extent of foaming, anti-foaming agents are often incorporated in formulations to give the relatively low foaming detergents required in most cleaning operations in the food industry.

1. Anionic surface active agents. The anionics are by far and away the largest group of surfactants but with the development of such a wide variety of these compounds it would be impossible to cover more than the major categories. Soap is an example of an anionic surfactant but, as already mentioned, it has a number of undesirable properties which severely restrict its use.

The principal anionic surfactants used today are alkyl (i.e. a chain of typically 12 to 18 carbon atoms with hydrogen atoms attached) sulphates or alkylbenzene sulphonates; examples of both are illustrated in Fig. 9.2. The hydrophobic portions of the molecules are represented, respectively, by the alkyl (i.e. lauryl) and benzene groups whilst the hydrophilic portions are represented by sulphate and sulphonate; the cations are commonly sodium or potassium. Another type of anionic surfactant is the sulphated ethoxylate range (e.g. sodium lauryl ethoxylate sulphate) also illustrated in Fig. 9.2. Virtually all anionic surfactants are high foam producers and none can be combined with a cationic detergent. The principal properties of major surface active agents are shown in Table 9.2.

2. Non-ionic surface active agents. The non-ionic surfactants are not dissociated in solution and can be used in conjunction with either

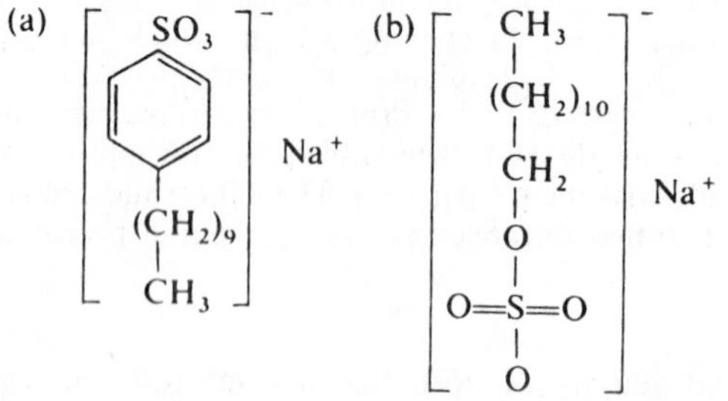

(c)

$$[C_{12}H_{25}-(OCH_2CH_2)_n-O-SO_3]^- Na^+$$

$$n = 1 \text{ to } 3$$

Fig. 9.2. Some common anionic surfactants: (a) sodium alkylbenzene sulphonate; (b) sodium lauryl sulphate; (c) sodium lauryl ethoxylate sulphate.

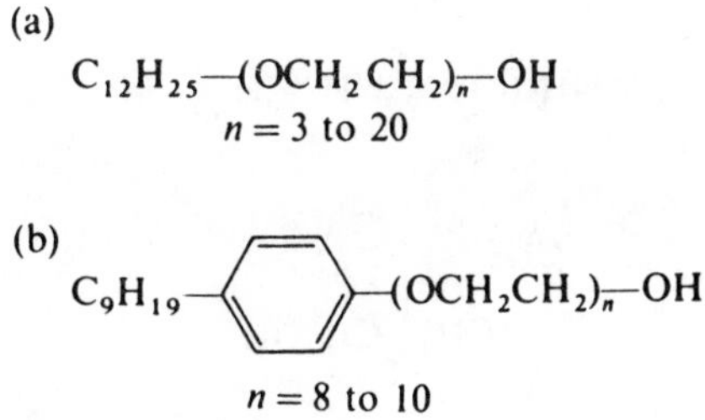

(a)

$$C_{12}H_{25}\text{---}(OCH_2CH_2)_{\overline{n}}\text{---}OH$$
$$n = 3 \text{ to } 20$$

(b)

$$C_9H_{19}\text{---}\langle\text{benzene ring}\rangle\text{---}(OCH_2CH_2)_{\overline{n}}\text{---}OH$$
$$n = 8 \text{ to } 10$$

Fig. 9.3. Some common non-ionic surfactants: (a) lauryl alcohol ethoxylate; (b) nonyl phenol ethoxylate.

anionic or cationic agents. They are powerful emulsifying agents, are unaffected by hard water and vary considerably in their foaming characteristics.

Many are highly soluble in water and are used mainly in liquid detergents. However, some of the non-ionic surfactants exhibit unusual solubility characteristics in water in that when they are heated they tend to separate out of the solution which becomes cloudy; the temperature at which this occurs is called the 'cloud point'. Practical use can be made of this phenomenon by operating the non-ionic above the cloud point temperature when anti-foaming characteristics, so important in high pressure cleaning with detergents, are induced; at lower temperatures the surfactant is redissolved and this change assists when subsequently cold rinsing.

As with the anionic agents, the list of non-ionic surfactants is an extensive one. Two of the major categories are based on products formed by condensation reactions between ethylene oxide and synthetic long chain alcohols (e.g. lauryl alcohol ethoxylate) or between ethylene oxide and alkyl phenols (e.g. nonyl phenol ethoxylate); their structures are illustrated in Fig. 9.3. The former represent the major group of low foaming surfactants in the UK.

3. Amphoteric surface active agents. Amphoteric surfactants can exist in solution in either the cationic or the anionic form depending on the pH. They are based on amino acids and have the general formula R-NH-CH$_2$-COOH where R is usually an alkyl radical. An example is dodecyl diaminoethyl glycine (Fig. 9.4) and it is in the anionic state that

(a)
$$C_{12}H_{25} \cdot NH \cdot (CH_2)_2 \cdot NH \cdot (CH_2)_2 \cdot NH \cdot CH_2 \cdot COOH$$

(b)

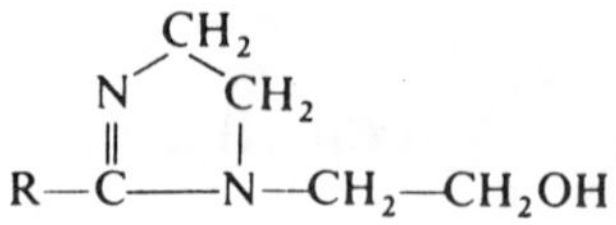

$$\begin{array}{c} CH_2 \\ N \diagdown CH_2 \\ \parallel \quad | \\ R-C-\!\!-\!\!-N-CH_2-CH_2OH \end{array}$$

Fig. 9.4. Typical amphoteric compounds: (a) dodecyl diaminoethyl glycine; (b) ethyl β-oxypropionic acid imidazoline.

detergency is active. The amphoteric agents are relatively good emulsifiers, are stable in both acids and alkalis and show a reasonable tolerance to hard water. Furthermore, those that are used as detergents do have some bactericidal activity; they are, however, relatively expensive, lack some of the desirable properties (see Table 9.2) and hence have a limited range of uses although, because of their mild nature, they are often incorporated in hand cleansers and hair shampoos.

9.5.2.4. Sequestering Agents
A truly soft water has inherent cleaning properties which are progressively weakened as more calcium and magnesium salts are dissolved in it; that is, the water gets harder and this is associated with an increasing tendency for salts to precipitate out forming scale. Sequestering agents are added to detergents to prevent salt precipitation although in the long term it is often far cheaper to soften a hard water supply than to add high concentrations of sequestrants to detergents. Obviously the amount of sequestrant that is added depends on the extent of water hardness and the overall detergent formulation.

1. Inorganic sequestering agents. The sodium polyphosphates are widely used as sequestrants but, in addition, they often provide other desirable properties to a detergent; many are good emulsifiers, dissolving and dispersing agents and they generally enhance rinsability (Table 9.2).

Of the polyphosphates *tetrasodium pyrophosphate* is cheap and commonly used but it functions primarily as a precipitant builder and tends to be a better sequestrant of magnesium than calcium. *Sodium tripolyphosphate* and *sodium tetraphosphate* are true complexing builders.

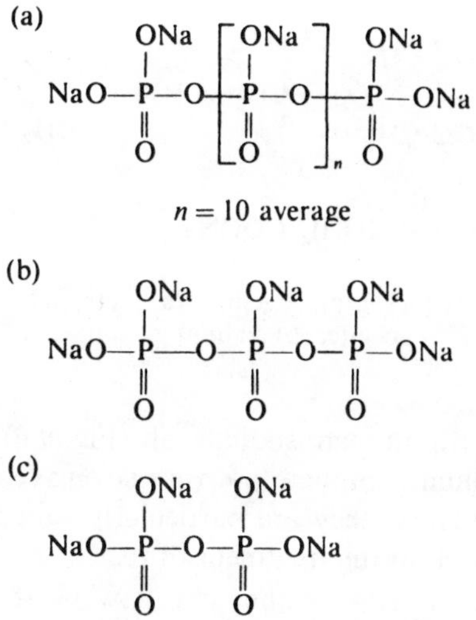

(a)

$$NaO-\underset{\underset{O}{\|}}{\overset{\overset{ONa}{|}}{P}}-O \left[-\underset{\underset{O}{\|}}{\overset{\overset{ONa}{|}}{P}}-O- \right]_n \underset{\underset{O}{\|}}{\overset{\overset{ONa}{|}}{P}}-ONa$$

$n = 10$ average

(b)

$$NaO-\underset{\underset{O}{\|}}{\overset{\overset{ONa}{|}}{P}}-O-\underset{\underset{O}{\|}}{\overset{\overset{ONa}{|}}{P}}-O-\underset{\underset{O}{\|}}{\overset{\overset{ONa}{|}}{P}}-ONa$$

(c)

$$NaO-\underset{\underset{O}{\|}}{\overset{\overset{ONa}{|}}{P}}-O-\underset{\underset{O}{\|}}{\overset{\overset{ONa}{|}}{P}}-ONa$$

Fig. 9.5. Inorganic sequestering agents: (a) sodium hexametaphosphate; (b) sodium tripolyphosphate; (c) tetrasodium pyrophosphate.

That is to say that they remove calcium and magnesium ions from water by complex formation rather than by undesirable precipitation of calcium and magnesium phosphates. With incorrectly formulated detergents there is a tendency, particularly at high temperatures, for the polyphosphates generally to lose their sequestering power due to their conversion, in solution, to the simpler orthophosphates which are weak sequestrants. *Sodium hexametaphosphate* is the least stable of the polyphosphates and is rather expensive; it is the best sequestering agent for calcium but is much less active against magnesium. Structures of typical polyphosphates are shown in Fig. 9.5.

2. Organic sequestering agents. The principal organic sequestering agents, also called chelating agents, are *ethylene diamine tetraacetic acid* (EDTA), *nitrilo triacetic acid* (NTA), their sodium and potassium salts, and the sodium salts of *gluconic* and *heptonic acids*. In spite of their relative expense they are fairly widely used, because of their high solubility in liquid detergent formulations.

NTA is generally considered the most cost-effective organic sequestrant. EDTA is even more efficient but is relatively expensive even in its

(a)

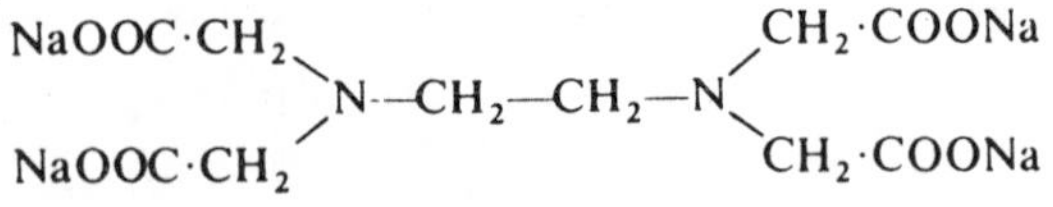

$$NaOOC \cdot CH_2 \diagdown \qquad \diagup CH_2 \cdot COONa$$
$$N\text{--}CH_2\text{---}CH_2\text{---}N$$
$$NaOOC \cdot CH_2 \diagup \qquad \diagdown CH_2 \cdot COONa$$

(b)

$$CH_2OH \cdot (CHOH)_4 \cdot COONa$$

Fig. 9.6. Organic sequestering agents: (a) tetrasodium ethylene diamine acetate; (b) sodium gluconate.

commonly used form, the tetrasodium salt (Fig. 9.6). Sodium gluconate (Fig. 9.6) and sodium heptonate are very strong sequestering agents at high levels of alkalinity; they are particularly suitable for sequestering iron and may be used in the treatment of rust.

9.5.3. Detergent Formulation

Modern detergent formulations are carefully blended mixtures of different chemicals each contributing to the desired properties of the detergent. Cost is an important factor so that detergents, wherever possible, should contain no wasted ingredients and they should be used at concentrations which most economically accomplish the required degree of cleaning. Thus each different soil to be removed and surface to be cleaned may ideally require a different detergent but, in practice, probably three or four different formulations will suffice.

Detergents are available as either powders or liquids. The former have the advantage of generally being more concentrated, there is less likelihood of wastage in the preparation of solutions but they do have to be dissolved to form those solutions. Liquids are, perhaps, more easily and accurately measured but, in practice, there is often considerable wastage of the concentrated detergent due to excessively strong solutions being prepared.

A general-purpose cleaner may contain alkali builders to break up grease, surfactants to improve wetting, dispersion and rinsability, and sequestrants to stabilize magnesium and calcium; the level of the sequestrants should be carefully adjusted depending on the degree of water hardness and in-use concentration. Sodium metasilicate may also be added as a cleaning agent with the added benefit of being a corrosion inhibitor especially if susceptible metals such as aluminium are to be cleaned. A better quality powder formulation for a general-purpose cleaner could be: sodium carbonate, 30%; sodium metasilicate, 35%; alkyl aryl sulphonate, 5%; sodium tripolyphosphate, 30%. Between 3 and 5 parts of powder would be added to 1000 parts of water to make up the detergent. For a strong alkali cleaner for heat-treated fats and proteins the formulation could contain sodium hydroxide, an alkali-stable sequestering agent and a non-ionic detergent. Thus this formulation could be: sodium hydroxide, 12 %; sodium carbonate, 20%; sodium metasilicate, 45%; sodium tripolyphosphate, 20%; non-ionic surfactant, 3%. For an acidic cleaner required for a light milkstone deposit or as a once weekly back-up to a more general-purpose cleaner, phosphoric acid may be used together with a low foaming non-ionic surfactant and possibly a corrosion inhibitor. A typical liquid formulation could be: phosphoric acid, 35%; surfactant, 1%; water, 64%. Other formulations have been listed by Tamplin (1980).

In the final analysis, however, the proof of the efficiency of any detergent is measured by the standard of cleanliness achieved in practice, and the time, effort and cost put in to achieve that standard. Modifications to formulations may be necessary if difficulties are being experienced but this should only be done in consultation with the detergent manufacturers.

9.5.4. Factors Affecting Efficiency of Detergents

The importance of the control of hardness by water softening or by the addition of sequestering agents to detergents has already been stressed but other factors also affect the efficiency of detergents. These include the concentration and temperature of the detergent solution, the time that solution is applied and the force with which it is applied.

For any detergent there is a minimum concentration necessary for effective cleaning under a given set of conditions; increasing the concentration above this minimum improves the cleaning action but at ever-decreasing rates and at increasing cost so that, on balance, there is an optimum concentration which should be aimed for under commercial conditions.

As the temperature is increased the reaction rate between the soil and the detergent is also increased as is the solubility of soluble materials and this means that soils can be more readily removed from surfaces; another advantage of higher temperatures is that viscosities are generally decreased resulting in increased turbulence, a feature of especial importance in CIP. Time effects are similar to concentration effects in that there are minimum and optimum times for contact between soils and detergents.

Obviously it is possible to effect cleaning merely by applying force (e.g. using brushes in manual cleaning) but this is extremely inefficient. Detergents are employed partly to lessen the need for force although, in practice, the two components are often combined. A good example of this is CIP where turbulence provides the cleaning force, the turbulence created being markedly influenced by the velocity of the cleaning fluid in the pipework.

9.6. CHEMICAL DISINFECTANTS

When disinfecting food contact surfaces of equipment and pipework it is rarely necessary to achieve absolute sterility. The usual aim is to so reduce the numbers of microorganisms that those remaining (i.e. some bacterial spores and possibly a few highly resistant vegetative cells) could not affect the microbiological quality of the food coming into contact with these surfaces. For this reason the absolute term 'sterilize' is really inappropriate and the popularly used term 'disinfect' will be employed in this discussion instead.

Either heat or chemicals can be used for plant disinfection operations but it is important to stress that these operations must be preceded by thorough cleaning. The efficiency of chemical disinfectants is undermined by the presence of soil and the cleaner the surface the more effective will be the disinfectant. Disinfection should follow immediately after cleaning although it is often advisable, if not essential, to disinfect surfaces a second time immediately before a process run if the equipment has been left unused for a lengthy period.

9.6.1. Desirable Properties

Disinfectants for use on food contact surfaces should ideally have the following properties:

1. Capable of rapidly killing microorganisms and, in particular, equally effective against both Gram positive and Gram negative bacteria. The majority of mould spores should be killed and the destruction of bacterial spores would be an added advantage.

2. Reasonably stable in the presence of organic residues and, if necessary, effective in the presence of hard water salts.
3. Non-corrosive and non-staining to plant surfaces of whatever type.
4. Odourless or have an inoffensive odour.
5. Non-toxic and non-irritating to the skin and eyes.
6. Readily soluble in water and readily rinsable.
7. Stable during prolonged storage in concentrated form and stable during short-term storage in dilute form.
8. Competitively priced and cost-effective in use.

9.6.2. Classification of Disinfectants

Whilst many chemicals, e.g. inorganic and organic compounds of mercury, are bactericidal very few have all the properties listed above. Those used in the food industry are generally restricted to four groups:

1. Chlorine-releasing compounds.
2. Quaternary ammonium compounds.
3. Iodophors.
4. Amphoteric compounds.

These are now considered in turn.

9.6.2.1. Chlorine-Releasing Compounds

The hypochlorites are the most widely used of all disinfectants in the food industry but there are a number of other chlorine-releasing compounds which are also used on a more limited scale. Amongst the latter are gaseous chlorine and chlorinated trisodium phosphate as well as the organic chloramines, derivatives of isocyanuric acid and dichloro-dimethylhydantoin.

In general, chlorine-releasing compounds are powerful disinfectants with a broad spectrum of activity, Gram positive and Gram negative bacteria being equally susceptible; furthermore, these compounds show some activity against bacterial spores. Many chlorine-releasing compounds are inexpensive; all are easy to use and unaffected by hard water. However, it is essential to maintain a high pH to prevent corrosion effects even though, as a consequence, some loss of bactericidal activity is experienced. Perhaps the main disadvantage of chlorine-releasing agents is that they are rapidly inactivated by the presence of organic matter although an additional weakness is that they must be carefully rinsed off to prevent corrosion.

Chlorine-releasing compounds available in powdered form are often assumed to be more stable than their liquid counterparts. However, the powders absorb water readily to become unstable so that desiccants need to be incorporated to enhance stability.

1. Hypochlorites. Hypochlorous acid (HOCl) is itself unstable but many of its salts can be prepared and these are invariably more stable. In solution these salts dissociate to form OCl — which is the ion that is responsible for the bactericidal properties of the hypochlorites. The most widely used salt is *sodium hypochlorite* (NaOCl) which is available in commercial form as a concentrated liquid containing *ca* 10–14% available chlorine. Rather interestingly it has been suggested that if this concentrated liquid is diluted with distilled water (1:1 or even 1:9) available chlorine levels fall more slowly during storage (Hoffman *et al.*, 1981). Also used is *calcium hypochlorite* ($Ca(OCl)_2$) which is available in powdered form and contains *ca* 30% available chlorine. In more diluted form aqueous solutions of sodium hypochlorite are widely used in the food industry as general disinfectants and in CIP systems; solutions should be freshly prepared and handled with care due to their skin irritating properties. In commercial formulations surfactants and stabilizers are sometimes added, the former to aid wetting and penetration characteristics, the latter to improve activity during prolonged storage. Hypochlorite solutions should always be stored in the dark or in opaque containers; stability is also improved if chill storage is employed. Solutions are more stable above pH 9·5 whereas germicidal activity is maximal between pH 4 and pH 5; at the latter pH corrosion effects are also maximal. Because of these corrosion problems, solutions of pH 10–11 are used and operating temperatures are kept relatively low since at higher temperatures corrosivity and loss of stability of the disinfectant occur. In-use concentrations typically vary between 50 and 200 ppm available chlorine and contact times of between 3 and 30 min are normal; it should be remembered that in each specific situation the minimum concentration and time to effect the necessary kill-off should be employed in order to avoid possible corrosion of susceptible surfaces.

2. Chlorine gas. Chlorine gas is commonly used for the disinfection of water supplies but it also has some application in the food industry. Where employed it must be fed into the water supply at a constant rate by means of a chlorinator. It is necessary to chlorinate above the 'break point' of water, that is the level at which the chlorine demand of the water, a variable factor depending primarily on the amounts of suspended solids and organic matter, has been satisfied. In fact chloramines (see later) are formed if ammonia-releasing compounds are present in the water and at higher chlorine doses these are oxidized. Only after this is the 'break point' reached so that thereafter any additional chlorine creates a residuum of free chlorine. A residual chlorine level of between 1 and 5 ppm is suitable for most continuous plant chlorination systems such as sprays for belts and elevators; higher levels (10–20 ppm) may be required for end of shift disinfection or for can cooling water.

3. Chlorine dioxide. The main advantages of chlorine dioxide (ClO_2) over chlorine are that it retains much of its bactericidal activity in the presence of organic matter and that it is most active at higher pHs; in fact it is effective over the pH range 3–13. Chlorine dioxide has become far more popular in recent years where water recycling systems are employed. Such systems, introduced as restrictions on the use of water increase, encourage slime development on product contact surfaces. Chlorine dioxide is far more effective than the more traditional chlorine-releasing compounds in controlling this problem which is mainly associated with vegetable processing plants.

4. Chlorinated trisodium phosphate. Commercially prepared chlorinated trisodium phosphate (CTSP; $4[Na_3PO_4.11H_2O]NaOCl$) gives a buffered solution of hypochlorite when dissolved in water. This relatively expensive compound is often incorporated in powdered formulations. The available chlorine content is low (4%) and it is somewhat inactivated in the presence of organic matter. Bromine-releasing compounds (e.g. sodium bromide) can be added to give enhanced bactericidal activity.

5. Chloramines. The chloramines, e.g. *chloramine T* (Fig. 9.7), *chloramine B* and *dichloramine T*, are very much more stable than the hypochlorites in the presence of organic matter, they are less irritating and toxic but their expense has undoubtedly limited their use. Furthermore, in spite of an available chlorine content of 25–30%, they are weaker bactericides except at high pH values (>10) where they are more active than the hypochlorites. Chloramines release chlorine slowly and are often used where equipment and utensils can be soaked for long periods since they are only weakly corrosive; rinsing after their application is necessary, however. They are often combined with alkali detergents to form detergent-sterilizers (see p. 368).

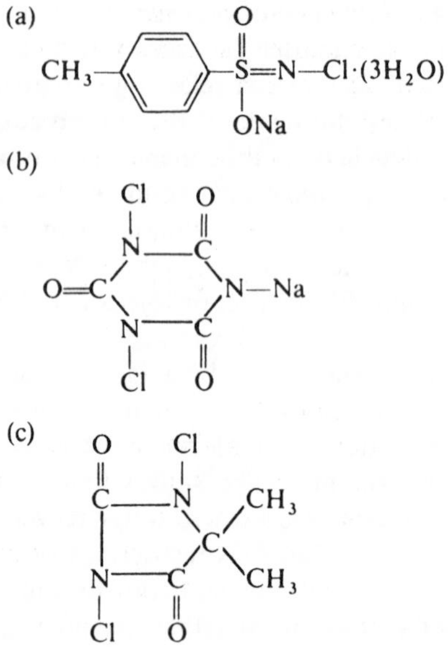

(a)

(b)

(c)

Fig. 9.7. Organic chlorine releasing disinfectants: (a) chloramine T; (b) sodium dichloroisocyanurate; (c) dichlorodimethylhydantoin.

6. Isocyanuric acid derivatives. Dichloroisocyanuric and *trichloroiso-cyanuric acids* have very high available chlorine levels but because of the low solubility of the acids in water their sodium salts are generally used for disinfection; these are available in powdered form and have somewhat lower available chlorine contents (e.g. sodium dichloroisocyanurate, 60%; Fig. 9.7). These compounds are, like the chloramines, relatively expensive, they are stable when stored under dry conditions, non-irritating and release chlorine slowly; unlike the chloramines, they retain their activity over a wide pH range (6–10). They are again used in alkali detergent-sterilizer preparations.

7. Dichlorodimethylhydantoin. When pure this compound (Fig. 9.7) is rather insoluble in water so that a technical grade powder of about 25% purity is employed which gives about 16% available chlorine. In many ways dichloromethylhydantoin is similar to the other organic chlorine-releasing compounds but it has greatest activity in acid conditions.

9.6.2.2. *Quaternary Ammonium Compounds*

Quaternary ammonium compounds, known as 'quaternaries', 'quats' or 'QACs', are essentially ammonium salts with some or all of the hydrogen atoms in the $[NH_4]^+$ ion substituted by alkyl or aryl groups; the anion is usually a chloride or bromide. The general formula is thus:

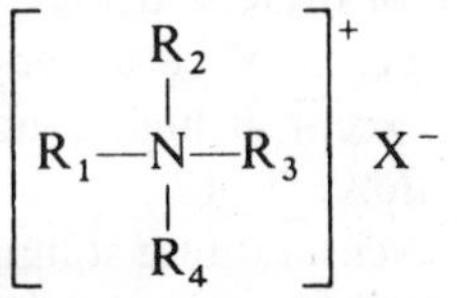

$$\left[R_1 - \underset{\underset{R_4}{|}}{\overset{\overset{R_2}{|}}{N}} - R_3 \right]^+ X^-$$

where R_1, R_2, R_3 and R_4 represent one or more alkyl or aryl groups substituting for hydrogen and X^- represents a halide, either Cl^- or Br^-. The large cation is the active part of the molecule whilst the anion is important only in that it can affect the solubility of the QAC. Examples of commonly used QAC disinfectants are cetyltrimethyl ammonium bromide and lauryldimethylbenzyl ammonium chloride; these are illustrated in Fig. 9.8. Many substitutions are possible theoretically but for maximum activity the alkyl chain must contain between 8 and 18 carbon atoms; in practice relatively few QACs are marketed.

The QACs are very active bactericides against Gram positive bacteria but are less effective against Gram negative forms unless sequestrants are present; bacterial spores are relatively resistant although their outgrowth may be prevented. After disinfection surfaces treated with QACs retain a bacteriostatic film due to the adsorption of the disinfectant on the surface; this film prevents the subsequent growth of residual bacteria. When required, rinsability can be improved by adding a small quan-

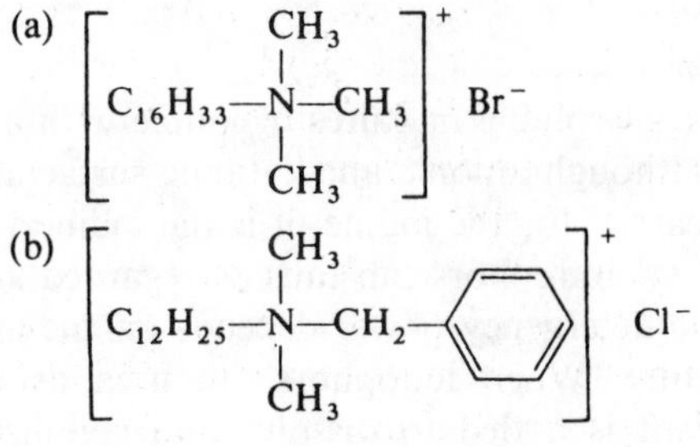

(a)
$$\left[C_{16}H_{33} - \underset{\underset{CH_3}{|}}{\overset{\overset{CH_3}{|}}{N}} - CH_3 \right]^+ Br^-$$

(b)
$$\left[C_{12}H_{25} - \underset{\underset{CH_3}{|}}{\overset{\overset{CH_3}{|}}{N}} - CH_2 - \bigcirc \right]^+ Cl^-$$

Fig. 9.8. QAC disinfectants: (a) cetyltrimethyl ammonium bromide; (b) lauryldimethylbenzyl ammonium chloride.

tity of a non-ionic surfactant to the disinfectant. QACs retain their activity over a fairly wide pH range although they are most active in slightly alkaline conditions and activity falls off rapidly below pH 5.

In comparison with the hypochlorites the QACs are more expensive but they have many desirable properties. Thus QACs are largely unaffected by the presence of organic debris, they are non-corrosive, although some types of rubber may be adversely affected, and they are non-irritating to the skin, except at high concentrations, so they can be handled with reasonable safety.

Quaternaries are stable even in dilute solution and when concentrated they can be safely stored for long periods with no loss of activity. Since QACs are cationic surfactants they do have some powers of detergency but they cannot, of course, be used in conjunction with anionic surfactants or even with certain non-ionic surfactants. Hard water salts tend to reduce the activity of QACs, the extent depending upon the length of the alkyl chain in the QAC; if the correct sequestering agents are employed activity is restored. Care must be taken with the selection of the sequestrant since some are incompatible with some QACs and cause their precipitation. Strong alkalis induce a similar effect and cannot be used with many QACs; in general, detergents containing such materials must be carefully rinsed off before the QAC is added.

QACs often foam vigorously in solution so they are generally unsuitable for CIP or spray systems. They are customarily used at concentrations of between 50 and 500 ppm, at temperatures in excess of 40°C and with contact times varying between 1 and 30 min.

Biguanides are other types of cationic disinfectant in use on a limited scale; they have the advantage of being more active against Gram negative bacteria, of not producing foam and of being unaffected by hard water.

9.6.2.3. *Iodophors*

Iodophors consist of soluble mixtures of iodine with a surfactant (typically non-ionic, although anionic and cationic surfactants may be used) which acts as a carrier for the iodine; it is the iodine which imparts the bactericidal activity. Iodophors can thus be regarded as detergent-sterilizers although the detergency power depends on the amount of surfactant in the mixture. When iodophors are used as disinfectants just sufficient surfactant is added to dissolve and stabilize the iodine but when used as detergent-sterilizers much more surfactant is added to improve detergency. Although iodophors are even less affected by pH changes than QACs, in practice an acidic component, usually phosphoric acid, is added to the iodophor in order to lower the pH of the solution. This is because iodophors are most active in a pH range of 3–5 and phosphoric acid buffers in this range.

Iodophors effect the rapid killing of a broad spectrum of bacteria and resemble hypochlorites in this respect but they also maintain a reasonable activity in the presence of organic wastes provided the pH is not above 4 and the quantity of waste is not excessive; iodophors are, however, much less active against spores than hypochlorites.

Iodophors are rather expensive and, in consequence, are not used widely; they are essentially non-corrosive, non-irritating, non-toxic and have little smell but they must be thoroughly rinsed off after use. Some plastic materials may absorb the iodine and become discoloured when exposed to these compounds; rubber also tends to absorb the iodine so that long contact times with the iodophor are to be avoided to prevent possible tainting of foods. Iodophors are unsuitable for cleaning equipment used for processing starch-containing foods since iodine forms a purple complex with starch. One advantage of the iodophors is that they are unaffected by hard water salts; they are also stable in concentrated form although over long periods of storage at high ambient temperatures some loss of activity is possible.

Iodophors are used mainly in dairies where, in addition to their bactericidal powers, the phosphoric acid is useful in the control of milkstone; iodophors are also used in the brewing industry. In CIP systems foaming is possible so that a low foaming surfactant should be incorporated in formulations used for this purpose. Operating temperatures of up to 50°C can be employed with iodine concentrations varying between 10 and 100 ppm.

9.6.2.4. *Amphoteric Compounds*
Whilst some of the amphoteric surfactants are primarily detergents with weak bactericidal powers there are others, the imidazoline derivatives, which are relatively stronger bactericides and weaker detergents; an example is ethyl β-oxypropionic imidazole (Fig. 9.4). As stated previously, amphoterics can exist as cations or anions depending on the pH of the solution and it is in the cationic state that these compounds are bactericidally active. They are generally more expensive than other disinfectants and are not particularly powerful bactericides although they can be blended with QACs to improve their efficiency. Amphoteric disinfectants are not markedly affected by organic matter or by water hardness,

they are non-corrosive, non-toxic and odourless, and they are stable, even in dilute form, for lengthy periods. However, they do tend to foam and, because of their expense and limited activity, amphoteric disinfectants have not been widely used in the food industry.

9.6.2.5. Phenolic Compounds

Many phenolic compounds have strong bactericidal powers and they are widely used as general-purpose disinfectants. Phenolics are not used in disinfecting operations on food plant because of their strong odours and because of the possibility of transmitting off-flavours to foods.

9.6.2.6. Detergent-Sterilizers

Detergent-sterilizers, popularly known as detergent-sanitizers, are essentially combinations of compatible and complementary ingredients; they contain a detergent and a separate disinfectant, so that cleaning and disinfection can be performed in a single operation. Many of the aforementioned ingredients have been used in one combination or another to produce detergent-sterilizers of varying efficiency; popularly used groupings are given in Table 9.3. In practice, the detergent-sterilizer formulations are likely to contain other components such as sequestering agents and buffers, and two surfactants are often included in a single formulation provided they are compatible.

Whatever its formulation a good detergent-sterilizer should ideally be effective against a variety of soils and a broad spectrum of microorganisms; it should be possible to use the compound in a variety of situations if its use is to be justified on economic grounds. In fact detergent-sterilizers are generally rather more expensive and less effective than their

Table 9.3
Commonly Used Detergent-Sterilizer Combinations

Detergent		Disinfectant
Inorganic alkalis	+	Hypochlorites
	+	Organic chlorine-releasing compounds
	+	QACs
Inorganic acids	+	Non-ionic surfactants
	+	Iodophors
Anionic surfactants	+	Organic chlorine-releasing compounds
Non-ionic surfactants	+	QACs
	+	Iodophors

separate counterparts although with the non-ionic surfactant + QAC combination somewhat greater efficiency is obtained. All detergent-sterilizer combinations can usefully be employed where soiling is light and where low temperature cleaning is desired. In addition there is no doubt that savings can be made in time and convenience if a single application of detergent-sterilizer proves adequate; this is reflected by the ever-increasing use of these compounds which are constantly being improved. One further advantage claimed for detergent-sterilizers is that hazardous bacteria should be killed when the agent is applied whereas in conventional cleaning viable bacteria may be discharged in the detergent waste.

Whatever cleaning or disinfecting agents are employed it is important to introduce alternative formulations at intervals in order to ensure that resistant food residues or bacteria do not accumulate.

9.6.3. Evaluation of Disinfectants

There are a number of tests that can be employed in the laboratory to evaluate the killing power of a disinfectant against microorganisms. These tests measure the rate of kill of selected bacteria or other microorganisms under specified conditions and a few of the more important tests, all of which are non-mandatory, are now reviewed.

1. The Rideal-Walker test. The Rideal-Walker test, first introduced in 1903, has been much modified over the years and is described in British Standard 541 (1934); today, after further amendments to the test, it is still extensively used for routine quality control of production batches and for the preliminary screening of new formulations. Its principal feature is that it was the first test introduced in which the test disinfectant was compared to a standard reference disinfectant (phenol) so that relative performances (i.e. phenol coefficients) could be established.

In the test the bactericidal powers of suitable dilutions of the disinfectant are compared with specified dilutions of phenol using a standard 24 h broth culture of the test organism (*Salmonella typhi*). The procedure involves adding 0·2 ml quantities of test culture to 5 ml quantities of dilutions of the disinfectant or phenol held at 17–18°C. After contact times of 2·5, 5, 7·5 and 10 min specified amounts (a standard loopful) of the mixture are inoculated into sterile broths; these broths are then incubated at 37°C for 48–72 h after which the presence or

Table 9.4
Typical Rideal-Walker Test Result

Disinfectant	Dilution	Contact time (min) of culture + disinfectant[a]			
		2·5	5	7·5	10
'X'	1:1000	—	—	—	—
	1:1100	+	—	—	—
	1:1200	+	+	—	—
	1:1300	+	+	+	—
Phenol	1:110	+	+	—	—

[a]+ = growth, — = no growth.

absence of growth is recorded. A typical test result is shown in Table 9.4. The Rideal-Walker or phenol coefficient is calculated by dividing the dilution of disinfectant which shows growth only at 2·5 and 5 min by the dilution of phenol giving the same result. From Table 9.4 the Rideal-Walker coefficient for disinfectant 'X' is therefore 1200/110 = 10·9 (approximately).

The Rideal-Walker test is simple to perform but it suffers from a number of disadvantages (Croshaw, 1981). Amongst these are: (1) there is an insignificant amount of organic matter present in the disinfectant: test organism contact mixture and this does not reflect the conditions normally found in commercial practice; and (2) the results are specific for the test organism used, i.e. *S. typhi*. Clearly the test would be far more relevant to the food industry if a wide range of test organisms were used, particularly if the range included the principal bacteria encountered on specified process lines.

2. The Chick-Martin test. Perhaps the greatest weakness of the Rideal-Walker test is the virtual absence of organic material in the test mixture, a weakness that is corrected in the Chick-Martin test. This test was originally introduced in 1908 and after modification became incorporated in British Standard 808(1938). As now constituted a sterilized yeast suspension is added as the organic soil to give a final concentration of 2·5 % in the contact mixture; a further difference from the Rideal-Walker test is that there is only one contact time (30 min). A typical test result is shown in Table 9.5 and it shows that dilutions are prepared in 10% steps and that a pair of broths is inoculated with the reaction mixture. The Chick-Martin

Table 9.5
Typical Chick-Martin Test Result

Disinfectant 'X'			Phenol		
Concentration (%)	Tube[a]		Concentration (%)	Tube[a]	
	1	2		1	2
0·411	—	—	2·00	—	—
0·370	—	—	1·80	—	—
0·333	+	+	1·62	+	+
0·300	+	+	1·46	+	+

[a]+ = growth, — = no growth.

coefficient is obtained by dividing the mean of the highest phenol concentration of the test organism (*S. typhi*) permitting growth in both tubes and the lowest phenol concentration showing no growth in both tubes by the corresponding mean concentration of the test disinfectant. Thus from Table 9.5 the Chick-Martin coefficient for disinfectant 'X' is:

$$\text{Mean concentration of phenol} \qquad \frac{1·62 + 1·80}{2} = 1·71$$

$$\text{Mean concentration of 'X'} \qquad \frac{0·333 + 0·370}{2} = 0·35$$

$$\text{Thus coefficient} = \frac{1·71}{0·35} = 4·9 \text{ (approximately)}$$

3. The Improved Kelsey-Sykes test. Phenol coefficient tests of whatever type suffer from being artificial in concept, giving results that show poor reproducibility, and comparing the test disinfectant with an unreliable control disinfectant (i.e. phenol). Any meaningful test design should include a more reliable measure of killing efficiency than the simple 'growth' or 'no growth' of *S. typhi* in the recovery medium, there should be a wider choice of test organisms and there should be neutralizers of the disinfectant in the recovery medium to prevent 'carry-over' effects. In answer to these criticisms Kelsey & Sykes (1969) introduced a new test for the assessment of disinfectants which, following further modifications, became the Improved Kelsey-Sykes test (Kelsey & Maurer, 1974). This test, which is an example of a 'use-dilution' test and can be used to test all types of disinfectant, is briefly described below.

Four test organisms (viz. *Escherichia coli, Staphylococcus aureus, Pseudomonas aeruginosa* and *Proteus vulgaris*) are conventionally included in preliminary screening tests to determine which is the most resistant to the disinfectant; other organisms including yeasts or moulds may be screened if required. The most resistant organism is normally selected for the detailed test which is performed at 20–22°C (other temperatures may be used) using three different concentrations of the disinfectant (viz. the concentration which is expected to pass the test, together with concentrations of + and −50% of that concentration). The disinfectant dilutions are prepared in 'standard' hard water (World Health Organization) as is the standard suspension ($10^8–10^{10}$/ml) of the organism which can be tested in the presence or absence of organic matter (i.e. with or without the addition of 2% sterile yeast added to the above suspension). Recovery broths (each 3 ml and normally containing Tween 80 as the disinfectant inactivator) are prepared in three sets of five tubes for each disinfectant concentration examined.

Thus, in the test, 1 ml of the bacterial suspension is added to 3 ml of the appropriate disinfectant concentration (time zero); after 8 min 0·02 ml of this disinfectant/test organism contact mixture is added to each tube of the first set of five recovery broths. Two minutes later (i.e. 10 min after time zero) a second 1 ml of bacterial suspension is added to the disinfectant/test organism mixture and after a further 8 min (i.e. 18 min from time zero) the second set of five tubes of broth is inoculated in the same manner as the first set. Finally, after a further 2 min a third 1 ml of suspension is added to the disinfectant/test organism mixture, left for a further 8 min (total time 28 min) and then the final set of broths is inoculated. Broths are then incubated at 32°C for 48 h when the presence or absence of growth is recorded (other incubation conditions may be used to suit the test organism selected). A typical test result is shown in Table 9.6. No growth of the test organism in at least two of the five

Table 9.6
Typical Kelsey-Sykes Test Result

Disinfectant 'X' concentration (%)	Broth set number[a]			Result
	1	*2*	*3*	
0·6	− − + + +	+ + + + +	+ + + + +	Fail
1·2	− − − − +	− − + + +	+ + + + +	Pass
1·8	− − − − −	− − − − −	− − − − +	Pass

[a] + = growth, − = no growth.

broths after disinfectant:test organism contact times of 8 and 18 min (i.e. columns 1 and 2) indicates that the disinfectant is satisfactory for use at the initial concentration (1·2 %) employed in the test; obviously as the test proceeds the disinfectant is progressively diluted by the addition of the suspension resulting in enhanced growth in the second and third sets of recovery broths. Since slight variations are found to occur in replicated tests it is recommended that the procedure is repeated on three successive days making the whole test relatively complex and time-consuming.

Cowen (1978) in a critical review of the Kelsey-Sykes test suggested that it was unsuited to many types of disinfectant but in spite of this and other criticisms the test has been widely adopted by official bodies throughout the world for the testing of disinfectants regardless of type or usage (Croshaw, 1981).

4. European Suspension test. Recommended test methods for disinfectants vary considerably amongst Member States within the European Community and with the removal of trade barriers it was deemed necessary to harmonize on disinfectant evaluation methods. As a result of collaborative studies a common test method, the European Suspension test (EST) has been agreed (Council of Europe, 1987).

In the EST the activity of the disinfectant is assessed at the lowest concentration recommended by the manufacturer against five test organisms (viz. *Staphylococcus aureus, Streptococcus faecium, Proteus mirabilis, Pseudomonas aeruginosa* and *Saccharomyces cerevisiae*). Bovine albumin is included in the test solutions as an organic soil at two concentrations (i.e. 0·03% to mimic 'clean' conditions and 1% for 'dirty' conditions) and the test is performed at 20°C.

The four specified bacterial strains are added to duplicate bovine albumin solutions to give a final cell concentration of 10^8/ml (10^7/ml for the yeast, *S. cerevisiae*). After 2 min contact the disinfectant is added to one suspension and standard hard water (Council of Europe, 1987) to the other, this acting as a control. After a further 5 min contact aliquots of each suspension mixture are transferred to tubes of inactivating liquid where a further 5 min contact time is prescribed.

To determine the fraction of surviving cells, viable counts are then performed on the two suspensions, using either the 'spread plate' or 'pour plate' method (see Section 4.3.1); plates are incubated at 30°C for 48 h. For the disinfectant to pass the test there must be a difference of at least 5 logarithmic reductions (i.e. 10^5 cells) in the viabilities of all test organisms in the control and test solutions respectively, and in both albumin concentrations; the test must also be performed in total on two separate days.

   5. *Surface tests.* There are a number of disinfectant tests which involve the use of an air-dried film of microorganisms which is then acted upon by the disinfectant; the surface film is prepared as a standard, with or without organic matter included. Stainless steel or glass surfaces are most commonly used but many others have been employed. One such method (Lisboa, 1959) originally introduced for dairy equipment employs short lengths (33 cm) of stainless steel tubing which are capped at both ends. Suitable suspensions or soils are introduced into the tube for the required time and then poured off. The test disinfectant is then added and after contact (1 min) is poured off and immediately replaced by a neutralizing solution to inactivate any residual disinfectant. The number of viable organisms remaining on the surface film is obtained by swabbing or rinsing the inside of the tube and performing counts on appropriate dilutions of the rinse fluids. Modifications to this method, suitable for use in the food industry, have been described by Blood *et al.* (1981).

9.7.   USE OF HEAT

Heat alone can be used to disinfect food plant and it can be supplied in the form of steam, hot water or hot air. Heat in the form of pressurized steam is the most effective method of sterilization; moist heat kills microorganisms at relatively low temperatures by denaturation of the protein but proteins are far more stable in dry conditions so that far higher temperatures and/or longer times are necessary to effect a kill using hot air. Thus, whenever possible, moist forms of heat are used to sterilize equipment but obviously the efficiency is also dependent on the temperature and time employed. Moist heat is a favoured disinfecting or sterilizing agent because it is non-corrosive, economical, has excellent penetration powers, leaves no residue and is active against the large majority of microorganisms.

9.7.1.   Steam

Saturated steam is a good disinfecting agent and has the potential to destroy all but the most heat-resistant bacterial spores. However, it is essential that heat is applied to surfaces, both internal and external, long enough to heat them to a temperature of *ca* 85°C for 1 min; under these conditions all but bacterial spores should be killed. To achieve these conditions it may be necessary to preheat with steam for many minutes before the treatment temperature is reached; longer preheating will obviously be required when larger items of equipment are being treated. Unfortunately steam guns of one form or another are frequently misused and the short-term blowing of steam against equipment may do more harm than good by supplying warmth and moisture to bacteria enabling them to thrive, especially in the presence of food debris. Steam guns and hoses can also be misused by applying them to inadequately waterproofed equipment (e.g. motors) so that electrical and lubrication faults develop; it must always be remembered that steam *per se* is not a cleaning agent and it should therefore only be used to treat clean surfaces.

9.7.2. Hot Water

Hot water at 80–90°C can be used to disinfect food plant. It is used at these temperatures in cleaning-in-place (CIP) systems where it usually acts as a detergent carrier rather than as the disinfecting agent; contact times of 5–15 min are desirable when it is used in the latter capacity. Small items of equipment can be disinfected by immersing them in water which is then heated to, say, 80°C for 10 min when all but bacterial spores should be killed. High pressure cold water jets and hoses cannot be used for disinfecting operations; the same can be said of water at temperatures below a minimum of 60°C and even at this temperature a contact time of *ca* 30 min will be necessary if the majority of vegetative bacteria are to be destroyed.

9.8.   DRY CLEANING

When dry foods have been processed cleaning is best achieved by using vacuum cleaners; these cleaners can either be portable or attached to a fixed vacuum cleaning system (see Section 9.10.6). The latter system can be exploited to clean floors, walls, ceilings and equipment, and can also be activated readily to clear unexpected spillages occurring during processing. There are many types of portable vacuum cleaners available (see Section 9.10.5) but whichever method is used difficulties may arise with dry cleaning. Dry foods tend to cake on surfaces and manual scraping or brushing may be necessary. Water and even liquid detergents should only be used with care in dry foods' areas since they may encourage caking of materials; the added moisture may also permit microbial growth.

9.9.   CLEANING-IN-PLACE (CIP)

CIP has largely replaced manual cleaning for equipment used to process liquid foods, i.e. in the dairy, brewing and soft drinks industries. The form CIP takes can vary considerably from the cleaning of individual items of equipment to sophisticated operations involving whole processing lines. The cleaning principle is similar in all cases and involves the sequential circulation of water, detergents and disinfectants through pipelines and processing equipment which remain assembled. The basic sequence of operations is: (1) a pre-rinse with cold water to remove gross soil; (2) the circulation of detergent to remove residual soil; (3) an intermediate cold water rinse to flush out detergent; (4) the circulation of disinfectant to kill any residual microorganisms; and (5) a final cold water rinse to flush out disinfectant. This can be varied, for example, by the use of detergent-sterilizers to replace stages (2) to (4), but the fundamental principles involved in conventional cleaning are strictly adhered to. Apart from the chemical effects of the detergents and disinfectants, the mechanical force generated by the flow of fluids through pipes or from spray heads assists in the removal of soil from food contact surfaces; in the case of pipes a velocity of *ca* 1·5 m/s is required to obtain the desired turbulence.

9.9.1.   CIP Systems

CIP may be practised where a closed circuit system is involved, this usually consisting of pipework, valves, pumps and a suitable vessel or reservoir; in the simplest systems the food processing vessel can act as the detergent reservoir. CIP is commonly used for pipelines, tanks, vats, heat exchangers and homogenizers but other food processing equipment can be suitably designed to enable CIP systems to be introduced.

There are two basic CIP systems which can be employed, designated 'single use' and 'recovery', although elements of these two systems are often combined; the relative merits of the systems are discussed by

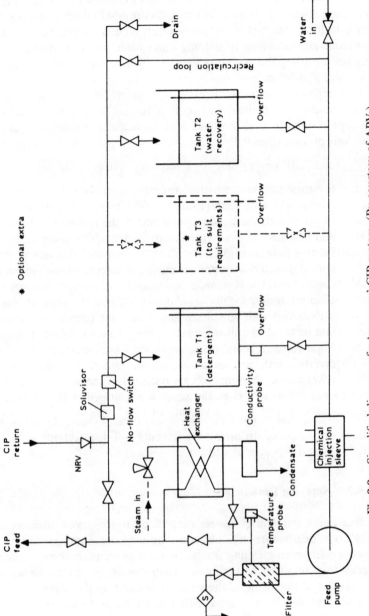

Fig 9.9. Simplified diagram of a 'recovery' CIP system. (By courtesy of APV.)

Tamplin (1980). In the 'single use' system the liquids are used only once and then discarded; this system may be operated either manually or automatically. In the 'recovery' procedure which is operated automatically solutions are recovered to holding tanks for re-use in subsequent cleaning operations. This system is best used where the degree of soil is not excessive and for more complex installations. In addition to the essential CIP equipment previously mentioned, such recovery systems should include timing and temperature-control equipment, filters, dosing units, storage vessels for both hot and cold fluids and the necessary recording instrumentation (Fig. 9.9).

A typical CIP programme for a recovery system could be:

1. Pre-rinse (5 min) with cold water from recovery tank.
2. Alkali detergent wash (15 min at 80°C); the first part is discarded to drain with the residual rinse water, the remainder is circulated and returned to a detergent recovery tank for further use.
3. Intermediate rinse (3 min) with cold water from the mains; the first part is diverted with any remaining detergent to the detergent recovery tank, the remainder is returned to the water recovery tank.
4. Cold sodium hypochlorite solution circulated (10 min); the first part is discarded with remaining rinse water, the remainder is circulated and returned to a disinfectant recovery tank for further use.
5. Final rinse (3 min) with cold water from mains; the first part is diverted with any remaining disinfectant to the disinfectant recovery tank, the remainder is returned to the water recovery tank until full when the excess water is discharged to the drain.

It is obvious that in these systems a very careful check should be maintained on the concentrations of detergents and disinfectants in the holding tanks and this is provided for.

9.9.2. Spraying Devices

Large tanks and similar vessels are difficult to clean and disinfect manually and, furthermore, it is obviously uneconomical and inefficient to fill such vessels with cleaning fluids. In these cases permanent or portable sprays can be fitted to the vessel when the design should be such that every part of the inside of the vessel is reached by the spray.

Spraying devices can have either fixed heads ('spray balls') or can revolve ('rotating jets') (see Fig. 9.10). The two types are compared

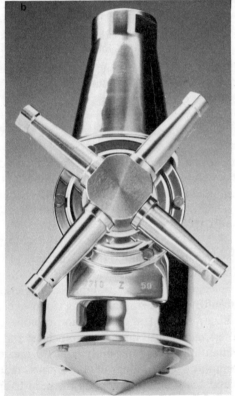

Fig. 9.10. Spraying devices. (a) A range of stainless steel spray balls. (By courtesy of APV.) (b) Hygienically designed rotating jet made of stainless steel and teflon. (By courtesy of Toftejorg Ltd.)

Food Microbiology and Hygiene

Table 9.7
Comparison of Spray Balls and Rotating Jets[a]

Spray balls	Rotating jets
Cheap	Much more expensive
Durable, no moving parts	Reliability variable and maintenance sometimes complex
Unaffected by high temperatures	Some cannot withstand high temperatures
Self-cleaning and self-draining	Not always self cleaning and self-draining
Can be made completely of stainless steel	Non-stainless steel parts
Continuous coverage in operation	Intermittent coverage in operation
Negligible effect when single hole blocked	With some nozzles blocking has serious effect
Radial throw limited and low direct impingement action	Long radial throw and higher direct impingement action
High flow rates	Lower flow rates

[a]From Tamplin (1980).

in Table 9.7 and, as can be seen, both types have their advantages; however, whatever the nature of the installation it is important to ensure good drainage in order to avoid the accumulation of CIP fluids in the vessel.

9.9.3. Benefits of CIP

In comparison with manual cleaning, the advantages of CIP include: (1) reduced labour costs; (2) more economic operation with optimum use of cleaning and disinfection solutions; (3) improved hygiene standards with cleaning and disinfection programme being followed exactly; (4) increased plant utilization with fast clean-up and rapid re-use possible; (5) fewer leaks and less mechanical damage to pipelines and equipment due to continual dismantling and re-assembly; and (6) greater safety by (a) reducing the amount of handling of hazardous materials such as stronger alkalis and acids, and (b) avoiding the need to enter and manually clean larger vessels. However, the installation of a fully automatic CIP system entails considerable capital investment although the capital will undoubtedly be recovered provided a suitable system is installed.

9.10. MECHANICAL AIDS FOR CLEANING

Today much food processing machinery and factory walls and floors are cleaned manually but there are many mechanical cleaning aids available which lighten the task. Personnel involved in cleaning should be able to complete the task as quickly as possible and with the minimum of effort; a full range of cleaning equipment should be provided therefore so that cleaning is facilitated by using the ideal tool for the job.

9.10.1. Pressurized Steam

As already mentioned, steam itself is not a cleaning agent but is used primarily for disinfecting metal surfaces. However, if sufficient pressure can be provided (i.e. at least 100 psi, or, more correctly, *ca* 700 kPa where 1 psi = 6·9 kPa approximately) steam guns can be employed to blow off debris. Unfortunately the mist created makes it difficult for the operative to assess the efficacy of the procedure; furthermore, aerosols may be created which might be potentially hazardous if pathogenic bacteria are contained within the droplets. Steam guns can also be dangerous to personnel and machinery if used improperly whilst low pressure steam hoses should not be used.

Steam guns can often be adjusted to operate as steam lances and are available in forms which mix steam with detergent solution or water at the required concentrations; these are widely used in the food industry and have the advantages of low maintenance costs and durability.

9.10.2. Hydraulic Devices

Low pressure water hoses are of limited value and their use should, at best, be restricted to floor cleaning. High pressure jets with nozzle pressures of 300–1200 psi are extensively used and abused. Although they are able to remove soil by physical force from otherwise inaccessible parts of machinery it is inevitable that some debris remains undisturbed; it would be far better if machinery were designed without inaccessible parts! High pressure water jets are usefully employed in the cleaning of floors, some wall surfaces and the external parts of certain types of suitable equipment, but problems with aerosols can limit their use. If required, all high pressure equipment should have easily interchangeable nozzles suitable for the various tasks to be performed.

9.10.3. Compressed Air

Compressed air can be used as the pressure source for water jets but its principal use in cleaning is as compressed air jets which are employed for removing dry powders and soil from equipment surfaces. Such jets have the advantage of low maintenance costs and durability but their use is restricted by the limited availability of compressed air and they have the disadvantage of spreading rather than eliminating dry materials.

9.10.4. Ultrasonics

Ultrasonic cleaning is a rather expensive and noisy technique that is sometimes used for smaller and more delicate items of equipment including plastic ware which would otherwise be hard to clean or which would be damaged by conventional cleaning techniques. Items to be cleaned are immersed in tanks containing a detergent solution at 60–70°C. An ultrasonic generator converts mains electricity to electrical energy at a high frequency (30 000–40 000 cycles/s) and transducers convert the energy into ultrasonic mechanical vibrations. These vibrations cause millions of microscopic vacuum bubbles to form and implode in the detergent solution. This process, known as 'cavitation', is responsible for the cleaning action.

9.10.5. Portable Cleaning Machines

There are many types of electrically driven portable machines available which perform a variety of tasks. Care must be taken when purchasing these machines as they are often expensive; trial runs should be arranged to assess their suitability. Machines should be mechanically and electrically robust but it is important to ensure that back-up maintenance is guaranteed and spares are readily available.

Vacuum cleaners which can operate under wet or dry conditions are used extensively and they are generally preferable to compressed air jets since, as mentioned earlier, the powdered soils and dust are contained. Other machines combine vacuuming with floor brushing, wet scrubbing or polishing (Fig. 9.11). Brooms, squeegees and rubber scrapers can still be used in the clearing up of soils during processing but only limited areas can be cleaned in this way. Portable pressure cleaners are still widely used; they can apply hot or cold water or detergent solution through a variety of nozzles and some can be used for foam cleaning.

Larger motor-driven vehicles can be used as sweepers in the transport reception and dispatch areas and in the surrounds generally.

9.10.6. Fixed Cleaning Systems

As with a fixed vacuum cleaning system, permanently sited cleaning stations can be installed at strategic points in the factory to provide supplies of pressurized steam or water as well as cleaning and disinfecting

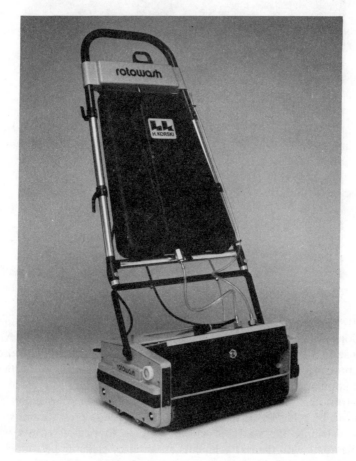

Fig. 9.11. A wet scrubber-vacuum combination machine suitable for use in smaller areas or confined spaces. (By courtesy of Rotowash UK Ltd.)

Fig. 9.12. Variable pressure cleaning unit. (By courtesy of KEW Industry Ltd.)

solutions which can be utilized by attaching the appropriate devices. Because of the possible deleterious effects of the indiscriminate use of high pressure cleaning systems variable pressure systems are becoming more popular. By interchanging the spray lance or other suitable attachment (not illustrated) a range of water pressure: water volume combinations can be achieved. Fixed pipes attached to these pressure units (Fig. 9.12) distribute the water or cleaning fluids to run-off points where flexible hoses with their attachments are used by cleaning personnel.

An extension of this system is one (i.e. the ring main system) where cleaning fluids are pumped from a central point to all parts of the factory; compressed air, steam and even a variable pressure system, can be provided in a similar way. Outlets are supplied at strategic points to which cleaning equipment can be linked.

9.11. FOAM CLEANING

Foam cleaning has become popular in recent years for walls, floors, inaccessible areas, vehicles and equipment with large food contact surface areas. In this form of cleaning a foaming agent is added to the detergent formulation to produce a long-lasting thick foam which enables the cleaning agents to have an extensive contact time with the soil; this is enhanced by the adhesive qualities of the foam which is even retained on vertical surfaces provided they are dirty. Thus once the foam cleaner has been applied the operative can move on, leaving the foam to work over a long period without further attention; in this way considerable time can be saved. Areas of 25 m²/min can be covered, the foam being left for 10–20 min, depending on the extent of soiling, before being rinsed off. Bactericidal agents are usually included in the rinse solution, QACs being normally used for this purpose. Other advantages of foam cleaning are that treated areas are easily recognized and that less cleaning solution is required since 1 part of water becomes 10 parts of foam. Thus foam cleaning can be usefully employed in dry foods factories, with the foam being removed by vacuum rather than by rinsing.

Limitations of this form of cleaning are the need for a pressure generating system, the necessity to ensure that the foam remains wet and the need to use somewhat higher detergent concentrations since the foams are used at room temperature.

Gel cleaners work on the same principle as foam cleaners but here a gelling agent replaces the foaming agent in the formulation.

9.12. CLEANING SMALL EQUIPMENT

Cleaning of small equipment or easily handled components of larger equipment should again follow the five basic steps of cleaning (see p. 344). Thus in manual cleaning a five-tank wash system (gross debris removal—warm detergent solution—rinse—disinfectant solution—final rinse) is justified if the disinfection stage is deemed necessary; at the very least the first three stages should be performed. Automatic washing equipment can be installed as an alternative to manual washing but the same basic cleaning sequence must still be observed.

9.13. PAPER AND FABRIC WIPERS

Wipers have become more popular as cleaning aids in the last few years partly as a consequence of the increasing costs of water supplies. Disposable wipers, typically made of recycled paper, are far more robust than previously and have a relatively high wet strength; they should only be used once so that cross-contamination problems are avoided. There is an ever increasing range of fabric reusable wipers becoming available which are made of tough, long-lasting materials such as polypropylene, cellulose and rayon; all exhibit a high degree of absorbancy and such wipers can be machine washed or sterilized. Wipers can be impregnated with antimicrobial agents (e.g. QACs) or detergents (e.g. non-ionics) to facilitate cleaning, although these wipers are normally disposed of at the end of the working shift. Typically wipers are colour coded so that specific tasks or working areas can be defined, thus minimizing the risk of cross-contamination.

BIBLIOGRAPHY

Cremieux, A. & Fleurette, J. (1983). Methods of testing disinfectants. In: *Disinfection, Sterilization and Preservation*, 3rd edn. Ed. S. S. Block, Philadelphia, Lea and Febiger.

Davis, J. G. (1968). Chemical sterilization. *Progress in Industrial Microbiology*, **8**, 141–208.

Elliott, R. P. (1980). Cleaning and sanitizing. In: *Principles of Food Processing Sanitation*. Eds A. M. Katsuyama and J. P. Strachan, Washington, Food Processors Institute.

Graham-Rack, B. & Binsted, R. (1973). *Hygiene in Food Manufacturing and Handling*, 2nd edn. London, Food Trade Press Ltd.

Parker, M. E. & Litchfield, J. H. (1962). *Food Plant Sanitation*. New York, Reinhold Publishing Corp.

Thompson, R. (1971). Detergents and biocides. In: *Hygiene and Food Production*. Ed. A. Fox, Edinburgh, Churchill Livingstone.

Troller, J. A. (1983). *Sanitation in Food Processing*. New York, Academic Press.

REFERENCES

Blood, R. M., Abbiss, J. S. & Jarvis, B. (1981). Assessment of two methods for testing disinfectants and sanitizers for use in the meat processing industry. In: *Disinfectants: Their Use and Evaluation of Effectiveness.* Eds C. H. Collins, M. C. Allwood, S. F. Bloomfield and A. Fox, London, Academic Press.

British Standard 541 (1934). *Technique for Determining the Rideal-Walker Coefficient of Disinfectants.* London, British Standards Institution.

British Standard 808 (1938). *Modified Technique of the Chick-Martin Test for Disinfectants.* London, British Standards Institution.

Council of Europe (1987). *Test Methods for the Antimicrobial Activity of Disinfectants in Food Hygiene.* Strasbourg, Council of Europe Publication.

Cowen, R. (1978). Kelsey-Sykes capacity test: a critical review. *Pharmaceutical Journal,* **220**, 202–4.

Croshaw, B. (1981). Disinfectant testing with particular reference to the Rideal-Walker and Kelsey-Sykes tests. In: *Disinfectants: Their Use and Evaluation of Effectiveness.* Eds C. H. Collins, M. C. Allwood, S. F. Bloomfield and A. Fox, London, Academic Press.

Hoffman, P. N., Death, J. E. & Coates, D. (1981). The stability of sodium hypochlorite solutions. In: *Disinfectants: Their Use and Evaluation of Effectiveness.* Eds C. H. Collins, M. C. Allwood, S. F. Bloomfield and A. Fox, London, Academic Press.

Holah, J. T. & Thorpe, R. H. (1990). Cleanability in relation to bacterial retention on unused and abraded domestic sink materials. *Journal of Applied Bacteriology,* **69**, 599–608.

Kelsey, J. C. & Maurer, I. M. (1974). An improved (1974) Kelsey-Sykes test for disinfectants. *Pharmaceutical Journal,* **213**, 528–30.

Kelsey, J. C. & Sykes, G. (1969). A new test for the assessment of disinfectants with particular reference to their use in hospitals. *Pharmaceutical Journal,* **202**, 607–9.

Lisboa, N. P. (1959). A tube test for evaluating agents possessing both detergent and sterilizing properties. *Proceedings of the XVth International Dairy Congress, London,* **3**, 1816–20.

Schwach, T. S. & Zottola, E. A. (1984). Scanning electron microscopic study on some effects of sodium hypochlorite on attachment of bacteria to stainless steel. *Journal of Food Protection,* **47**, 756–9.

Tamplin, T. C. (1980). CIP technology, detergents and sanitizers. In: *Hygienic Design and Operation of Food Plant.* Ed. R. Jowitt, Chichester, Ellis Horwood.

Chapter 10

CLEANING AND DISINFECTION: PRACTICAL APPLICATION

10.1. AIMS

In practical terms the aim of an acceptable cleaning programme (including, where necessary, a disinfection element) must be to restore equipment and especially food contact surfaces to the required level of cleanliness. For non-food contact surfaces and floors, walls and ceilings the standards demanded when cleaning will be lower than for the food contact surfaces themselves.

Cleanliness as applied to food plant embraces physical, chemical and microbiological factors. Physical cleanliness implies the absence of *visible* soils on surfaces but for food contact surfaces chemical cleanliness (i.e. absence of residual chemicals including cleaning agents and disinfectants) and microbiological cleanliness (i.e. residual microorganisms so low in numbers that they could not significantly affect the quality of foods coming into contact with those surfaces) are also demanded.

Any standards laid down must be achieved unfailingly because if there is any deviation from the standards an insidious build up of bacteria and other microorganisms may occur over a period of time. In such a situation an increasing residuum of bacteria can be left unnoticed on food contact surfaces; as a result there is a tendency for an indigenous flora of bacteria to be created which is associated with items of equipment which have been inadequately cleaned. This, in turn, causes the bacteriological quality of the food to deteriorate progressively over a number of days from the commencement of substandard cleaning. Even with an inadequate cleaning programme bacterial numbers will eventu-

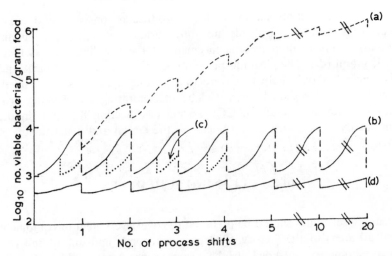

Fig. 10.1. Effect of different cleaning programmes on the bacteriological quality of food under process conditions. (a) Product with excessive build up of bacteria and with unsatisfactory cleaning programme at end of shift. (b) The same product showing excessive increase in bacterial numbers during shift but with efficient end of shift clean. (c) The same product with a mid-shift clean. (d) Relatively uniform product with little change in bacterial numbers.

ally stabilize in the food but the quality of the final product may be far inferior to that which can be achieved with an effective cleaning programme. This point is illustrated in Fig. 10.1, graph (a).

In certain more hazardous processes a build up of bacteria on food contact surfaces (e.g. cutting blades) is inevitable during production runs but such increases should be restricted by frequent cleaning of the affected surface so that the desired uniformity of the product, so important in quality assurance, can be maintained. Figure 10.1, graph (b), illustrates a situation where too great a change in the bacteriological quality of the product is found suggesting the need for additional cleaning during the shift, the benefits of which are illustrated in Fig. 10.1, graph (c). The reverse effect can sometimes be seen with liquid foods such as milk. Here the food itself can flush out bacteria from poorly cleaned surfaces and where this occurs the first milk samples down the line will be the most heavily contaminated; these will be followed by steadily improving samples until a satisfactory and stable product is obtained.

Thus the aim at all times must be to produce as consistent and as acceptable a product as possible, not only during a single production run but also daily from the commencement of the installed process (Fig. 10.1, graph (d)). The cleaning programme, which must be efficient and economical, should help to ensure that product quality is maintained; further control should be achieved by incorporating the cleaning programme into an overall HACCP system (see Section 8.2) but, in the final analysis, so much is dependent on the design and construction of food plant.

10.2. CLEANING SCHEDULES

Cleaning schedules should be prepared to cover every item of equipment and all areas of the factory. Toilets, cloakrooms, and dining and rest areas should be included in the cleaning programme. The cleaning schedules should contain the following essential information: (1) the method of cleaning which should include details (and possible hazards) of the cleaning and disinfecting agents to be used, together with the quantities and dilutions required and the exact method, time and temperature to be used when applying the solutions; (2) the sequence of cleaning which should be so arranged that recontamination of previously cleaned equipment is avoided: (3) the extent to which equipment should be dismantled and whether the assistance of a fitter is required—obviously equipment should be installed, where possible, that is readily dismantled; (4) details of likely 'trouble spots' where extra care should be taken due to equipment design defects; (5) the time which should be taken over individual cleaning tasks; (6) the frequency of cleaning; and (7) the person responsible for each cleaning task and the person whose responsibility it is to see that all the tasks have been properly performed.

Where non-automatic cleaning is used, proper measuring devices should be provided to prevent wastage of cleaning and disinfecting agents. Sufficient time must always be allowed for cleaning so that it can be performed diligently; no matter what pressures there are from production, the maintenance of product quality is crucial so that cleaning time must be regarded as sacrosanct. The same philosophy must apply when considering the frequency of cleaning which must again reflect the need for uniformity in quality; thus, as was mentioned in Section 10.1, certain items or parts of equipment may require extra cleaning during a shift, or a continuous cleaning system may have to be installed to prevent significant build up of bacteria.

10.3. USE OF LABOUR

Traditionally cleaning has been regarded as a menial occupation carried out by the lowest quality of available labour, often with inadequate cleaning aids. Fortunately, in the food industry at least, this view is changing although a certain stigma is often still unjustifiably attached to this important job; in fact, no facet of food production is any more important than the cleaning and disinfection of equipment, and cleaning should be regarded as a technology rather than a necessary chore.

Employees selected for cleaning duties can be from either of two sources: they may comprise a sanitation squad hired or trained specifically for that purpose, or they may be production workers in which case they are often kept on at the end of a shift to do their own cleaning. Both arrangements have their proponents in the food industry, and the pros and cons are briefly reviewed.

The advantages of a sanitation squad are that they come into work fresh and alert so that more thorough cleaning might be expected. Furthermore, they can be trained to strip the equipment where necessary rather than have to wait for a fitter; when operatives are responsible for their own cleaning it is often difficult to obtain a fitter where the equipment requires dismantling. Sanitation squad training should include an in-depth understanding of cleaning equipment and its maintenance, and squad members should have some understanding of the principles of microbiology and food hygiene. Sanitation squads can be employed either on a continuous basis, or to work during short breaks and at the end of a production run, or to work at the weekend when they may also be used for more general cleaning tasks. Perhaps the main disadvantage of a sanitation squad is that the time taken to complete the cleaning programme is protracted because the numbers employed must be limited on financial grounds. As a consequence, food debris may be left unattended for many hours and thus may become increasingly more difficult to remove. A second problem is that labour may be difficult to engage in times of relative labour shortage as the hours worked are often unsociable; this difficulty can be partially overcome by giving prestige to the sanitation squad backed up by financial rewards for maintaining high standards of cleanliness. At least in larger food processing companies it is likely that sanitation squads will be employed to clean floors, walls, ceilings and factory surrounds, and in these areas they must be regarded as indispensable.

One of the advantages of operatives being made primarily responsible for cleaning their own food processing equipment is that they know the equipment intimately and should be able to undertake the task efficiently as soon as production ends. They should also more readily recognize incipient equipment faults and, perhaps most important of all, they ought to take a pride in the appearance and cleanliness of the equipment and its contribution to a high standard of product. There is no doubt that this view is widely held in the food industry but clearly much depends on the quality of labour employed; furthermore, indifference to the task can develop due to tiredness at the end of a long shift or due to complacency and the question can be asked whether it is fair to expect employees to perform such an important task at the end of the working day. The system can be useful where two or more shifts are operating; here the operative is cleaning to a standard acceptable to the next employee on the machine who would quickly complain if faults were found. Again, where double shifts are operating with a minimum delay between them, it could be arranged that the incoming shift is responsible for cleaning before production commences.

If operatives are employed for cleaning it is important that they, too, are given some basic training in the use of cleaning equipment and are encouraged to appreciate the importance of good cleaning practices; they should also fully understand that food processing equipment is expensive and should not be mishandled. It is important that cleaning is not left to inadequately trained employees, whosoever is finally used.

10.4. THE ROLE OF MANAGEMENT

Senior management have a dominant role to play in relation to food hygiene since it is they who are ultimately responsible for budgeting decisions relating to the purchase of suitable food processing equipment and cleaning aids; they, too, are responsible for deciding the frequency of and time allowed for cleaning and for promoting general attitudes to hygiene within the company. If senior staff are seen to be genuinely concerned about hygiene this concern will undoubtedly be transmitted to all employees.

Final responsibility for hygiene should be assumed by a member of senior management such as the company quality controller or an equivalent; responsibility should not be delegated to junior management.

The organization of hygiene on a departmental basis will depend upon the size of the company but provided it is of sufficient size hygiene interests are best served by creating a separate department for this purpose; at very least hygiene should be accommodated as a distinct subsection within the quality control department. The company hygiene officer will be a member of junior or middle management but should be answerable directly to the company quality controller. The hygiene officer should be well versed in the relevant principles of microbiology, chemistry, food technology and food hygiene and, at least in larger companies, should be employed full-time on hygiene duties; all too often hygiene officers have diverse responsibilities which divert attention from their principal duty. One area where they can be usefully employed is in training in their subject so that all employees at least understand the basic principles of hygiene.

The cleaning of food equipment must always be monitored by bacteriological checks so that close liaison between the hygiene section and the quality control or microbiology laboratory is important. The results of the checks should be made available to all employees and where operatives are primarily responsible for cleaning their own equipment competitions could be introduced between shifts or sections, based on the results obtained. Food hygiene, even cleaning, should be made interesting to the operatives and better results will be achieved if all staff are kept well informed. Obviously management should familiarize themselves with the results of bacteriological line checks and should arrange regular meetings with all those involved, both in cleaning and production, so that problems can be fully aired and discussed.

It must be remembered that good hygiene extends to habits in toilets, cloakrooms, dining and rest areas and the cleaning of such areas; it should be the responsibility of the hygiene officer to check these facilities regularly.

10.5. MISCELLANEOUS PROBLEMS ENCOUNTERED

Efficient cleaning is fraught with a multitude of difficulties many of which have been discussed already. These include: (1) the faulty design and incorrect siting of food processing equipment; (2) recontamination of equipment that has been previously cleaned; (3) inadequate time for cleaning or cleaning too infrequently; (4) insufficient or poor-quality labour employed; (5) misuse of cleaning and disinfecting agents by gross variation from the recommended concentrations; and (6) wrong mental attitudes of management and operatives to an important operation.

Additional problems may be encountered if the factory hygiene officer is of insufficient status to enforce the recommended procedures or if cleaning supervisors or operatives are absent and nobody is detailed to replace them, i.e. there is a lack of communication amongst the various grades of employee. These problems are exacerbated when there is a high turn-over of labour resulting in new employees being given insufficient training.

With regard to the cleaning methods themselves, these are sometimes altered without full recognition of the possible long-term detrimental effects which the change may have on the physical state of the equipment or on the efficacy of action on the residual soils or microorganisms. Certain salesmen can be very persuasive and produce glossy handouts containing impressive figures extolling the virtues of their products; the figures may well represent results that were obtained under test conditions totally different from the environment in which the product is to be used commercially. It is important therefore to always test the cleaning or disinfecting agent under typical commercial conditions before introducing it on a large scale in the factory. On the other hand, it must never be assumed that cleaning or disinfecting methods are effective because they always have been in the past; checks should be made occasionally to confirm their continued reliability.

Finally, mechanical cleaning aids may break down and have to await repairs; when this happens it puts additional pressure on the remaining working equipment which is, in turn, more liable to fail. Repairs may in consequence be hurried or inadequate and the whole cleaning programme may be jeopardized; this situation is often associated with a reluctance of senior management to spend sufficient money on cleaning aids.

10.6. ASSESSMENT OF CLEANING EFFICIENCY

The efficiency of process line sanitation can be checked by visual inspection or by using microbiological techniques. Visual inspection is a simple but rough method which cannot determine the microbiological cleanliness achieved. Much depends on the care taken by the inspector. With experience the inspector may know where to look for signs of inadequate cleaning but residual soils do vary in their visual detectability (e.g. some soils are not visible if there is a film of water on the cleaned surface) and high intensity lighting must be directed on surfaces during inspection. In spite of these shortcomings visual inspections are worthwhile provided they are performed assiduously. They can be carried out during or immediately after cleaning or even shortly before the start of the next production run. Inspections should be performed randomly so that employees involved in cleaning are unaware when they are due. A check-list of various items of equipment should be prepared and items, once cleaned and inspected, can be given a cleanliness rating which can be compared with earlier data. All the findings should be recorded in an inspection report. Additionally, records of the cleaning materials, dilutions and application times used should be kept where applicable. If equipment has not been adequately cleaned there should be time available for corrective action before processing is restarted; particular care should be taken with the subsequent inspection of equipment given low cleanliness ratings.

Other visual tests for the cleanliness of surfaces are rather less subjective and these, together with more sophisticated tests, have been reviewed by Kulkarni *et al.* (1975).

The most commonly used tests are microbiological and these have been discussed in Chapter 4. When applied to surfaces, these tests typically involve estimations on given surface areas of the total numbers of viable bacteria at specific incubation temperatures, indicator organisms (i.e. coliforms, *E. coli*, etc.) and, where warranted, specific food spoilage, or food poisoning bacteria. In general, however, estimations are limited to 'total numbers' as this is the most sensitive guide to cleanliness. The bacteria are removed from surfaces by means of sterile swabs, by rinsing with a known volume of sterile diluent or by agar contact methods. Because microorganisms are often unevenly distributed on equipment surfaces as large an area as possible, but including less accessible points, should be sampled; for this reason swabbing methods have found particular favour. The inherent problem with standard microbiological methods is that test results are normally unavailable until after a 48 h period due to the need for incubation of the test samples. However, very rapid if rough results are obtainable if based on any residual food on contact surfaces. Thus it is possible to test for the presence of retained protein using special strips which change colour depending on the amount of residual protein, the colour being measured colorimetrically (Anderson *et al.*, 1986).

It should not be forgotten that the processed foods themselves may also be used to check cleaning efficiency. Foods first in contact with cleaned surfaces are most liable to pick up residual bacteria so that higher than expected counts on such foods may be indicative of inadequate cleaning. This principle is often used to monitor cleaning where liquid foods are being produced. Here samples are taken at regular intervals early in a production run and if there is a gradual reduction in the bacterial count from the first sample onwards, rather than a fairly constant set of results, cleaning has been inadequate; this is the result of the liquid flushing out the residual bacteria as has been mentioned earlier in this chapter.

In conclusion it is pertinent to stress that when monitoring techniques show up areas where cleaning proves to be troublesome it is vitally important to concentrate efforts in those areas and not to accept evidence of inefficient cleaning complacently.

BIBLIOGRAPHY

Dunsmore, D. G., Twomey, A., Whittlestone, W. G. & Morgan, H. W. (1981). Design and performance of systems for cleaning product-contact surfaces of food equipment: a review. *Journal of Food Protection*, **44**, 220–40.

REFERENCES

Anderson, M. E., Huff, H. E., Marshall, R. T. & Naumann, N. D. (1986). A simple qualitative method for detecting cleanliness of food contact surfaces. *Journal of Food Protection*, **49**, 342–6.
Kulkarni, S. M., Maxcy, R. B. & Arnold, R. G. (1975). Evaluation of soil deposition and removal processes. An interpretive review. *Journal of Dairy Science*, **58**, 1922–36.

Chapter 11

WATER AND WASTE TREATMENT

11.1. INTRODUCTION

It is essential in all food processing establishments to have an adequate supply of water which is utilized in a variety of operations. Much of the water required will have to be of potable quality (i.e. it is drinkable water and as such is free from bacteriological or chemical hazards and has no unpleasant taste or smell); this water is used, for example, in transporting foods and in cleaning equipment. In contrast, water used for heating and cooling, which does not come into direct contact with the food, need not be of such a high quality. For reasons of economy, water is often re-used within the factory in a sequence of operations so that when it is finally discharged it may be of a very poor quality and be very heavily contaminated with microorganisms, food wastes or soil. Such water may have to be purified before it can be accepted into drainage systems or rivers and thus the concept of waste disposal within the food industry can be of considerable importance.

Waste materials from food processing factories are very variable in composition and may include solids and liquids, foods high in protein or high in carbohydrates, fatty foods and chemical residues used in both processing and cleaning. Such materials are generally not hazardous and are biodegradable but their high organic content can pose special problems; in particular, such wastes can substantially increase the *biological oxygen demand* (BOD) of the water. The BOD can be defined briefly as the quantity of dissolved oxygen consumed by chemical or microbiological action when a sample of water is incubated in the dark for a given

397

time (usually 5 days) at a given temperature (20°C). Since oxygen is required for the oxidation of organic matter, the BOD gives a rough indication of the quantity of such matter present in the sample.

Effluent with a high BOD can create pollution problems if discharged directly into a river; the consequent removal of oxygen can readily upset the whole ecological balance and may even result in the death of fish over lengthy stretches of a stream or river. Clearly the standards for trade effluents required by regional or local water authorities will vary; much will depend upon where the effluent is to be discharged, the degree of dilution in the receiving waterway and whether municipal facilities are available for its reception and treatment. It is at least likely that screening of solid materials and removal of grease and fat may be required before wastes can be discharged into sewers. At the other extreme, where no municipal facilities are available, the food manufacturer will be required to set up a handling, treatment and disposal system. Such a system may be required to treat, either by chemical or biological methods, the wastes from the factory and this can prove to be very expensive. It may be possible to offset some of this expense by recovering products from the waste which can be sold as 'by-products'. One thing is certain, wastes of whatever description should be disposed of as rapidly as possible so that pests are not encouraged and offensive odours are not allowed to develop.

Many of the above issues are considered in more detail below.

11.2. WATER: ITS POLLUTION, PURIFICATION, DISTRIBUTION AND EXAMINATION

As stated previously, potable water should not contain harmful bacteria or chemicals, or have an unpleasant taste or smell. The necessary legal requirements are included in the Water Supply (Water Quality) Regulations (1989) which embody the legislation included in the 1980 EC Drinking Water Directive (80/778/EEC).

Rain water is relatively pure but by the time it is collected as surface water, in reservoirs or natural upland lakes fed by rivers, it will have picked up microorganisms, organic matter and other impurities (e.g. dissolved minerals and iron) making purification essential. Rain water may also be precipitated and collected as ground water which is recoverable from wells or springs. Such water, although normally containing a high concentration of dissolved minerals, is usually very much purer than surface water as it contains fewer microorganisms and less organic matter; however, purification may still be necessary. Where ground water supplies are available, economic considerations usually justify their use especially where large quantities of water are required. It may be that such water can be used without purification but even if purification is necessary net costs may still be favourable.

Harmful microorganisms associated with polluted water supplies have included those causing cholera (not now a problem in developed countries), typhoid, salmonellosis, bacillary and amoebic dysentery, and many human enteric virus infections including those causing infective hepatitis and poliomyelitis. Fortunately standard water treatments should effectively eliminate these hazards although in the last few years cryptosporidiosis, a gastroenteritis caused by a protozoan parasite, has been emerging as a potential threat even where drinking water has been treated conventionally. Perhaps of greater concern is the presence in water of many of the chemical pollutants which are not removed by standard water treatments and which are toxic to man or aquatic life, and may be non-biodegradable. Amongst these chemical pollutants are fertilizers, insecticides, herbicides and detergents, although nowadays all of these organic compounds should be biodegradable; however, nitrate levels are now a cause of concern as nitrates may be converted in the body into nitrosamines which have been implicated as carcinogens (see p. 161). Other pollutants include mercury, cadmium and lead which are not harmful when consumed in very low concentrations but which are not metabolized and may accumulate to dangerous levels when consumed over long periods of time.

11.2.1. Water Purification

The treatment that water receives to render it potable depends upon the initial level of pollution and it is preferable therefore to obtain natural water which has as low and consistent a level of pollution as possible. Surface waters are collected and held in reservoirs before treatment and during this period of storage many beneficial changes occur in the quality of the water. The microbial population is substantially reduced as is the extent of general organic pollution; much of this is due to natural sedimentation and flocculation (see below), which finally induces a partial clarification of the water. However, recontamination can occur due to the presence of birds, seagulls being a particular nuisance as they are frequently carriers of human intestinal pathogenic bacteria; algae, too, can encourage the growth of bacteria if allowed to flourish in the warm summer months, the bacteria thriving on decomposing algal cells.

For relatively clean waters the treatments following initial storage may be limited to quickly filtering the water through coarse sand and chlorination. For most waters more complex treatments are necessary and these usually involve *coagulation* and *flocculation*, and *slow sand filtration*; additional pretreatments are used for the poorest quality waters.

11.2.1.1. Pretreatments
Following initial storage, pretreatments may include *micro-straining, rapid sand filtration, biological sedimentation and prechlorination*. The object of the first three processes is to remove larger particulate matter including tastes and odours whilst prechlorination reduces the numbers of viable bacteria to the levels found in better quality natural waters.

11.2.1.2. Coagulation and Flocculation
When coagulants such as aluminium sulphate, sodium aluminate, ferrous and ferric sulphate, and certain polyelectrolytes are added to water, suspended materials are destabilized and become attracted to each other to form agglomerations of fine particles. The coagulated particles subsequently build up into flocs (*flocculation*) which are large enough to settle out fairly quickly. The floc provides a trap for smaller particles and these also settle out so that most of the suspended matter (e.g. clay, silt, microorganisms, general organic debris, etc.) can be removed by sedimentation, the clarified water being run off from the upper part of the tank.

11.2.1.3. Slow Sand Filtration
It is customary to combine the above processes with some form of sand filtration. All that is necessary at this stage is probably a fairly coarse sand filter so that filtering is rapid but the most widely used form is the slow sand filter. This employs a fine sand (0·2–0·4 mm diameter) together with anthracite and gravel in layers. The crucial area is limited to the top few centimetres of the sand layer where a zone of active bacteria, fungi and protozoa ('schmutzedecke') provides the filtering action; this necessitates the slow passage of water through the filter so that large filter beds are required. Suspended matter and bacteria are removed by the browsing microorganisms, protozoa being particularly important in this respect; in addition, natural tastes and odours are removed, and ammonia is oxidized to nitrate.

It is sometimes deemed necessary, for aesthetic reasons, to supplement the above processes by passing water through activated carbon filters to remove the last traces of organic materials including those responsible for undesirable tastes and odours; no improvement in the biological purity of the water is effected by this process.

11.2.1.4. Disinfection

This is the final stage in the water purification process and the one that is absolutely essential. The previous treatments merely improve the quality of the water to a standard where disinfection can be performed efficiently. The method of disinfection used is normally chlorination (see p. 362) which should kill any remaining cells and provide residual chlorine for the water distribution system. Ozone has also been widely used as a disinfectant in European countries; it suffers from the disadvantage that it is difficult to retain residual ozone in the distribution system.

11.2.2. Water Distribution

Water is of the highest quality after disinfection and before it enters the distribution system. Some deleterious changes, microbiological or chemical, may occur during distribution before the run-off point is reached. Bacteria can gain access to the system by various means, e.g. during pipeline repairs, from faulty jointing materials, from tap washers or by back siphonage caused by lowered pressures. Common sources of bacterial recontamination are static water lines where chlorine levels are reduced or non-existent, such recontamination manifesting itself in the form of biofilms developing along pipelines (Geldreich, 1989). Since many of the bacterial types developing are often derived from sewage or soil, it is essential that frequent bacteriological testing is carried out on samples taken from a variety of points in the distribution system.

11.2.3. Legionnaires' Disease

Legionnaires' disease is so named because of its association with an outbreak of severe pneumonia which occurred at the 1976 Annual Convention of the Pennsylvania American Legion. Although the exact cause of that outbreak was never proved conclusively it is now known that this infection is caused by a bacterium, *Legionella pneumophila*; this Gram negative rod can be readily isolated from natural waters and hot and cold water systems. The organism can survive normal water purification treatments to enter the mains supply and it can be readily isolated from tap washers and water. *L. pneumophila* apparently thrives in the presence of soluble iron and it has been suggested that corroded pipelines aid its spread (Mepham, 1989).

There is mounting evidence to indicate that legionellas are able to grow within free-living amoebas and other microorganisms (Lee & West, 1991). Hundreds of legionella cells can be formed within a single amoeba whilst cysts formed by the amoebas as part of their life-cycle have been shown to protect the legionellas from high concentrations of free chlorine (Kilvington & Price, 1990); thus the efficacy of water purification can be undermined.

Legionellas are carried in fine droplet aerosols and only in this form can they cause an infection by entering the lungs where, after an incubation period of 2–14 days, the pneumonia-like symptoms develop; Legionnaires' disease cannot be contracted by drinking contaminated water. Aerosols may emanate from water cooling systems, showers, flush toilets and even taps.

Typically larger outbreaks occur when the contaminated aerosol produced by a cooling tower enters the air-conditioning system. Thus two such outbreaks were reported in Bolton in late 1988 (Mitchell *et al.*, 1990). Thirty-seven cases of Legionnaires' disease were diagnosed whilst a further 23 patients had milder 'flu-like symptoms associated with a *L. pneumophila* infection. Of the 37 patients with Legionnaires' disease 33 worked or had contact with an engineering factory or its vicinity, whilst 21/23 with the milder symptoms worked at the factory. Ventilation methods varied at the factory but it was proved that it was possible for aerosol from one specific cooling tower (from which *L. pneumophila* was isolated at a concentration of 30 000 cells per ml water) to be spread by prevailing winds throughout the main factory.

In the above incidents it appeared that the cooling tower was treated regularly with biocides but this was clearly inadequate. To prevent such incidents it is important to erect cooling towers well away from ventilation ducts; disinfection of the whole cooling system must be performed and monitored regularly but, in addition, growth of protozoa and algae must be prevented. Finally, the importance of the water temperature in a distribution system should be stressed. Cold water lines should be insulated so as to maintain a temperature of below 20°C whilst hot water lines should maintain a temperature of above 50°C; hot water in storage tanks should be held at 60°C.

11.2.4. Bacteriological Examination of Water

Contamination by sewage is the greatest potential hazard for drinking water since sewage is the source of most enteric pathogens. As with foods it is impracticable to examine for a wide range of pathogenic microorganisms which may require sophisticated and lengthy isolation procedures; in addition, the numbers present in polluted water will be a fraction of the numbers of normal intestinal organisms present. Again, as with foods, 'indicator organisms' (see Section 4.3.7) are looked for and, if present, faecal contamination of the water must be presumed with the consequent possibility of intestinal pathogens being present.

The organisms used as indicators of faecal pollution in water are coliforms (and *Escherichia coli* in particular) and, where anomalous results have been obtained, faecal streptococci (enterococci) and *Clostridium perfringens* (DHSS, 1983). The examination methods used in food testing for the first two groups of organisms have been described previously (see Section 4.3.7); with water, membrane filtration has largely superseded multiple tube methods. In the filtration technique a measured volume of water (normally 100 ml), or a suitable dilution of it if heavily contaminated, is drawn by vacuum through a sterile membrane composed of cellulose esters, the membrane being supported on a sintered glass disc in an aluminium funnel (Fig. 11.1). The bacteria are retained on the grid-marked membrane which is removed from the filtration apparatus, placed face upwards on a suitable selective medium (e.g. MacConkey agar for *E. coli*) and incubated. The individual bacteria are nurtured by the underlying medium and form colonies on the surface of the membrane; by counting the colonies the number of bacteria in the water sample can be found.

If *E. coli* is recovered from a sample it means that pollution by sewage of human or animal origin has occurred. Since *E. coli* dies off in water, high recoveries indicate recent and heavy pollution whilst low recoveries indicate either recent and slight pollution or heavier but more distant pollution. Even the presence of coliforms in l00 ml chlorinated water is unacceptable, although in practice slightly lower standards are permitted in samples taken from the distribution system (DHSS, 1983). These include:

1. No sample of 100 ml should contain more that 3 coliforms.
2. No two consecutive samples to contain coliforms in 100 ml.
3. Coliforms should not occur in more than 5% of samples, provided that at least 50 samples per year are examined.

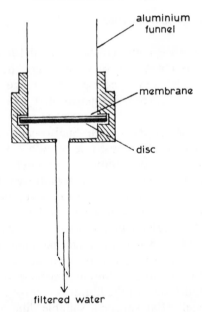

Fig. 11.1. Membrane filtration apparatus.

Clostridium perfringens is a normal inhabitant of the gut of man and many other animals and is consequently voided in small numbers in faeces. Examination for this organism in water is carried out where a considerable time may have elapsed between possible pollution and analysis since the spores produced by *C. perfringens* can survive in water for long periods; furthermore, they are relatively resistant to toxic materials so that waters containing residual chlorine may well contain such spores in the absence of vegetative cells.

In testing for *C. perfringens* it is customary to preheat the water at 75°C for 10 min to destroy non-sporing bacteria before adding aliquots of the water to appropriate quantities of the test medium (i.e. 50 ml water to 50 ml double strength medium, 5 × 10 ml water to 5 × 10 ml medium and 5 × 1 ml water to 25 ml single strength medium). The medium used is Differential Reinforced *Clostridium* medium (Gibbs & Freame, 1965). Samples of water containing a wide variety of *Clostridium* spp. will turn the medium black due to the reduction of sodium sulphite and the precipitation of ferrous sulphide. A loopful of any culture showing such a reaction after 48 h incubation at 37°C is inoculated into freshly boiled and cooled litmus milk (the heating drives off any dissolved oxygen). If *C. per-*

fringens is present the classical 'stormy clot' reaction will be observed after incubation as above; this reaction is typified by acidification and co-agulation of the milk, the clot being disrupted by gas production and possibly blown to the top of the test-tube.

11.2.5. Virological Examination of Water

The routine examination of water for viruses can only be attempted by the largest and better equipped water authorities at the present time as the techniques involved are complicated and time-consuming whilst results may be delayed for up to three weeks (Grabow, 1986). As yet, no standards have been laid down within the EC for virus numbers in drinking water although these will, no doubt, be introduced in due course since discussions are now taking place within the Community.

The correlation between counts of faecal indicator bacteria and their virus counterparts (i.e. the enteric viruses) is variable since water samples sometimes contain the latter in the absence of the former. This is due to the far greater stability of viruses in water; they may remain viable in water for up to 6 months and yet their actual numbers in human faeces ($< 1 \times 10^6$ per gram) are only a fraction of that attained by *E. coli* ($1{-}10 \times 10^8$ per gram).

The most dangerous viruses associated with the alimentary tract are the enteroviruses (including polio, cocksackie and echo viruses, the latter two causing viral meningitis and intestinal upsets), infectious hepatitis virus, rota-viruses, reoviruses, adenoviruses and norwalk virus; the last four groups cause fevers, respiratory infections and two forms of intestinal upsets respectively. Clearly all these viruses should be absent from potable waters.

Cultivation of viruses in the laboratory involves the inoculation of concentrated water samples into susceptible living cells. Chick embryos can be used for this purpose but more widely used are the tissue culture techniques in which viable cells, extracted from living tissues, are resuspended in a complex nutrient medium and placed in a suitable container. Typically monkey kidney cells are used in water analysis but they suffer from the weakness that they cannot detect some of the viruses of primary concern such as hepatitis A and rotaviruses. The kidney cells become attached to the sides of the container and form a monolayer which can be readily infected with viruses. The numbers of virus particles in a given sample can be estimated by measuring the number of plaques that develop after the sample has been incubated with the cell film for one to several days. The plaques represent the dead cells produced as a result of their infection by the virus ('cytopathic effect') and are shown up when the cells are stained to differentiate between those that are living and those that are dead. Thus, provided a correct sample concentration has been assayed, the number of virus particles in the sample can be determined. However, since many important water-borne viruses cannot be cultivated on cells, newer techniques are now being exploited for virus detection. These include immunofluorescence, ELISA and the use of gene probes (Gerba *et al.*, 1989). The last named detects viral nucleic acid, is highly sensitive and does not require cultivation of the virus; recently, the use of the polymerase chain reaction (PCR) to detect even a single virus particle in water has been considered (West, 1991).

As indicated above, concentration of the water sample is normally necessary due to the low density of viruses especially in drinking water. Many techniques for concentrating the samples have been tried (reviewed by Hurst *et al.*, 1989) including: (1) virus collection in gauze pads or other suitable filter material which when subsequently eluted provide the samples for tissue culture inoculation; (2) ultrafiltration; (3) flocculation; and (4) affinity chromatography. The first named technique is perhaps the most widely used to-day in the form of 'directed adsorption'. In this viruses are adsorbed onto either filters (e.g. cellulose esters) or granular solids (e.g. powdered glass); the viruses are recovered for tissue culturing either by dissolving the filter material or by elution followed by filtration.

It has been suggested that bacterial viruses (i.e. bacteriophages) and, in particular, coliphages can be used as faecal indicators since there is a correlation between their numbers and the numbers of faecal bacteria they are associated with; in fact, as with other viruses, phages remain viable for longer periods than their host bacteria so that more remote pollution can be monitored; again the complex methodology required for phage assay has limited the use of this technique.

11.2.6. Cryptosporidiosis

The protozoan parasite *Cryptosporidium parvum* has been responsible for an increasing number of cases of cryptosporidiosis in recent years. Infection from certain foods such as raw milk or undercooked sausages is possible as is infection from livestock; in this case young children handling calves or lambs are most at risk (Casemore, 1990). Recently water has emerged as an important source of the protozoan and this is accentuated when mains water becomes contaminated because so many people are then at risk (Badenoch, 1990).

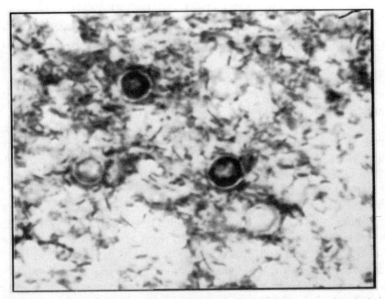

Fig. 11.2. Oocysts of *Cryptosporidium parvum* (× 1200). (By courtesy of R. Jackson.)

Infection typically follows the ingestion of water contaminated with resistant oocysts (Fig. 11.2), a stage in the complex life-cycle of *C. parvum*. The incubation period is 3–7 days and the illness is characterized by gastroenteritis with severe and often prolonged diarrhoea accompanied by mild 'flu-like symptoms. Cryptosporidiosis is far more severe in immunocompromised patients (e.g. those suffering from AIDS) and such people may be faced with a life-threatening illness.

One of the largest outbreaks of cryptosporidiosis occurred in 1988 in Carroltown, Georgia, US, where some 13 000 cases were recorded. In the UK an outbreak in the Swindon area in early 1989 possibly affected as many as 5000 persons (Badenoch, 1990). In both these incidents conventional water treatments were shown to be inadequate, primarily due to shortcomings at the filtration stage which probably allowed the *C. parvum* oocysts through; if this occurs it has been shown that the oocysts are totally resistant to chlorine-based disinfectants at the levels used in water treatment (West, 1991). In the UK alone there were over 21 000 cases recorded in the 1983–9 period stressing the importance of this illness to-day; the fact that current diagnostic methods are tedious and rather inefficient only exacerbates the problem.

11.3.   LIQUID WASTE AND ITS TREATMENT

11.3.1.   Requirements Imposed on Effluent Discharge

Today increasing emphasis is being placed on controlling pollution by reducing or eliminating waste and by encouraging better environmental standards generally. At the present time the 1975 EC Directive on Waste Management is being strengthened and Member States are being required to encourage recycling and to use waste as an energy source. Thus in the UK the newly introduced Environmental Protection Act (1990) is far tighter in pollution control legislation than the earlier Control of Pollution Act (1974).

In the UK this legislation is essentially administered by a newly constituted body, the National Rivers Authority (NRA). The NRA is thus responsible for the overall management of rivers and the control of effluent discharge into them and thus acts as a regulatory body over the private water companies. Any food manufacturer discharging waste into rivers should normally be required to ensure that it complies with the so-called Royal Commission 30:20 Standard (cited by Report of the Working Party on Sewage Disposal, 1970); in practice, however, consent conditions may be relaxed by a water authority if it is deemed to be environmentally safe.

The 30:20 Standard is based on a total suspended solids (SS) maximum of 30 mg/litre of waste water (the SS is obtained by filtering a known volume of water and weighing the residue consisting of both inorganic and organic matter after drying) and a BOD of 20 mg/litre (see p. 397).The SS and BOD are sometimes loosely combined to indicate the 'organic load' but a more widely used test as a measure of the total load is the chemical oxygen demand (COD; the amount of oxygen used in the chemical oxidation of the matter present by potassium dichromate under specified conditions). Other criteria used in pollution control include pH (which should be between pH 6 and 9, although standards may be tighter), temperature (not above 30°C), toxic metals (2 mg per litre) and a dilution requirement that effluent should be discharged at a maximum of 1 part to 8 parts of the receiving water; this latter requirement is often relaxed significantly however.

In many cases liquid effluents are discharged into sewers for treatment at regional water authority sewage works. These effluents are often untreated by the food manufacturer, it being generally believed that municipal treatment is cheaper than on-site treatment by the manufacturer.

Whilst this practice is widely recommended because of the high capital outlay required for the installation of a suitable effluent treatment plant, doubts have been expressed that biodegradable wastes are, in the long run, most effectively and cheaply treated in this way (e.g. Grant, 1975); furthermore, since the charges levied by water companies for waste treatment are still increasing, it is becoming more advantageous to integrate treatment plant with the processing side of the factory. The system adopted will depend on factors such as the volume and strength of the effluent to be treated.

Standards may be applied by water companies for effluents discharged to sewage works and figures in the 400–500 mg/litre range for both SS and BOD are typical. Since much untreated effluent from food processing factories may be in the 1000 to > 5000 mg/litre BOD range, it is clearly necessary that such effluents will require at least partial treatment before being discharged.

A further difficulty with effluents from food factories is that they may fluctuate widely in both volume and composition at different times of the day or week. Such variations are frequently associated with plants operating a batch processing system; obviously the most dramatic change occurs at the end of production when cleaning fluids are being used.

The more domestic type of sewage, consisting mainly of human excreta and other welfare area waste, is usually disposed of through the normal sewage system and does not merit further attention here.

11.3.2. Liquid Waste Management

Whilst food processing waste is inevitably non-toxic and biodegradable a considerable amount of waste is inevitable—in fact, few industries create such quantities. It is essential therefore to reduce the volume and strength of the waste to a minimum. Screening of effluent removes larger solids and even this simple treatment can have a beneficial effect on BOD. Perhaps as much as 25% of the effluent load can be eliminated by greater care being taken in processing and cleaning (e.g. by avoiding spillages or over-filling, by using dry-cleaning aids and by careful monitoring of liquid volumes used in wet cleaning, and by re-use of water).

Management of waste involves many other considerations. Information should be obtained on all relevant existing legislation and local by-laws and any likely future controls. It is necessary to ensure that the authority's sewage works has the capacity to treat the factory effluent and, if so, the likely cost of using those services; these costs should be compared with those for installing effluent treatment plant on site. It may be advantageous to classify the types of waste according to their pollution levels and on any special characteristics; such wastes may then be treated separately, some passing directly to the sewer whilst others are given a partial or complete treatment on site. On-site treatment becomes far more profitable where by-products can be recovered to be converted into marketable commodities, a situation which is much encouraged to-day. If on-site treatment is decided upon it must be capable of producing a consistent end-product, in terms of both the quality of effluent and the recovered material(s). It is clearly advantageous to use plant that is simple to operate and does not create malodours. Another possibility is for the company to make a direct grant to the water company so that the latter may increase the capacity and range of treatments available at the local sewage works to allow for the treatment of the food wastes.

Whatever system is decided upon the primary aims must be to: (1) minimize non-utilizable wastes; (2) maximize the recovery of wastes that can be converted to profitable by-products; and (3) recycle or re-use materials such as water, cleaning agents, etc., which can be employed in a sequence of operations.

Under no circumstances is the indiscriminate dumping of waste allowed; if such action is contemplated it should be borne in mind that heavy charges can be levied on companies prosecuted.

11.4. LIQUID WASTE TREATMENT METHODS

Food effluents may be treated physically, biologically or chemically. Physical treatments usually represent the 'primary' stage of treatment although in some cases they may be the only treatment required. Biological treatments are normally used at the 'secondary' stage although they may be used as primary or 'tertiary' treatments with certain types of effluent. Tertiary treatments are often necessary where high standards of effluent purity are required and where heavily polluted wastes are being treated.

11.4.1. Primary Treatment

Effluent is usually subjected to a primary treatment in order to remove the coarser suspended and other insoluble solids which might otherwise interfere with the efficiency of subsequent treatments. Such material is conventionally removed by one or more of three systems: *screens, settling tanks* and/or *cyclones*. Screening is widely used, often in conjunction with settling tanks, and has proved a very economical method. Screens of different mesh size are available and may be of the stationary, rotary, vibrating or centrifugal type; care must be taken to see that the screens are cleaned and serviced regularly. Settling tanks, or *clarifiers*, allow more finely dispersed solids to be removed by gravity. The clarified liquor rises to the top to be run off, whilst the solids accumulate as a 'primary sludge' which is often treated by anaerobic biological systems (see Section 11.4.2.4); reductions in effluent BOD of up to 35% can be effected by such treatment. *Grit chambers*, essentially preliminary settling tanks where the flow of liquid is sufficiently rapid that only the denser solids are allowed to settle, are often used where effluents contain high levels of inert solids (e.g. sugar beet factories). Cyclonic treatments involve effluent being forced down a cone under high pressure; the swirling action created forces the heavier solids to the side and then downwards to be discharged at the bottom whilst the treated water passes out at the top of the cyclone.

Effluents from abattoirs and meat processors may contain fats and oils which, as well as contributing to a high BOD, can also agglomerate and form scums. Such materials can be readily removed in grease traps, the accumulating fats being skimmed off and subsequently incorporated into animal feedingstuffs. Wastes containing fluids with very high or low pH values (e.g. cleaning fluids) or containing high salt concentrations (e.g. brining solutions) may have to be diluted before discharge; pre-adjustment of pH may be necessary in extreme cases.

As has been mentioned earlier, effluent characteristics can vary dramatically, both quantitatively and qualitatively over a production sequence. In order to minimize the effect of these variations and produce a more stable effluent flow balancing in holding tanks is widely used. It is often required by the water authority and may be virtually essential where treatment plant is installed at the factory. By this means 'shock loads' which may seriously affect the treatment efficiency are largely avoided and secondary treatments can be applied to wastes of a more constant quality.

11.4.2. Biological Treatment

Because of its high organic content food waste is more effectively treated by biological means at the secondary stage; in fact, conventional treatments at sewage works employ similar systems. Biological treatments, which may be aerobic or anaerobic, essentially involve the conversion of colloidal and suspended organic matter into CO_2 and water (aerobic digestion), or into methane (CH_4) and hydrogen sulphide (H_2S) (anaerobic digestion); in addition to these end-products, solids in the form of cellular material are also produced. A wide variety of microorganisms is involved in these processes which also remove dissolved solids to a certain extent. Full treatment may be achieved by single-, two- or even multi-stage biological systems. These traditional systems suffer from the weakness that they convert potentially useful waste materials into products of little value; however, many effluent treatment systems are now adapted to recover utilizable by-products. A further weakness is that, for aerobic treatment, wastes should preferably have a nutritional BOD:nitrogen:phosphorus ratio of *ca* 100:5:1. Wastes from meat processing factories, for example, are very high in nitrogen and may require its partial removal before treatment whereas other wastes (e.g. from breweries) are very low in nitrogen and phosphorus and cannot be effectively treated unless these elements are added in some form; one easy way of doing this is to mix brewery effluents with domestic sewage which is high in nitrogen and phosphorus.

The three main aerobic biological treatment systems are: (1) lagoons; (2) activated sludge systems, and (3) trickling filters. These are considered in turn together with anaerobic digestion methods.

11.4.2.1. Lagoons

In their simplest form lagoons, also known as stabilization or oxidation ponds, are very shallow (< 2 m) and permit sunlight to penetrate to the base. This encourages the growth of algae which supply much of the oxygen required for BOD reduction, although oxygen can also be supplied by mechanical means. These lagoons have been used for cannery and brewery wastes and for the final 'polishing' of effluents (i.e. removing the last traces of pollutants) at the tertiary stage. They produce a stable effluent but require a relatively large area of land. In addition there is a tendency for anaerobic conditions to be created, even in these shallow ponds, by an excess of organic matter or by temperature changes; such conditions can give rise to odour problems but these can be controlled by the addition of oxidizing agents. Another problem is that these lagoons are normally designed to treat effluent over periods of several weeks to months; however, it has been found that some lagoons pass the effluent through in a day or even less so that effluent is passing out virtually untreated.

Aerated lagoons are somewhat deeper basins (2–3 m) which rely upon surface aerators or compressed air diffusers to supply the necessary oxygen. They are often divided into sections, some of which may be anaerobic, the effluent flowing from one compartment to another with the solids in suspension at all times. This flow is accompanied by a progressive lowering of effluent BOD with 80–90% removal being achieved for abattoir and meat processing waste. If correctly designed with different compartments the 'short circuiting' mentioned above should not occur with this type of lagoon.

11.4.2.2. *Activated Sludge Systems*

These systems are very widely used and are particularly useful where larger volumes of high organic load wastes, such as in meat processing, are being treated and where an effluent of a high and consistent standard is required. Reductions in organic loadings of over 90% can be achieved. The system relies on the activity of very dense microbial populations in the form of flocs (i.e. activated sludge—see below) held in suspension in the waste waters in a highly aerated state to form the 'mixed liquor'. The floc organisms utilize the organic matter to produce CO_2, water and new cellular material. The flocs are composed of a mixture of microorganisms; the proportions of the types comprising the mixture vary according to the nature of the waste being treated. *Zoogloea* spp. (essentially slime producing, filamentous pseudomonads) are very important aids in the formation and maintenance of flocs whilst other bacterial genera involved in the process include *Pseudomonas, Flavobacterium* and *Alcaligenes*. Many different types of protozoa also play a part in the purification process. They ingest bacteria and have an important role in determining the amount of sediment (i.e. sludge) produced; they also utilize organic matter in general and thus assist in reducing BOD levels.

Aeration of the effluent is a continuous process with retention times normally being a few hours in the aeration tank; treatment is aided mechanically by the use of aerators or compressed air diffusers. The treated liquor passes to settling tanks where the mass of new cellular material should settle out relatively quickly as sludge, allowing the now clarified effluent to be run off. Part of the sludge is recirculated as activated sludge, it being mixed with the incoming effluent in the aeration tank to form the 'mixed liquor' mentioned earlier. The balance of the 'mixed liquor' is critical in this process and nutrients must be added as needed to maintain the optimum activity of the microorganisms. Because

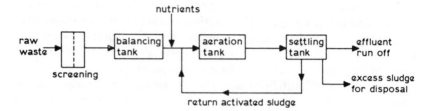

Fig. 11.3.   Flow diagram of a simple activated sludge system.

excessive cellular material is produced, portions of the activated sludge must be removed and disposed of at intervals (see later). A diagram of the activated sludge system is given in Fig. 11.3.

On occasions the activated sludge fails to settle efficiently and can pass over with the 'clarified' effluent to raise its BOD. This problem, known as 'bulking', has many causes. It is frequently associated with the presence of filamentous bacteria, particularly *Sphaerotilus natens*, but may also be caused by insufficient oxygen or an excess of carbohydrate. Although activated sludge systems can be set up quickly and are readily adaptable they are rather expensive to install, require skilled operation and are rather sensitive to shock loadings. For this last reason in particular, many modifications to the conventional system have been introduced.

Amongst the more interesting of the modifications is the 'oxygen activated sludge' system which has many benefits and is likely to be more widely used in the future. In this process pure oxygen is supplied to the reactor tank which is of gas-tight construction to facilitate good contact between the gas and effluent; by using oxygen the respiratory activities of the activated sludge bacteria are considerably enhanced. Far less mechanical power is needed with this system and this should balance the additional cost of the oxygen supplied. This process has been successfully introduced for dairy effluents and has overcome sludge bulking faults often associated with this type of effluent. Furthermore, the oxygen activated sludge system is better able to cope with shock loadings as such loadings automatically trigger an increased supply of oxygen; conversely a reduction in demand results in a diminution in the flow of oxygen to the system.

Other variations of the activated sludge process in common use attempt to overcome the inherent weakness of the conventional system in that there is a very high oxygen demand at the point where the effluent enters the aeration tank. In the 'extended aeration' system, for example, only low organic loadings are accepted but such systems cope better with shock loadings and produce very low BOD effluents. A similar 'extended aeration' principle is the basis of the 'Pasveer ditch', a popular form of oxidation ditch used mainly by the dairy industry. In this system waste is circulated around an oval ditch by means of angled aerator blades and then removed at a suitable point where partial settling has occurred; the ditch is about 1 m in depth and is trapezoidal in cross-section. Amongst other modifications in use is the 'step aeration' activated sludge system where waste is introduced at intervals along the length of the aeration tank to even out the oxygen demand.

11.4.2.3. Trickling Filters

These filters (Fig. 11.4), also known as percolating or biological filters, typically consist of a cylindrical concrete tank some 2 m in depth which is packed with inert material such as stones, clinker, gravel or plastics. The effluent is distributed over the uppermost layers of the bed by means of a rotating arm distributor and the effluent trickles down over the materials where oxidation of the organic waste occurs. This is initiated by the action of a succession of balanced biological communities which become established as slime layers at different levels in the bed on

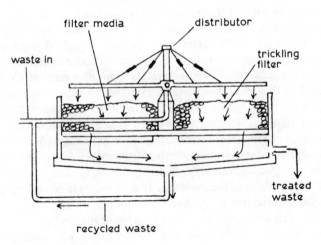

Fig. 11.4. Diagram of a trickling filter.

the surface of the packing materials. The slime, or biological film, contains a complex community of bacteria, protozoa, fungi, worms and fly larvae which all play a part in the digestion process; of necessity, such a balanced community may take several months to develop.

The slime is sloughed off periodically and replaced by fresh organisms. It sometimes increases in thickness to a point where anaerobic conditions are created and the efficiency of the system is thereby undermined. Over-loading with high BOD effluents is conducive to this condition and can result in 'ponding' which is manifested by pools of water developing on the surface of the filter bed; this is essentially created by blockage of the waste channels. Packing materials should have a large surface area:volume ratio and be of such a size as to ensure good aeration and drainage; if these conditions are observed shock loadings should be readily accommodated with BOD reductions of up to 95% possible. If excessive biological growth persists a widely used system is alternate double filtration which employs two trickling filters arranged in sequence. The incoming high BOD effluent is treated by the first filter creating a biomass build-up in it; the partially treated effluent then passes to the second filter where biological growth will be much reduced due to the low organic load. After a period the filters switch positions so that control of growth can be maintained.

Many of the difficulties experienced with the traditional low rate trickling filters are overcome when high rate plastics filled bio-filters are used. The plastic packing may be made from polystyrene, polypropylene or polyvinyl chloride. 'Flocor' is a widely used example of such materials; they are geometrically designed to give the maximum surface area: volume ratio and at the same time permit a high flow rate to be achieved in conjunction with efficient ventilation thereby eliminating blockages. Because these plastics are so light they can be installed in towers and in this form can be used as a pretreatment stage in biological oxidation where highly polluted effluents are handled; BOD reductions of 60–70% can be achieved. This high rate biological filtration can be followed, perhaps after recirculation through a second bio-tower, by conventional low rate treatments; the latter may take the form of activated sludge or trickling filter systems although other treatments, in different combinations, are possible. It is important to ensure that the system chosen will achieve the desired quality of effluent and is economic to install and run; the high rate biological filter has been successfully used to treat a variety of food wastes including those from fruit, vegetable, meat and poultry processing establishments.

11.4.2.4. Anaerobic Digestion

In the anaerobic digestion of organic materials bacteria produce a mixture of carbon dioxide and methane as end-products. A variety of bacteria convert larger organic molecules (e.g. proteins, fats and starch) into smaller molecules (e.g. amino acids, long chain fatty acids and sugars) which, in turn, are converted into short chain fatty acids (e.g. acetic acid), carbon dioxide and hydrogen. A specialized group of methanogenic bacteria (e.g. genus *Methanobacterium*) which only grow in the absence of oxygen convert, in separate reactions, the acetic acid, and the carbon dioxide and hydrogen, into methane. The advantage of this system is that, in the anaerobic environment, bacterial growth is far slower and thus a marked reduction in the volume of sludge is achieved. Balanced against this is that treatment periods are far longer and, because bacterial growth is also retarded at low temperatures, warm wastes at 25°C or over are conventionally treated to make the process economically viable.

The anaerobic digestion process has been applied to the treatment of primary sludge and sludge from biological treatment processes. The sludge is digested for several weeks in closed tanks maintained at 30–35°C by combustion of part of the fermentation gases produced; the sludge gives rise to an inert and odourless humus. This humus can be easily dewatered before disposal as waste or used for agricultural purposes; at the same time the methane produced can be used commercially as a fuel. Anaerobic biological processes are also suitable for the treatment of warm wastes with very high (>10 000 mg/litre) BOD and SS levels, e.g. from slaughter-houses, canning plants and distilleries. One variation, which has been used for highly polluted carbohydrate wastes, is the anaerobic filter (Fig. 11.5); here the effluent passes into the base of the fermenting vessel and is run off towards the top. The methanobacteria are particularly active in the lower layers of the gravel which fills the chamber and the gases produced by these organisms are drawn off from the top of the chamber. It is possible to have an anaerobic activated sludge system where the normal aeration tank (Fig. 11.3) is replaced by a closed fermentation vessel fitted with a gas outlet and coupled to a closed settling tank. An advantage of both the latter systems is that important slow-growing microorganisms are retained in the reactors and are not flushed out with the treated effluent as is the case with the conventional activated sludge process; because of the retained indigenous flora effluent can be treated relatively quickly. These systems are particularly useful in the treatment of low strength:high volume

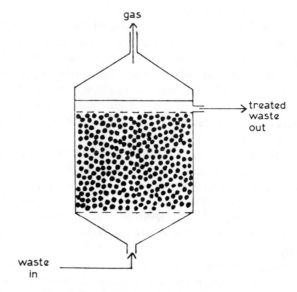

Fig. 11.5. Diagram of an anaerobic filter.

effluent which could not be treated practicably by normal methods and this has greatly increased interest in anaerobic digestion in the last few years.

Anaerobic lagoons have depths of 3–6 m and are also able to deal with heavy organic loads; they are often used as a preliminary to aerobic treatments where the heavier loads may create difficulties.

11.4.2.5. *Sludge Disposal*

A vast amount of sludge is produced in waste treatments and much of its volume is due to the high water content (95–99%), the exact amount depending on the process used; this water is difficult to remove. As has been mentioned in the previous section, anaerobic digestion is widely used for sludge treatment, one of its great advantages being that, after treatment, the sludge is more readily dewatered by air drying, heating, filtration or centrifugation. With these conditioning treatments the cost of final disposal is much reduced. Of the sewage sludge produced in the UK up to about 50% has been traditionally dumped at sea, some 25% is used as agricultural fertilizer and 25% is disposed of on land in other ways such as by dumping in disused mines. Clearly the massive dumping of sludge at sea is no longer acceptable environmentally so that other means of disposal must be found. Incineration is a potentially attractive alternative, the resultant ash being a safer end-product to dump than sludge where methane production in land fills is always a potential hazard; a further attraction of incineration is that the energy created can be utilized in, for example, the generation of electricity. The most profitable use of sludge is as an additive in animal feedingstuffs although possible future legislation enforcing its pre-sterilization could make this outlet less viable economically.

11.4.2.6. Spray Irrigation
Dispersal of suitable liquid wastes by spraying them on land is a very economical method and has proved popular in the UK. Ideally large tracts of porous soil, preferably sown with grass, are required. If available, this method of disposal may even be used to treat effluent directly after screening; however, spray irrigation is also effective as a tertiary stage treatment. The grasses absorb much of the water which is subsequently released into the atmosphere as vapour whilst the organic waste is digested aerobically by soil microorganisms. Spray irrigation therefore largely eliminates any chance of river pollution and, at the same time, supplies nutrients to the soil. It has been successfully employed, for example, in the treatment of distillery, dairy and cannery wastes.

11.4.2.7. Production of Single Cell Protein
The rapidly expanding world population together with acute food shortages in many parts of the world have heightened awareness of the need to utilize waste, both liquid and solid, to formulate novel foods. Waste from the food processing industry is potentially of great value and its efficient utilization would restrict environmental pollution. Ideally the aim should be for virtually wasteless food processing operations and such concepts have been suggested (Filka, 1986).

Wastes, mainly from the fruit and vegetable processing industries are rich in carbohydrates, vitamins and other nutrients whilst the dairy industry generates a vast amount of whey which is high in carbohydrates and proteins. Microorganisms can be grown on such wastes to produce microbial protein (i.e. single cell protein) which contains all the essential amino acids. Unfortunately its high nucleic acid content, if undegraded, creates a high concentration of uric acid in people fed on the protein resulting in gout so that single cell protein is used as a nutritional supplement in foods or animal feedingstuffs.

The technology for single cell protein was developed in Germany during World War 1 but it is only in the last 20 years or so that there has been a major expansion in the field. Many processes have been designed for a wide variety of food industry wastes but few have reached full-scale production.

Yeasts are most widely used as a single cell protein source but moulds and a few bacteria can also be utilized. Amongst the better known ventures was the Symba process which employed yeasts in the treatment of waste effluents from potato processing plants. Two types of yeast were grown together in the waste; the first, a saccharomycete produced an enzyme amylase, which breaks down starch into sugars whilst the second yeast, a candida, utilized the sugars to produce large amounts of cellular material. The final product contained nearly 50% protein and also had a high vitamin content. Unfortunately the process was discontinued due to the variable availability of the potato wastewater; many processes involving the fermentation of lactose in whey, a by-product of cheese making, have been running successfully to produce yeast single cell protein (Goldberg, 1985).

11.4.3. Chemical and Physical Treatment Methods

11.4.3.1. Dissolved Air Flotation
Dissolved air flotation is used for the removal of suspended solids with consequent BOD reductions of up to 90% being possible. Although running costs are rather high, its success can be measured by the variety of total treatment systems in which it has been incorporated. In the air flotation method air is dissolved in water under high pressure to produce a stable bubble-free solution which is suddenly released into an open flotation tank. Here it comes into contact with the waste liquid but with the marked drop in pressure the air comes out of solution as a mass of tiny bubbles; these air bubbles adhere to the particles of suspended matter and carry them to the surface. A scum of solids is thus formed on the surface which can be skimmed off whilst the clarified liquid can be run off at the bottom. Coagulating agents such as ferric chloride or aluminium sulphate are often added to the waste liquid to improve the efficiency of the air flotation technique. This technique has been used for the recovery of fats and protein from the meat, poultry, fish and dairy processing industries.

11.4.3.2. Coagulation and Flocculation
This technique, the basis of which has been described previously (see

Section 11.2.1.2), can be used in the treatment of food waste, often at the tertiary stage or as an aid to dissolved air flotation. There is a tendency for increased amounts of sludge to be produced if the coagulating agents (i.e. inorganic metal salts such as aluminium sulphate or commercial polyelectrolytes) are not carefully measured; these agents can be rather expensive but flocculation must still be regarded as a useful supportive or polishing technique.

A technique that has grown in popularity in comparatively recent years is the use of lignosulphonates to precipitate protein from wastes from meat, poultry, fish and dairy processing plants. This treatment is usually followed by dissolved air flotation, the flotated sludge being finally dewatered and dried; in this form the sludge can be used for animal feedingstuffs, lignosulphonates having been approved as an additive by both EC and US authorities (Hopwood, 1980). Additional advantages of lignosulphonates are that they are bacteriostatic so that microbial counts tend to be reduced in the effluent which is also largely fat free: furthermore, because of the bacteriostatic effects the animal feed itself should have reduced counts thereby possibly enhancing its safety.

11.4.3.3. Ion Exchange

Oppositely charged particles attract each other to differing extents depending upon their charge densities. Use is made of this principle in ion exchange where the more highly charged particle (either cation or anion) replaces a less highly but similarly charged particle in a suitable exchange bed. Ion exchange resins may be acidic (i.e. loaded with cations such as Na^+ which are exchanged for, say, Ca^{2+}) or basic (i.e. loaded with anions such as Cl^-). They are typically composed of cross-linked polystyrene or polyacrylic acid molecules to which either negatively fixed (for cation exchange) or positively fixed (for anion exchange) functional groups must be attached together with the mobile exchangeable counter ions—it is possible to have a mixed bed of cationic and anionic exchange resins.

Many organic waste materials carry a net +ve or −ve charge so they can be treated and removed from the effluent by adsorption onto the exchange resin. Unfortunately many of the conventional ion exchange resins have a poor adsorptive capacity for larger molecules like proteins; however, resins based on cross-linked regenerated cellulose have a highly porous structure into which protein molecules can easily penetrate so that these materials have considerable potential in waste treatment and by-product recovery. By varying the degree of cross-linking it is possible to alter the permeability characteristics so that resins can be prepared for specific tasks.

As will be appreciated, the resins will become loaded and lose their efficiency for ion exchange so that a regeneration process becomes necessary. This involves desorption of the entrapped charged molecules or ions and their final replacement by the original exchangeable counter ions. This process involves treating the resin beds with regeneration solutions which may have to be recycled many times before washing off excess counter ions with water. Fouling of the resin beds can occur due to an accumulation of suspended solids, growth of bacteria or algae or due to the build up of miscellaneous inorganic or organic compounds; such fouling can be treated with non-ionic detergents, sterilizing agents or dilute acids.

Ion exchange is used principally for the demineralization of water (i.e. removing, for example, calcium, magnesium and phosphate ions); it is also used in the tertiary treatment of whey and brine solutions, in by-product recovery and to remove nitrate from water.

Electrodialysis produces similar effects to ion exchange but relies on electrical forces to induce movement of molecules through chemically treated plastic membranes of different permeabilities. The membranes allow the passage of smaller molecules but are impermeable to larger molecules.

11.4.3.4. *Miscellaneous Techniques*

There are several other techniques, used mainly at the tertiary stage or in by-product recovery, which are worthy of note. Mention has already been made of centrifugation and evaporation for the dewatering of sludge but these concentrating techniques can also be used as a stage in by-product recovery. Other techniques include *activated carbon adsorption* and the separation of molecules by *ultrafiltration* or *reverse osmosis*.

Activated carbon adsorption is a relatively cheap tertiary treatment technique for removing organic waste materials which are difficult to remove by other means. The effluent is passed over beds of the activated carbon which, because of its large surface area, adsorbs dissolved organic molecules. It is possible to regenerate the loaded activated carbon by dewatering and then heating it to high temperatures when the adsorbed organic molecules are oxidized and volatilized.

Ultrafiltration and reverse osmosis are techniques, used singly or in combination, which are particularly useful in the removal of the last traces of colloidal or dissolved solids. Both, like electrodialysis, rely on plastic membranes to concentrate and separate soluble ions, molecules, and colloidal and particulate materials. The force inducing movement across the membrane is supplied by a high pressure differential across the membrane itself. With ultrafiltration the solution is separated into two parts, one containing larger molecules and the other smaller, their relative sizes depending upon the pore dimensions of the membrane. Reverse osmosis also separates the solution into two parts one of which is virtually pure water. A pressure is applied that is greater than the osmotic pressure to drive the water from the solution, the retained solution becoming more and more concentrated. The materials in it can, if necessary, be treated subsequently by ultrafiltration. Much depends on the quality of the membranes which require careful cleaning and maintenance.

These two techniques have been applied successfully to the recovery and concentration of protein from whey discarded in cheese manufacturing, in the preparation of vegetable protein and in the recovery of sugar from confectionery wastewater.

11.4.4. Examples of Effluent Treatment Systems

11.4.4.1. Poultry Processing Effluent

The principal wastes emanating from poultry processing factories are feathers, offal, blood, general proteins and fats, in the dissolved, colloidal or particulate form. The solids can be screened and, to reduce an otherwise excessively high BOD, it is common practice to recover the blood in a separate operation after bleeding the birds; the blood is often incorporated into animal feedingstuffs. Grease can be removed in traps after which the effluent passes to a flow balancing tank, a particularly important stage where biological treatments are being used.

If effluent with a very high BOD has to be fully treated a two-stage biofiltration system can be employed (Fig. 11.6; Summers, 1972). As can be seen, the effluent passes from the flow balancing tank to the first stage biofiltration tower (high rate, plastic packed) from which solids can be removed as required. Recycling of part of the effluent is possible to maintain maximum efficiency of the biological system, the remaining effluent passing to the second stage biofiltration tower (again high rate) where a further reduction in BOD is effected; again, recycling of effluent at this stage is possible. The effluent then passes to a conventional biological treatment system (e.g. activated sludge) for final treatment and disposal.

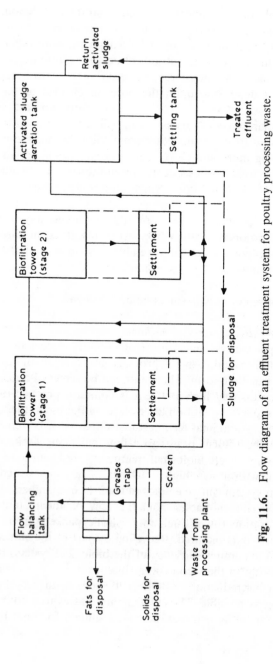

Fig. 11.6. Flow diagram of an effluent treatment system for poultry processing waste.

Similar systems to this have been successfully employed in the treatment of wastes from many other food processing plants including those producing dairy products, and frozen and canned foods.

11.4.4.2. Meat Processing Effluent

A typical treatment system for a factory handling meat products is illustrated in Fig. 11.7 and has been described more fully by Stracey (1975). After initial screening, fat trapping and flow balancing, effluent passes to the primary sedimentation tank which is fitted with scrapers and sludge removal devices. After these preliminary treatments the effluent passes at a controlled rate to an activated sludge aeration tank, aeration being provided by rotors. The number of paddles on the rotors decreases along the length of the tank thus providing, by its 'tapered aeration' action, another mechanism by which oxygen demand may be equilibrated throughout the activated sludge system. The treated waste then passes to a conventional settling tank, the sludge from which can be returned to the aeration tank or disposed of as waste. The clarified liquor is pumped to a nitrifying filter, clarified in a further settling process and finally discharged into a river or stream.

Some nitrification does occur in most biological waste treatment processes but where the wastes contain high levels of protein it is normal to find high concentrations of reduced nitrogen compounds such as ammonia. In such circumstances an additional nitrification stage may be required in order to oxidize these reduced compounds which would otherwise adversely affect the BOD levels in the receiving waters. Nitrification involves the oxidation of ammonia, by aerobic nitrifying bacteria, through nitrite to nitrate; this process is best achieved in a submerged filter with air or oxygen being bubbled through to supply the considerable oxygen demand.

However, as nitrates are themselves now a cause of concern (see Section 3.14.2.2), their denitrification to nitrogen has become an increasingly important process; this denitrification is effected by certain types of bacteria under anaerobic conditions. Much of the nitrogen thus formed escapes into the atmosphere in the gaseous form and this reduces the levels of nitrate in the waste liquor and thus helps to prevent the process known as 'eutrophication'. In this process the ecological balance of the receiving water is upset by the excessive growth of algae and other aquatic plants which utilize the large amounts of nitrogen compounds present, i.e. these nitrogen compounds act as fertilizers.

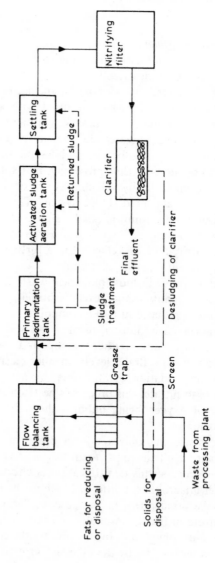

Fig. 11.7. Flow diagram of a treatment system for waste from a meat products factory.

11.4.4.3. Sugar Processing Effluent

The manufacture of sugar from sugar beet uses large volumes of water and anaerobic treatment of the wastewater is commonly employed. Shore *et al.* (1984) have described a system where anaerobic digestion is employed both to treat anaerobically stabilized lagoon water and excess sludge from an activated sludge plant which is also used in effluent treatment.

Since sucrose is leached readily from cut beets bacteria attack the sugar producing acids and there is a consequent fall in pH. This is controlled by the addition of lime (calcium oxide) which maintains a pH of *ca* 7·0 thus making the effluent more suitable for biological treatment. The treated lagoon water (or wastewaters generally could be used) is fed to an Anamet anaerobic digester (Fig. 11.8) via heat exchangers where the temperature of the water is raised to 35–40°C. The wastewater is held in the digester for about 4 days whilst being continuously stirred. After discharge the mixture of water and biomass is separated during passage through a flocculation tank and a special clarifier where the flocs are separated by gravity. The settled solids (i.e. concentrated sludge) are returned to the digester while the overflowing clear effluent can either be discharged, if of the required standard, or be 'polished' if required. The methane gas which collects in the digester can be used as fuel for drying beet pulp in the factory. There are other economic advantages in the system described including reduced energy and nutrient costs and, in addition, BOD reductions of up to 90% should be achievable.

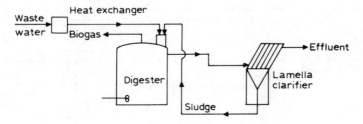

Fig. 11.8. Flow diagram of Anamet anaerobic digestion system. (After Shore *et al.*, 1984.)

11.5. SOLID WASTE DISPOSAL

11.5.1. Organic Solids

The screening of waste liquors to effect the removal of organic solids is a prerequisite in the treatment of liquid food effluents. Food solids removed in this way include bones and other meat scraps, feathers, pea vines, pods, cores, peelings, trimmings and the shells of nuts and eggs. Many of these can be treated and incorporated into animal feedingstuffs and/or composted to produce agricultural fertilizers.

Examples of by-products produced from solid wastes include fats from meat wastes (often incorporated into commercial soaps) and gelatin from bones, whilst bones are collected mainly by animal by-products processors and rendered into bone tallow and bone meal. However, there are many other uses for bone by-products. For example, protein from bones is incorporated into comminuted meat products (e.g. sausages, beefburgers and pies) and is also used in soups, sauces and gravies (Jobling & Jobling, 1983).

Fruit and vegetable solid waste can be used immediately as an animal feed but its low protein content makes it unattractive nutritionally. However, by using the solid waste as a substrate for the growth of the mould *Aspergillus niger*, a nutritionally rich single cell protein is produced; Davy (1981) has described such a process (Fig. 11.9). After a preliminary washing stage the solid waste is cut up to increase the surface area:volume ratio thus making the fermentable sugars more readily available. The moisture content is reduced but still allows for mould

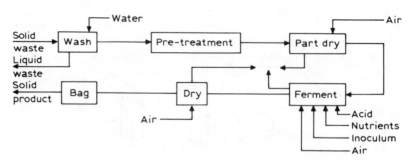

Fig. 11.9. Process flow diagram for fruit and vegetable waste treatment. (After Davy, 1981.)

growth but the lowered a_w prevents the growth of contaminating bacteria and yeasts. Fermentation, which is performed at 28–30°C for 36 h in a low pH (*ca* 3) environment, is initiated by the addition of a spore suspension of *A. niger* together with nitrogen and phosphates as nutrient supplements; aeration during fermentation is essential and this is aided by agitation of the solids in the fermenter. Many fruits and vegetables could be treated by this process to provide a dry protein-rich product which should be economically attractive as an animal feed. Fruit effluents may also be fermented to produce ethyl alcohol, and citrus fruit wastes may be used as sources of citric acid, citrus oil and gelling agents or may be used as feed supplements due to their high vitamin C content; charcoal may be produced from nut shells.

In spite of the wide range of recovery processes which have found application in the food industry there is still a great deal of solid food waste which must be disposed of directly. Such waste should be collected in bins or hoppers and regularly transferred to skips or trucks for final disposal (see below); particular care should be taken to see that these various containers are washed out after emptying or they may attract insects and other pests.

The final disposal of organic waste can be by composting (discussed earlier), incineration or some form of dumping (Bridgwater & Mumford, 1979). Incineration at high temperatures (600–800°C) involves the complete destruction of all organic material, the main products being carbon dioxide and water; any subsidiary gases are deodorized at these temperatures and electrostatic precipitators can be incorporated to clean up the exhaust fumes.

Properly controlled 'land filling' is the most acceptable alternative form of disposal. If disused mines or quarries are available these should be utilized. With land filling techniques it is important to ensure that adequate layers of soil are sandwiched between the waste to absorb excess moisture. Failure to do this can result in the reconstitution of the waste as a slurry after heavy rain, etc., and this can give rise to offensive odours as a result of digestion by anaerobic bacteria. The disposal of solid food waste by tipping or by depositing it in shallow trenches should be avoided as waste left exposed in this way will cause odours and may also create a nuisance by attracting rodents, insects and scavenger birds.

11.5.2. Inert Solids

Large inert solid wastes include glass bottles, metal cans, plastic containers and cardboard boxes. These wastes are often just compressed in volume by powerful waste compaction systems and disposed of at tipping sites, especially selected to cause the minimum damage to the environment, and which in the long run will be land filled. However, alternative treatment methods are available and some of these are described below.

11.5.2.1. Glass

Glass is non-degradable but it is chemically inert and does not pollute the environment. Identical glass is used for both returnable and non-returnable bottles, and, whilst there may be economic reasons for not encouraging a 'deposit-and-return' system, the vast increase in non-returnable bottles has added to the litter problem in the UK. Since glass can also be recycled, far greater effort should be made to collect discarded and broken bottles; the glass can then be used in the manufacture of new bottles but it may also be incorporated into other materials such as glass wool, bricks and even street paving materials (Woodroof & Phillips, 1974).

11.5.2.2. Metal Cans

Cans are made of either tinplate or aluminium, and in either form can be readily crushed. Their recycling would appear desirable but recycling of traditional cans is complicated by the presence of tin which adversely affects steel recovery. Tin from cans has been used in the production of stannous fluoride which is often incorporated in toothpastes. Tin-free coatings have been developed for soft drinks and beer cans but cost has limited their use. It is more economical to recycle aluminium cans and the wider use of this metal should aid recycling programmes.

11.5.2.3. Plastics

Plastic materials are very widely used by the food industry but it is possible to dispose of them since they can be easily crushed. Most plastics can be safely incinerated as carbon dioxide and water are the main products; toxic fumes are not given off by those plastics used in the food industry.

One answer to the problem of the disposal of plastics is to use plastics, now being developed, which are degraded by UV light from the sun; such plastics can be degraded to a particulate form in a matter of a few days to several months and when particulate they are then biodegradable (Woodroof, 1975). Another interesting development has been the introduction of PET bottles. PET (polyethylene terephthalate) is becoming widely used in the bottling of carbonated drinks. The plastic is an excellent packaging material being inert, pure and physically strong so that a large volume (2–3 litres) of beverage can be safely bottled. A further advantage is that the PET can be recycled and treated in different ways to produce a wide range of useful by-products. Such by-products can be used, for example, as paint additives, pipework, fabrics, tail-light lenses for cars and fibreglass reinforcement. Obviously any recycling scheme depends upon the efficiency of bottle recovery which can be either voluntary or mandatory (i.e. a deposit system). A voluntary scheme was introduced in the UK in the Leeds/Bradford area in 1977 since when it has expanded successfully to become a landmark in waste recycling schemes; many similar voluntary or mandatory schemes have been introduced in Europe based on the Leeds system. One of the great advantages of these schemes is their contribution towards litter reduction, a publicity angle which could be encouraged by food manufacturers.

11.5.2.4. *Cardboard and Paper*
Packaging material is generally disposed of by compression and tipping but recycling can be a viable proposition. Cardboard can be pulped on site with water which can be removed and recycled; the dried pulp is then transported to a paper recycling plant. In view of the vast amounts of cellulose contained in cardboard it would be impossible to utilize more than a fraction in normal biodegradation processes but cellulose breakdown by microbial fermentation is feasible in the presence of additional nutrients and can yield fuels and animal feedingstuffs as by-products.

BIBLIOGRAPHY

Armitt, J. D. G. (1981). Brewery and malthouse effluents and their management. In: *Brewing Science*, Vol. 2. Ed. J. R. A. Pollock, London, Academic Press.

Bellamy, W. D. (1983). The use of microbiological agents in upgrading waste for feed and food. In: *Upgrading Waste for Feeds and Food*. Eds D. A. Ledward, A. J. Taylor and R. A. Lawrie, London, Butterworths.

Grainger, J. M. (1987). Microbiology of waste disposal systems. In: *Essays in Agricultural and Food Microbiology*. Eds J. R. Norris and G. L. Pettipher, Chichester, John Wiley and Sons.

Green, J. H. & Kramer, A. (1979). *Food Processing Waste Management*. Westport, Avi Publishing Co.

Hutchinson, M. & Ridgway, J. W. (1978). Microbiological aspects of drinking water supplies. In: *Aquatic Microbiology*. Eds. F. A. Skinner and J. M. Shewan, London, Academic Press.

Ilett, K. J. (1979). Dissolved air flotation and hydrocyclones for wastewater treatment and by-product recovery. *Food Processing Industry*, **48** (June), 38–41.

McCarty, P. L. & Haug, R. T. (1971). Nitrogen removal from wastewaters by biological nitrification and denitrification. In: *Microbial Aspects of Pollution*. Eds G. Sykes and F. A. Skinner, London, Academic Press.

McCaull, J. & Crossland, J. (1974). *Water Pollution*. New York, Jovanovich.

Melnick, J. L. (1978). Water as a reservoir of virus in nature and means for control. In: *Viruses and Environment*. Eds E. Kurstak and K. Maramorosch, New York. Academic Press.

Mosey, F. E. (1981). Anaerobic biological treatment of food industry waste waters. *Water Pollution Control*, **80**, 273–89.

Nyns, E.-J. (1986). Biomethanation processes. In: *Biotechnology,* Vol. 8 (Microbial Degradations). Eds H.-J. Rehm and G. Reed, Weinheim, Germany, VCH Verlagsgesellschaft mbH.

Pepper, D. (1973). Reverse osmosis and ultrafiltration techniques. *Effluent and Water Treatment Journal*, **13**, 779–84.

Solt, G. S. (1973). Waste treatment by ion exchange. *Effluent and Water Treatment Journal*, **13**, 768–73.

Thomas, W. M. & Berryman, P. M. (1986). Water and wastewater. In: *Quality Control in the Food Industry*, 2nd edn, Vol 2. Ed. S. M. Herschdoerfer, London, Academic Press.

REFERENCES

Badenoch, J. (1990). *Cryptosporidium*—a water-borne hazard. *Letters in Applied Microbiology*, **11**, 269–70.

Bridgwater, A. V. & Mumford, C. J. (1979). *Waste Recycling and Pollution Control Handbook*. London, George Godwin Ltd.

Casemore, D. P. (1990). Epidemiological aspects of human cryptosporidiosis. *Epidemiology and Infection*, **104**, 1–28.

Davy, C. A. E. (1981). Recovery of fruit and vegetable waste. In: *Food Industry Wastes: Disposal and Recovery*. Eds A. Herzka and R. G. Booth, London, Applied Science Publishers.

DHSS (1983). *The Bacteriological Examination of Drinking Water Supplies 1982*. Reports on Public Health and Medical Subjects No. 71. London, HMSO.

Filka, P. (1986). Design of wasteless food processing plants. In: *Food Engineering and Process Applications*, Vol. 2 (Unit Operations). Eds M. Le Maguer and P. Jelen, London, Elsevier Applied Science Publishers.

Geldreich, E. E. (1989). Drinking water microbiology—new directions toward water quality enhancement. *International Journal of Food Microbiology*, **9**, 295–312.

Gerba, C. P., Margolin, A. B. & Hewlett, M. J. (1989). Application of gene probes to virus detection in water. *Water Science and Technology*, **18**, 147–54.

Gibbs, B. M. & Freame, B. (1965). Methods for the recovery of clostridia from foods. *Journal of Applied Bacteriology*, **28**, 95–111.

Goldberg, I. (1985). *Single Cell Protein*. (Biotechnology Monographs, Vol 1). Berlin, Springer-Verlag.

Grabow, W. O. K. (1986). Indicator systems for assessment of the virological safety of treated drinking water. *Water Science and Technology*, **18**, 159–65.

Grant, R. A. (1975). Protein recovery as an effluent treatment process. *Effluent and Water Treatment Journal*, **15**, 616–21.

Hopwood, A. P. (1980). Recovery of protein and fat from food industry waste waters. *Water Pollution Control*, **79**, 225–32.

Hurst, C. J., Benton, W. H. & Stetler, R. E. (1989). Detecting viruses in water. *Journal of the American Water Works Association*, **81**(9), 71–80.

Jobling, A. & Jobling, C. A. (1983). Conversion of bone to edible products. In: *Upgrading Waste for Feeds and Food*. Eds D. A. Ledward, A. J. Taylor and R. A. Lawrie, London, Butterworths.

Kilvington, S. & Price, J. (1990). Survival of *Legionella pneumophila* within the cysts of *Acanthamoeba polyphaga* following chlorine exposure. *Journal of Applied Bacteriology*, **68**, 519–25.

Lee, J. V. & West, A. A. (1991). Survival and growth of *Legionella* species in the environment. In: *Pathogens in the Environment*. Ed. B. Austin. Journal of Applied Bacteriology Symposium Supplement No. 20, 121S–9S.

Mepham, P. (1989). Legionnaires' disease. *Environmental Health*, **97**, 74–9.

Mitchell, E. *et al.* (1990). Two outbreaks of Legionnaires' disease in Bolton Health District. *Epidemiology and Infection*, **104**, 159–70.

Report of the Working Party on Sewage Disposal. (1970). *Taken for Granted*. London, HMSO.

Shore, M., Broughton, M. W. & Bumstead, N. (1984). Anaerobic treatment of waste waters in the beet sugar industry. *Water Pollution Control*, **83**, 499–506.

Stracey, I. G. (1975). Treatment of abattoir waste and meat-processing effluent at Haverhill Meat Products Ltd. *Water Pollution Control*, **74**, 101–8.

Summers, T. H. (1972). Wastes from poultry processing origins and treatment. *Effluent and Water Treatment Journal*, **12**, 299–305.

West, P. A. (1991). Human pathogenic viruses and parasites: emerging pathogens in the water cycle. In: *Pathogens in the Environment*. Ed. B. Austin. Journal of Applied Bacteriology Symposium Supplement No. 20, 107S–14S.

Woodroof, J. G. (1975). Plant sanitation and waste disposal. In: *Commercial Fruit Processing*. Eds J. G. Woodroof and B. S. Luh, Westport, Avi Publishing Co.

Woodroof, J. G. & Phillips, G. F. (1974). *Beverages: Carbonated and Noncarbonated*. Westport, Avi Publishing Co.

Chapter 12

HYGIENE AND TRAINING OF PERSONNEL

All employees working in or, where allowed, passing through any area handling food ingredients, preparing or processing foods, or storing final products, should comply with the hygienic practices laid down by the company. These practices may vary somewhat from area to area but they should reflect the need for personal cleanliness and suitable clothing and footwear; the need to guard against contamination of food materials by foreign matter must also be ensured; such contaminants may be human hair or other items associated with operatives, e.g. buttons, jewelry.

For persons working in direct contact with foods or food contact surfaces further rules may have to be applied but, in general, these rules should not differ in principle from those laid down for all employees present in food processing areas. Important factors to be considered are discussed in the following sections.

12.1. HAND WASHING AND CARE OF THE HANDS

In the case of food equipment contact surfaces it is always possible to so reduce the numbers of contaminating microorganisms that any remaining will not undermine the quality of the processed food (i.e. disinfection of food contact surfaces is possible). Unfortunately it is impossible to disinfect skin to the same degree and hands are therefore a potentially important means of distributing microorganisms. Such distribution may involve the transfer of organisms from hands to food or their transfer from food to food via the hands; especial care must therefore be taken to ensure that these transmission routes are minimized.

Conventional *but thorough* hand washing using soap and water removes the transient flora, that is microorganisms, mainly bacteria, picked up temporarily from the environment. Thus enteric bacteria, including *Escherichia coli* in profusion and salmonellas if present, may pass through toilet paper on to the hands but such organisms should be removed by thorough washing. Tablet soap is not recommended although it has been shown that such soaps, with or without added antibacterials, only harbour a light, transient and rapidly changing flora (McBride, 1984).

It is virtually impossible to remove the indigenous flora on the hands although some reduction can be effected by using bactericidal soaps and creams. This resident flora is a particular problem when employees are carriers of *Staphylococcus aureus* in their nasal passages; many of these employees inevitably will also carry the organism on their hands as part of the established flora (see p. 45). It would clearly be impracticable to debar some 30–40% (the likely *S. aureus* carrier rate) of employees from contact with foods and therefore other means of restricting the dissemination of *S. aureus* must be adopted; incidentally steps taken to limit the spread and growth of *S. aureus* in foods will also serve to similarly limit other organisms carried on the skin. Efficient methods of food processing, as have been described in previous chapters, constitute an essential element in controlling such contamination but personal hygiene and the use of a variety of other aids also play a prominent part.

12.1.1. Personal Hygiene

Personnel must be encouraged to develop an attitude where hand washing becomes a virtually automatic response to certain situations. Thus hands, and wrists, should be thoroughly washed: (1) before starting work; (2) before and after lunch and tea breaks; (3) after using the toilet; (4) when leaving or returning to the processing area for any other reason; (5) when changing jobs within the process area; (6) when the hands become unexpectedly soiled or contaminated in any way such as after handling equipment or food which may be of a sub-standard quality.

As well as washing of hands, fingernails should also be scrubbed after visiting the toilet or when hands are heavily soiled; to facilitate this, fingernails should be kept short and clean. Nail brushes should be provided which should be in good condition and kept as clean and as dry as possible when not in use; daily disinfection by boiling or by rinsing in detergent followed by soaking in hypochlorite is also recommended (Tebbutt, 1984).

General advice on the care of hands and fingernails should be available; nail varnish should not be permitted and nail-biting employees should be discouraged to the extent of redeploying them in areas where they cannot come into direct contact with food.

12.1.2. Bactericidal Soaps and Creams

The regular washing of hands in warm water with non-medicated soaps, which should, of course, be dispensed as a liquid or powder, may be perfectly adequate in many situations. However, it is often preferable, particularly where more hazardous foods are being prepared, to include a bactericide in the cleaning solution and/or to introduce a hand dip disinfectant to be used at the end of the wash; by these means even the resident bacterial flora can be substantially reduced.

Many germicides which have been shown to reduce progressively the resident skin flora have been incorporated into soaps, hand wash solutions and 'barrier' hand creams; one such germicide, hexachlorophane, which was very popular some years ago has been found to have toxic side-effects and is no longer used. A germicide that is widely used today in hand washes is chlorhexidene ('Hibitane') which is prepared as a solution in ethyl alcohol and can induce an immediate *ca* 98% reduction in bacterial counts on the hands following a conventional light wash; chlorhexidene also has a considerable residual effect on the bacterial flora and it becomes even more effective after it has been applied several times (Lowbury *et al.*, 1974). Other germicides include trichloro hydroxydiphenyl ether (Irgasan DP 300), chloroxylenol and iodophors. All are commonly used although only iodophor (with 0·75% available iodine) can be compared to chlorhexidene in its effectiveness (Sheena & Stiles, 1983*a*).

Barrier creams are widely used for skin protection and if correctly formulated are surprisingly effective in reducing the release of bacteria from fingers (Sheena & Stiles, 1983*b*); some creams incorporate QACs as disinfectants which can further enhance their effectiveness.

Liquid soaps can occasionally become infected by *Pseudomonas* spp. particularly when the dispensers are 'topped-up' rather than cleaned out and disinfected between runs; this practice should be avoided but a further protection against infection is afforded by the incorporation of disinfectant in the soap.

The final drying of hands should be by means of disposable paper towels or warm-air driers, although with the latter regular maintenance is necessary.

12.1.3. Use of Gloves

Rubber gloves have been widely used as a means of restricting the spread of bacteria. When new and in good condition gloves are a help but, all too frequently, they are worn until the glove surfaces become roughened, porous and even split; in this state they are more a hazard than a help since they may harbour large numbers of bacteria on their damaged surfaces. Where gloves are re-used they should be washed and disinfected regularly, both inside and out, and discarded when they begin to deteriorate. Disposable gloves have rightly become more popular and, provided they are disposed of at the end of the shift, they are recommended where direct handling of food cannot be avoided as with, for example, the stripping of flesh from cooked chicken carcasses.

12.2. PRACTICES, GOOD AND BAD

Employees must be trained in good personal practices which must be respected at all times and especially in the vicinity of foods being processed. As with personal hygiene mentioned previously, good practices must become almost instinctive so that there is less likelihood of friction arising between enforcing supervisory staff and employees.

12.2.1. Clothing and Jewelry

Outdoor clothing should be housed in lockers and should not come into contact with protective clothing used in the factory. Protective clothing should be housed separately, correctly worn, clean and in a good state of repair; no temporary repairs involving the use of safety pins, for example, should be permitted. Loose items such as pens, pencils, paper clips and drawing pins should not be carried in open pockets, even by management. Suitable footwear should also be provided for employees working in food processing areas; rubber boots are recommended for most situations provided suitable non-slip floor surfacing is used.

Whilst a plain ring may be worn on the hand all other jewelry including stoned rings, bangles, necklaces and pendants must be removed before entering the processing area; similarly watches should be removed as should ear-rings unless of the 'sleeper' type. False eyelashes and the use of excessive make-up or perfume should not be permitted.

12.2.2. Hair

Human hair is often responsible for foreign matter complaints which can be virtually eliminated by the careful of hair restraints. It should be borne in mind that humans lose about 100 head hairs every day so that effective covering is essential. Hair nets, worn in conjunction with hats, caps or scarves, are to be preferred: the mere placing of a cap or tying of a head scarf on the head with the prime view of appearance is unfortunately totally inadequate. To be effective all head hair must be scrupulously contained and nothing short of this will suffice.

Where moustaches or beards are not closely trimmed some form of effective restraint is again required; well-trimmed beards and moustaches do not require covering since hair loss from the face is minimal.

12.2.3. Other Bad Practices

There are many bad practices or habits which should be avoided by personnel. Amongst the more obviously offensive and hazardous are spitting, smoking, sweet eating and gum chewing although these are, of course, totally prohibited or only allowed in approved areas. It is also well known that coughs and sneezes create aerosols of microorganisms which may be transmitted to foods or to other personnel. Other practices are equally undesirable, not only aesthetically but also because bacteria are possibly disseminated to foods or food contact surfaces; such practices include the licking of fingers, the scraping of teeth with fingernails and nose picking.

12.3. HEALTH SUPERVISION

12.3.1. Pre-Employment Medical Examination

All applicants seeking employment in the food industry, who are likely to come into direct or indirect contact with foods, should be examined medically to ensure fitness for work. The examination should include answering a questionnaire recording the past medical history of the applicant. Detailed information should be obtained on infections of the digestive tract including typhoid fever, paratyphoid, salmonellosis, bacillary and amoebic dysentery and gastroenteritis involving symptoms of sickness and diarrhoea. Information should also be obtained as to whether the applicant has suffered within, say, the last 12 months from skin rashes, boils, sores and infected wounds or from discharges from the eyes, nose and ears; these abnormalities are normally associated with the presence of large numbers of *S. aureus* and even though the person no longer manifests symptoms he or she may still continue to harbour the organisms and thus represent a hazard to the industry.

Common sense dictates that any applicant who has suffered from typhoid or paratyphoid should not work in a food factory in a job where there is any reasonable possibility of that person coming into contact with food materials; even though clearance from pathogens is required under the food hygiene legislation, the intermittent nature of salmonella excretion in faeces of carriers (see Section 2.3.1.2) implies that three, or even five, stool samples taken on successive days from a carrier may yield false negative results. Applicants suffering from any of the other complaints listed should again not be allowed in food processing areas until declared free of infection; this declaration should be supported by negative test results (to presence of salmonellas, shigellas and intestinal parasites) on stool specimens taken on a minimum of three successive days where intestinal complaints are involved.

12.3.2. Health Monitoring of Employees

Obviously the rationale of the medical examination and questionnaire for applicants should be applied to existing employees on a continuous basis. Thus no person who is suffering from the complaints mentioned in the previous section should be allowed into food processing areas. It is essential that management is informed immediately of any such complaints so that the correct remedial action may be initiated. In many instances this will mean that the employee affected will be sent home and only allowed back following clearance from the general practitioner supported, where necessary, by negative stool test results; however, as stated previously too much reliance cannot be placed on such results and, undoubtedly, the routine testing of stool samples taken from employees at regular intervals has little to commend it.

Where minor complaints are involved it may often be permissible to temporarily transfer the affected employee to a non-food handling job outside the food processing area. Management should also be informed where employees have been in close contact with people suffering from gastrointestinal disorders or have visited countries where such disorders are prevalent or endemic.

As is required by the Food Hygiene Regulations, all cuts, sores and other skin abrasions must be covered with approved waterproof dressings which prevent the transmission of skin bacteria on to other surfaces; waterproofing can be achieved by using fingerstalls where finger injuries are involved. Dressings must be replaced regularly and not allowed to become dirty. *Every precaution should be taken to see that dressings are not lost* but if the unlikely happens management must be informed *immediately.* Recovery of lost dressings may be aided by incorporating metal strips into brightly coloured dressings thus enabling them to be located by metal detectors as well as being conspicuous.

Health checks at the start of each working day are sometimes deemed necessary for employees working in 'high risk' areas. It would be preferable if employees appreciated that it was in their own interests, as well as a social responsibility, for them to report illness or injury voluntarily; such voluntary reporting is made easier when a good rapport exists between management and the workforce, something which, unfortunately, is too often lacking.

Lastly, conditions should be created to minimize the possibility of accidents. Most accidents result from falls due to slippery floors or steps, from collisions due to poor lighting or poor factory layout, from fatigue induced by too noisy an environment, or from poorly designed or unguarded equipment. All these aspects have been discussed in previous chapters; when correct designs and procedures have been established accidents, which undermine the health of employees just as much as illnesses, can be reduced to an absolute minimum.

12.4. INDUCTION AND IN-SERVICE TRAINING

Regardless of their grade, all new employees should be trained in the basic principles of food and personal hygiene detailed in the earlier part of this chapter. Training should also be given in the basic concepts of bacteriology. This training should stress the ubiquitous nature of bacteria, their means of dissemination, their rapid growth under certain conditions and their role in food poisoning and food spoilage. Finally, information should be given on the methods that are used in the cleaning and disinfection of equipment and working areas in general; the reasons justifying the adoption of these methods should be explained.

Company hygiene regulations should be set out in booklet form; all new employees should be provided with the booklet and be required to sign that they have read and understood the rules. Non-company labour, employed in contract work in food processing areas, must also be conversant with the regulations and should, if necessary, undergo special training.

Training, to be effective, must be continuous so that regular talks, discussion groups and refresher courses are required. Aspects of hygiene should be illustrated by posters displayed at suitable places and these posters should be changed regularly to provide greater visual impact. It is equally important to check that the hygiene regulations are being observed by employees in order to ensure that indifference does not develop over a period of time — familiarity can breed contempt!

Obviously the extent of training required varies with different categories of staff and training needs should be assessed carefully. How this should be done has been illustrated admirably by ICMSF (1988). Categories defined include food producers (e.g. farmers), line operators, quality control personnel, warehouse and transport staff, and management. As an example, the suggested knowledge required by line operators is given in Table 12.1. It is stressed that management should be fully conversant with HACCP principles and their application. Also of crucial importance is that top management give their full support to the training programme and are fully acquainted with its principles. There is no doubt that a well-constructed programme will reap many benefits such as reducing the level of supervision required, increasing productivity and reducing the level of consumer complaints which can result in adverse publicity and declining sales. Legislation now being introduced by the EC does require companies to introduce and maintain training schemes for all levels of staff. In the United Kingdom the Food Safety Act (1990) requires that personnel preparing or handling food must undergo training in food hygiene.

Any training programme should be expounded by a teacher with an enthusiasm for and an understanding of the subject and who is able to explain things at the level of understanding of the audience. Talks may be given by the training officer, hygiene officer, company bacteriologist,

TABLE 12.1
Basic Knowledge Required by Line Operators[a]

The major sources of microorganisms in the product for which they are responsible.

The role of microorganisms in disease and food spoilage.

Why good personal hygiene is required.

The importance of reporting illness, lesions and cuts to supervisory personnel.

The nature of the control required at their point in the process.

The proper procedures and frequency for cleaning equipment for which they are responsible.

The procedures necessary to report deviations from control specifications.

The characteristics of normal and abnormal product at their given step in the process (e.g. colour, texture, package integrity, odour).

The importance of maintaining proper records.

How to monitor CCP of operations within their responsibility.

[a]From ICMSF (1988).

a member of the quality assurance staff, a welfare nurse or any other responsible person. It may sometimes be necessary to involve outsiders such as Environmental Health Officers or other persons working in the Public Health field; they are usually very willing to help.

Teaching aids are a very important part of a training programme, visual demonstrations in the form of films, slides, wall charts and posters being very effective. Fortunately such aids are now readily obtainable from appropriate sources.

BIBLIOGRAPHY

Ayliffe, G. A. J., Babb, J. R. & Lilly, H. A. (1981). Hand hygiene disinfection. In: *Disinfectants: Their Use and Evaluation of Effectiveness.* Eds C. H. Collins, M. C. Allwood, S. F. Bloomfield and A. Fox, London, Academic Press.

REFERENCES

International Commission on Microbiological Specifications for Foods (ICMSF) (1988). *Microorganisms in Foods*, Vol. 4, (Application of the Hazard Analysis Critical Control Point (HACCP) System to Ensure Microbiological Safety and Quality). Oxford, Blackwell Scientific Publications.

Lowbury, E. J. L., Lilly, H. A. & Ayliffe, G. A. J. (1974). Pre-operative disinfection of surgeons' hands: use of alcoholic solutions and effects of gloves on skin flora. *British Medical Journal*, **4**, 369–72.

McBride, M. E. (1984). Microbial flora of in-use soap products. *Applied and Environmental Microbiology*, **48**, 338–41.

Sheena, A. Z. & Stiles, M. E. (1983a). Immediate and residual (substantive) efficacy of germicidal hand wash agents. *Journal of Food Protection*, **46**, 629–32.

Sheena, A. Z. & Stiles, M. E. (1983b). Comparison of barrier creams and germicides for hand hygiene. *Journal of Food Protection*, **46**, 943–6.

Tebbutt, G. M. (1984). A microbiological study of various food premises with an assessment of cleaning and disinfection practices. *Journal of Hygiene, Cambridge*, **92**, 365–75.

Chapter 13

LEGISLATION

13.1. INTRODUCTION

Legislation relating to foods was originally introduced in many countries to prevent the sale of fraudulent products and was concerned with compositional or weight defects. Only in more recent times has the legislation been expanded to include public health considerations such as those relating to the transmission of harmful bacteria in foods.

In the United Kingdom the philosophy has been to restrict the amount of legislation to the minimum and to place the onus of responsibility for food safety and hygiene in the hands of the food industry. Legislation has been introduced therefore only where a need for it has been demonstrated and, when introduced, the legislation has been kept as general as is practicable. More specific legislation may have been necessary where foods presented a more serious hazard; such legislation generally defined the processing standards required rather than quoting microbiological standards which must be attained.

In contrast, many European countries have introduced legislation to combat *potential* or *theoretical* hazards whether or not such hazards exist in practice. Such legislation often includes microbiological standards and, as a result, is far more detailed than is equivalent legislation in the United Kingdom. However when the United Kingdom joined the then termed European Economic Community (EEC) in 1973 there was a commitment to create a single internal market amongst the Member States. The ready movement of goods including foods from one community member to another had often been prevented by existing

444

national laws creating trade barriers. It was therefore necessary for legislation to be harmonized so that the free movement of goods between all Member States would be facilitated. This legislative process is to be completed by 31 December 1992 when the Single Market is finally established. Thus at the present time there is a massive legislative programme which must be completed by the end of 1992 although much of it has already been implemented by Member States. The impact of the European Commission's legislation on Member States' national legislation is therefore considerable and thus Community legislation must be discussed before considering, more briefly, aspects of the national legislation of individual countries.

13.2. FOOD LEGISLATION WITHIN THE EUROPEAN COMMUNITY

13.2.1. The Legislative Process

Initially community officials become aware of the need for European Community (EC) legislation for possibly many reasons but any legislation introduced must take into account the interests of Member States and, particularly, the development of the single internal market. Experts in the area of concern are consulted; the main food legislation advisory bodies are the Scientific Committee for Foods, the Advisory Committee on Foodstuffs and the Standing Committee for Foodstuffs.

The Scientific Committee is composed of members who are leading scientists able to give 'opinions' (i.e. advice encouraging good practices but not legally binding) on matters relating to the health and safety of foods. The Advisory Committee comprises permanent members together with co-opted experts; this committee deals with problems concerning the harmonization of legislation relating to foodstuffs. The final Standing Committee is made up of permanent government experts from Member States; as with the Advisory Committee it provides opinions on matters under consideration by the Commission.

Following their deliberations a working paper will be issued, followed by draft legislation drawn up by the Commission's Directorate General responsible for that policy area. Directorates General are part of the European Commission (i.e. are equivalent to ministers within a national government although they are civil servants) and draft proposals are submitted to the Commission proper. If the proposals are adopted they pass to Council which is the highest decision-making body within the EC. From Council the proposals pass, in turn, to both the European Parliament and to the Economic and Social Committee; the latter consists of employers' representatives, trade unions, consumer associations and other interested parties. Opinions from these two sources are considered in due course by Council which has the power to amend the adopted proposals in the light of the additional information received. After further discussions involving several stages within Council a decision is finally taken on the form of legislation, if any, to be introduced. Legislation can be made statutory provided a majority of Member States agree to it, although there is a clause that if 'vital national interest' or fiscal matters are involved unanimity is required.

Legislation appears in the following forms:

Regulations	— are mandatory and legally binding in their entirety on all Member States; no alteration of national law is required for their implementation.
Directives	— are again mandatory on Member States who are required to introduce national laws to comply with the Directive within a specified time limit. Define a result to be achieved but means of achieving it left to Member State.
Decisions	— specific rules addressed to Member States, companies or individuals; are binding in their entirety.
Recommendations and Opinions	— Not binding on Member States but give Community view on good practices.

13.2.2. Examples of EC Food Legislation

It would be quite impossible to cover all aspects of EC food legislation dealing, as it does, with labelling (e.g. ingredients, additives, colouring materials and emulsifiers), packaging and the materials used for packaging (covered here is the problem of the migration of constituents of plastic materials into foods), weights and volumes, and fiscal matters such as VAT. Even legislation dealing more obviously with foods, food safety and hygiene will have to be restricted to important aspects and material representative of the general aims and tenor of that legislation. Fortunately much of it is repetitive so that the examples quoted are illustrative of much additional legislation. However it is worth noting that legislation of relevance comes from two policy areas, viz. from the Directorates General responsible for Agriculture (DG6) and Internal Market Affairs (DG3) respectively. The areas covered by these two often overlap and, although their aims in terms of food safety and hygiene may be common, their attitude to the form of legislation necessary to achieve those aims is by no means harmonic! In general legislation emanating from DG3 is more flexible, and hence more practicable, than that from DG6 whose approach can be unduly authoritarian. Thus, at the present time there is some conflict between the two Directorates General about certain aspects of legislation in the pipeline; in addition, there is still a quantity of EC legislation in draft stages which may not be agreed by Member States in its present form for other reasons.

Regulations can be regarded as the most rigorous form of EC legislation and there are many Regulations dealing with foodstuffs. However, legislation of interest dealing with food safety and hygiene is overwhelmingly in the form of Directives although there are proposals in the pipeline which, if implemented, would see the introduction of Regulations in the field of food safety and hygiene.

One of the earliest Directives of interest was Council Directive 64/433/EEC (1964) which dealt with health problems affecting Community trade in fresh red meat; this contained much useful information and it certainly represented a standard for subsequent legislation. This Directive is therefore considered in some detail together with more recent legislation derived from it.

13.2.2.1. Fresh (Red) Meat Legislation

Council Directive 64/433/EEC, a DG6 document, deals with intra-Community trade in fresh red meat and requires that such meat (e.g. beef, lamb and pork) must be fresh when exported to another Member State. Detailed requirements relating to the slaughtering, dressing, storage and transport of the meats (restricted to large cuts such as half carcasses or quarters) attempt to achieve this aim.

Approved slaughterhouses are required to have adequate lairage facilities, separate rooms for slaughtering and dressing, sufficiently large chilling rooms and adequate separation between 'clean' and 'contaminated' parts of a building. Separate lockable premises must be provided both for sick animals and for their slaughter. Rooms where work on meat is undertaken must have slightly sloping, easily washable and waterproof flooring with adequate drainage; walls must be smooth with rounded angles and corners, and painted up to at least 3 m with coloured and washable paints. There must be adequate ventilation and lighting. In addition, cutting plants, which must also be approved, must have refrigeration equipment so that the butchered meat can be held at a constant temperature of not more than 7°C; equipment in these establishments must be constructed of non-corrodible materials which are easy to clean and disinfect, and tables should be provided with detachable cutting tops.

The Directive also requires that staff working in different areas must be provided with separate changing rooms which must have sufficient showers, lavatories, wash basins with hot and cold water, and disposable hand towels. Staff are expected to wear clean working clothes and headgear; they must be in possession of a medical certificate, renewed annually, showing that they are not harbouring harmful microorganisms, and any person who is a possible source of contamination is prohibited from working with meats.

Amongst the rules for *ante mortem* inspection included in the Directive are those stating that animals brought to the slaughterhouse should be killed immediately and subsequent bleeding of the animal must be complete; this must be followed by immediate skinning of the carcass (pigs exempted) and, thereafter, immediate evisceration. Amongst *post mortem* rules are those requiring carcasses to be cooled immediately and then held at not more than 7°C (3°C for offal) following inspection; temperatures are also laid down for the transport of meats which must be in sealed vehicles or containers. Requirements for vehicles include the need for inside surfaces coming into contact with meats to be non-corrodible, smooth, and easy to clean and disinfect. Vehicles used for transporting meats cannot be used for the transport of live animals, whilst no other product may be transported at the same time as meats. All meat must bear an official health mark and must be accompanied by a health certificate during transport to the country of destination.

Subsequent legislation, primarily in the form of Directives amending Council Directive 64/433/EEC, has further tightened and extended the legal requirements. Thus Council Directive 69/349/EEC extended the role of official veterinary officers to include the supervision of the cleanliness of buildings, facilities and instruments, staff hygiene, and sampling for microbiological tests, these to be performed in approved laboratories.

Other additions in this Directive were that taps must not be hand-operated and, at the time of cutting, the pH of the meat must be between 5·6 and 6·1. It is worth noting that the above and earlier legislation allowed Member States to prohibit the importing of fresh meat from other Member States if conditions governing approval were not being met; Member States were also able to prevent the movement of meat within the Community where disease had broken out. Council Directive 72/462/EEC expanded the above legislation to embrace 'third countries' (i.e. non-Community members) so that meat exported by these countries to Member States would be of the required EC standard. Council Directive 83/90/EEC was a major revision of Council Directive 64/433/EEC in the light of more modern processes and techniques. It introduced a general section on conditions required for approval of fresh meat establishments. (N.B. In the UK this is the responsibility of the Ministry of Agriculture, Fisheries and Foods (MAFF) but in most Member States it is in the hands of official veterinary organizations.) The impact on food hygiene and safety legislation was only minor however, e.g. hot water to be supplied at not less than 82°C, doors to be made of non-corrodible material, and demarcation by colouring of pipework carrying potable and non-potable water. Additional requirements for slaughterhouses included the use of overhead rails for the movement of meat carcasses and the suspension of meat from such rails during dressing. Cutting room temperatures must not exceed 12°C and all operations such as cutting and boning should be performed on meat with an internal temperature of 7°C or under; however permission was given for meat to be cut warm if transferred directly from the slaughterhouse.

Council Directive 85/323/EEC requires that any operator or proprietor handling fresh meat must carry out regular checks on the general hygienic conditions of the establishment, these checks to include microbiological analyses. It was appreciated that harmonized microbiological testing methods would be required but veterinary officers were held responsible for assessing results and for any other tests deemed necessary. Spot checks on fresh meat establishments were introduced in Commission Decision 85/446/EEC to verify that the Directives were being obeyed. Minor amendments of interest in Council Directive 88/288/EEC were, first, the inclusion of meat establishment ceilings in the legislation, the requirement being that they should be clean or easily cleaned, and second, arm-operable wash basin taps were no longer permissible.

With the mass of legislation which had accumulated on fresh meat it seemed prudent to update Council Directive 64/433/EEC once again and this has been done recently. Council Directive 91/497/EEC contains a lengthy Annex which is effectively a complete redraft of the original Directive even if much of it remains intact. Now included are poultry, rabbit, horse and mechanically recovered meats but retail outlets remain excluded from the legislation.

The microbiological tests to be performed are now much more detailed in that there must be checks on utensils, fittings and machinery at all stages of production; where necessary final products should be analysed. (N.B. This seems to imply the implementation of HACCP principles.)

General conditions are laid down for the approval of meat-handling establishments. There are also special requirements for slaughterhouses and cutting plants; in both the required layout and structure of premises to be approved is stressed. Staff hygiene is covered and it is noteworthy that annual medical certificates for staff working on and handling meats are no longer necessary. More precise information is given on meat-handling hygiene; for example, evisceration must be complete within 45 min of stunning. Requirements for the transport of fresh meat now include the use of hermetically closed systems or sealed transport so that the prescribed meat storage temperatures are not exceeded. During transport carcasses and smaller cuts of fresh meat must also be suspended from corrosion resistant fittings so that the meat does not touch the floor of the vehicle. Packaged and unpackaged meats may only be transported together where there is adequate physical separation.

Derogations are allowed for small slaughterhouses and cutting plants but they must still be approved nationally. The general conditions needed for approval are similar in principle to those for larger businesses except that there is a degree of flexibility over the need for separate rooms for different processing operations; instead, clearly separated areas are considered acceptable.

Finally, requirements are laid down for professional qualifications for auxiliaries. Their course of instruction should typically consist of 400 h theoretical work together with 200 h practical training under an official veterinarian.

One of the difficulties facing the UK is to establish the exact role of veterinary officers, who, in continental Member States, have long had an important part to play in monitoring health aspects of foods. Unfortunately in the UK relatively few veterinary officers have had adequate food hygiene training and experience, particularly at the local level and they have lacked the required knowledge of food processing operations. And yet they are answerable, through MAFF, to the

Standing Veterinary Committee in Brussels, a rather inflexible authority in terms of interpretation of legislation. There should clearly be a greater understanding of and sympathy in the application of the legislation but this can only come with experience of the food industry. A further complication in the UK is the position of Environmental Health Officers (see later) whose interests frequently overlap those of the veterinary officers, but whose training may be more suitable when dealing with food hygiene and safety matters.

13.2.2.2. *Meat Products and Minced Meat Legislation*

The main legislation on meat products is to be found in Council Directives 77/99/EEC and 88/658/EEC. Meat products were defined in the earlier Directive as having been prepared wholly or partly from meat which has undergone treatment to ensure a certain degree of preservation, i.e. heating, salting or drying or any combination of these treatments. Again there are detailed requirements relating to the structure of the premises, the facilities in the premises, to the equipment, to the correct operation of the premises and to working temperatures (cutting up and deboning rooms, curing rooms and slicing/cutting and wrapping rooms must all be at temperatures not exceeding 10°C); the Directive also covers personal habits and the exclusion of personnel from handling raw meats if they are carriers of or are suffering from an infectious disease. An element of final product testing is introduced into this legislation by stipulating that for meat products in hermetically sealed containers (i.e. primarily cans) the efficacy of the seals and the heat treatment received must be assessed at regular intervals. This assessment may involve the incubation of representative samples to confirm that the product and containers are sound. Again, it is the responsibility of Member States exporting foods to supply an EC health certificate confirming that sound raw materials were used and that production was in accordance with EC standards.

Council Directive 88/658/EEC extends the above legislation to parallel that seen in equivalent fresh meat legislation. Minor points of interest are that meat or meat preparation containers must not be placed directly on to floors and arm-operable taps are prohibited. There is however, a derogation allowing wood to be used in rooms where meat products are smoked, cured, matured, pickled, stored or dispatched, if the wood is essential for technological reasons. (N.B. The definition of meat product is changed in this Directive to meat that has undergone treatment to alter the characteristics of the cut surface from that of fresh meat.) There is a slight relaxation in the maximum allowable temperature (now 12°C) of rooms where cutting, curing, etc., is performed. There are additional requirements detailed for establishments manufacturing meat products; these are concerned primarily with the layout and structure of rooms used for different purposes. More stringent testing protocols are laid down for canned products which include regular checks of residual chlorine levels in the can cooling water. A section is introduced on prepared meals which must be produced under conditions of strict temperature control. Such meals must be packaged immediately before or after cooking, after which the meals must be cooled to an internal temperature of 10°C within 2 h.

Council Directive 83/201/EEC establishes exceptions from Council Directive 77/99/EEC for certain products containing only a small percentage (under 10%) of meat or meat products. In essence the derogations permitted are for the foodstuffs other than meat although further derogations may apply if the meat has been completely treated before receipt. Council Directive 89/227/EEC extends EC standards to meat products prepared in third countries exporting to Member States; the Directive further requires Member States to draw up a list of approved third countries from which Community Members may obtain meat and meat products. Finally Council Directive 89/397/EEC allows for inspection, without prior warning, of foods at all stages of preparation. Inspection includes the sampling and analysis of meats, staff hygiene, and the examination of microbiological and other relevant results and documentary material.

Turning to minced meat, and meat in pieces of less than 100 g, the main legislation covering these foods is to be found in Council Directive 88/657/EEC. The general composition and intent of the legislation is similar to that for fresh meat and meat products. Thus, in terms of control of temperatures, that of premises is restricted to 12°C whilst the internal temperature of meat may not exceed 7°C. However, where the duration of operations on minced meat exceeds 1 h, the meat may not exceed 4°C although this requirement can be relaxed if no hygiene problem is shown. Immediately after production meat should be stored at 2°C if chilled. (N.B. This temperature is regarded as too low since surface freezing is possible.) It is again incumbent on the owner to carry out regular checks on general hygiene and production conditions. Of particular interest in this Directive is the inclusion of specific microbiological tests which must be carried out, mainly on a daily basis, together with the imposition of bacteriological standards which must be adhered to.

Standards are laid down for 5 groups of bacteria (viz. aerobic mesophiles, the viable counts being performed at 30°C; *E. coli*; sulphite-reducing anaerobes, these are roughly equivalent to *Clostridium perfringens* and a number of closely related species; staphylococci, *S. aureus* is assumed here, and salmonellas; Table 13.1). Three categories of contamination are recognized for the first 4 groups of bacteria based on: (1) counts up to and including *m*; (2) counts between *m* and threshold *M*; and (3) counts above *M*. The quality of any consignment is considered *satisfactory* where all counts are equal to or less than 3 *m* (that is, taking *E. coli* as an example the maximum permitted would be 150 per g and not 50 per g); this refers to counts using a solid growth medium and where a multiple-tube liquid medium is used the equivalent count is equal to or less than 10 *m* (i.e. for *E. coli* a 167 per g maximum). Typically liquid media counting techniques give rather better recoveries but they are laborious to do. The apparent anomaly created is a good example of the sometimes illogical thinking of those responsible for this sort of detailed legislation. Thus, having clearly defined the 'satisfactory threshold' as *m* (i.e. 50 per g for *E. coli*), 3 *m* is then taken as 'satisfactory'. It is thus difficult to determine what the true threshold really is!

The second category of quality is *acceptable* where *M* is not exceeded in any count (i.e. 500 per g for *E. coli*) and where $^c/_n$ is equal to or less than $^2/_5$. An *unsatisfactory* quality of consignment is where *M* is exceeded in any count or where $^c/_n$ is greater than $^2/_5$. For salmonellas only 2 categories are recognized, *satisfactory* where no salmonellas are isolated and *unsatisfactory* where they are.

In addition to the above criticisms it could justifiably be questioned whether this degree of detail is necessary in microbiological standards. As much could be achieved by a simplified format with emphasis on salmonella testing; this would in no way undermine the quality and safety of meats.

Turning to more general matters, there is a proposal in draft form allowing derogations for food companies producing only limited amounts of minced meat and meat preparations; this would enable companies to reach EC standards over an agreed period without preventing them trading nationally. This example of flexibility is eminently sensible.

13.2.2.3. *Poultry and Poultry Meat Legislation*
Legislation concerning poultry and poultry meat is to be found principally in Council Directive 71/118/EEC and in the revised version, Council Directive 75/431/EEC. The former Directive, which covers

TABLE 13.1

Bacteriological Standards for Minced Meat and Meat Pieces of Less than 100 g or as an Ingredient in Meat Preparations[a]

Bacteria	Counts per g		Criteria	
	M^b	m^c		
Aerobic mesophiles (30°C)	5×10^6	5×10^5	$n^d = 5,$	$c^e = 2$
E. coli	5×10^2	50	$n = 5,$	$c = 2$
Staphylococci (S. aureus)	5×10^2	50	$n = 5,$	$c = 1$
SO_2 reducing anaerobes	100	10	$n = 5,$	$c = 1$
Salmonellas	Absent in 25 g		$n = 5,$	$c = 0$

[a] From Council Directive 88/657/EEC.

[b] M = acceptability threshold above which counts are no longer satisfactory where M equals $10m$ (count in solid medium) and M equals $30m$ (count in liquid medium using multiple tubes).

[c] m = threshold below which all counts are satisfactory

[d] n = number of samples tested.

[e] c = number of samples giving values between m and M.

chickens, turkeys, ducks, geese and guinea-fowls, again prescribes similar requirements to those for fresh red meat and meat products. This Directive covers both chilled and frozen poultry and poultry meat as well as any poultry which has not undergone a preservation process; poultry meat must not have been treated with antibiotics, preservatives or tenderizers. Member States must ensure that birds are killed and processed as quickly as possible in approved slaughterhouses, and that carcasses are handled hygienically during processing. Slaughter rooms must be large enough for stunning and bleeding, on the one hand, and plucking and scalding on the other, to be performed in separate places; a separate room is required for evisceration.

Essentially similar conditions are laid down for poultry meat (i.e. parts of carcasses and boned meat). Thus cutting premises must again be approved whilst with both poultry and poultry meat *ante mortem* and *post mortem* inspection must be performed by official veterinary officers or qualified assistants. Appropriate health certification and labelling is required and hygiene requirements for staff follow the normal pattern.

Temperature control for poultry and poultry meat is stringent in that meat must be stored and transported at or below 4°C. Transport is permitted in vehicles in which live animals have been transported, provided that the vehicle has been thoroughly cleansed, disinfected and deodorized where necessary; fresh meat may also be transported at the same time as substances likely to affect it provided suitable precautions are observed!

Perhaps the most interesting aspect of this legislation is the condemnation of the spin-chiller system (see Section 3.5.1) which, according to Council Directive 71/118/EEC, was to be prohibited from January 1976. Difficulties in enacting this requirement were already evident in Council Directive 75/431/EEC which postponed any change to January 1978. A further derogation in this Directive allows EC exemptions for farmers supplying poultry directly to the consumer although Member States are required to maintain suitable health controls. Spot checks were incorporated into the legislation and bacteriological sampling for harmful organisms was introduced.

Council Directive 78/50/EEC further extended the period of derogation on the use of spin-chillers but it was now stated that derogations would not be allowed beyond August 1982. This Directive stipulates the volume of water needed in the chillers for different sizes of poultry carcass and requires that carcasses must be propelled constantly, by mechanical means, through a counter-flow of water (see Section 3.5.1). The temperature of the water in the chiller tank(s) at entry and exit of birds must be no higher than 16°C and 4°C respectively. More recent Directives have further extended the derogation period on the use of spin-chillers, the replacement of which is, at least for the time being, apparently deemed impracticable; as a palliative detailed instructions for cleaning chillers were introduced.

Council Directive 85/324/EEC requires operators to perform regular checks on the general hygienic conditions of all stages of production. The formulation of a code of good hygiene practice for poultry meat is the same as in Council Directive 85/323/EEC for fresh red meat. More recent legislation in proposal form extends the scope of the requirements on poultry to include pigeons, farmed pheasants, quails and partridges.

13.2.2.4. Fish, Shellfish and Fishery Products Legislation

It is only very recently that substantive legislation concerned directly with fish and related foods has been ratified but there still remains only a fraction of the legislative material that there is for the meats discussed above.

Before the recent legislation Council Directive 79/923/EEC was concerned with the quality of waters required for shellfish (i.e. bivalve molluscs such as oysters, mussels, clams, cockles and scallops). It prescribed standards which were to be achieved within 6 years. During this period it was assumed that pollution levels would be reduced substantially so that the guideline standards quoted were seen primarily as a target to be achieved. A microbiological standard was proposed along with many other recognized pollution indices such as an acceptable pH range, dissolved oxygen level and minimal concentrations of various trace metals. The microbiological standard allowed a maximum of 300 faecal coliforms per 100 g shellfish flesh or per 100 ml intravalvular liquid; this guideline standard was mandatory for shellfish waters from which shellfish were to be eaten 'directly'.

Turning to the recent legislation Council Directive 91/493/EEC lays down health conditions for the production and placing on the market of fishery products. This Directive aims to harmonize measures for the handling and treatment of all fish and shellfish at all stages up to retailing. The general content and aims of the legislation are in line with that for meat and need not be reiterated. There is the usual detail on hygiene standards, this covering both factory vessels (i.e. larger fishing boats on which some processing is carried out) and fish processing establishments; smaller fishing vessels are not included in the legislation. Temperature control requirements for fish after unloading are specified; emphasis is placed on the importance of speed in handling unloaded fish and the need to keep the temperature of fish as low as possible by the use of ice is stressed. Where fish are not dispatched or processed within a 'reasonable time after unloading' they must further be stored in ice in a cold room.

Where fish are consumed raw, or where certain types of fish (e.g. herring and mackerel) are cold smoked or marinated, they must be examined for the presence of nematode larvae and frozen ($-20°C$) for at least 24 h. Further checks required include a sensory evaluation to confirm freshness and, where doubt exists, follow-up biochemical tests (e.g. on histamine levels) should be performed. Microbiological tests are not obligatory but they can be performed where public health may be at risk.

A further major recent Council Directive, 91/492/EEC, extends the legislation in Council Directive 79/923/EEC and lays down health conditions for the production and placing on the market of live bivalve molluscs. It again deals with conditions for harvesting areas together with requirements for purification plants, processing establishments and transport. Microbiological standards are included on this occasion. For bivalves intended for immediate consumption the standards are 300 and 230 per 100 g respectively for faecal coliforms and *E. coli*, together with a requirement that *Salmonella* spp. must be absent in 25 g samples of flesh; further standard maxima are stipulated for certain toxins associated with shellfish (see Section 2.6.2.1). Again it is recognized that there is a need for agreed analytical methods to be established whilst the effectiveness of faecal indicator bacteria counts is to be reviewed.

13.2.2.5. *Milk Legislation*

The major legislation concerning milk is to be found in Council Directive 85/397/EEC although there are proposals for Regulations awaiting agreement which extend and strengthen the legislation. The principles embodied in the Directive are that milk must come from healthy cows and be so treated that human health is not endangered. However it is recognized in the Directive that there are many problems to be resolved so that initial action is restricted to fixing general standards for heat-treated milk.

The Directive includes the usual criteria for approval of dairies and collection centres covering the premises and staff. There are special requirements for equipment covering items such as mechanical fillers and automatic sealers for containers, cooling equipment and storage facilities for heat-treated milk. It is specified that equipment surfaces coming into contact with milk must be made of smooth materials which resist corrosion and are easy to clean and disinfect.

After milking, milk must be stored immediately under hygienic conditions and, if not collected from the farm within 2 h the milk must be cooled to 8°C (daily collection) or 6°C (non-daily collection); in addition, the temperature of the milk must not exceed 10°C during transport to the depot. Raw milk should be heat treated within 4 h of arrival at the depot but, if not, it must be cooled to 5°C and maintained at that temperature until so treated. The total viable count (30°C incubation) of such milk should not exceed 300 000 per ml but tighter standards were envisaged at a later date. However if the raw milk is not treated within a maximum of 36 h further microbiological testing is necessary to ensure that the count does not exceed 600 000 per ml when normal processing may proceed. (N.B. It is interesting to speculate how this particular criterion can work in practice since a minimum incubation period of 24 h would be necessary before counting of colonies could commence; by this time there would, presumably, be a further increase in the total count of the raw milk.)

Pasteurized milk, required to be held at 71·7°C for 15 s or any other equivalent combination of temperature and time, must also satisfy certain

Table 13.2
Bacteriological Standards for Pasteurized Milk[a]

Bacteria	Step 1[b]	Step 2[c]
Pathogens	None	None
Coliforms (per ml)	<5	<1
Total viable count 30°C (per ml)	50 000 max	30 000 max
Milk incubated at 6°C for 5 days		
Total viable count 21°C (per ml)[d]	250 000 max	100 000 max

[a] From Council Directive 85/397/EEC.
[b] Milk for export.
[c] With effect from April 1990 for milk consumed directly without heating; from January 1993 for other milk unless postponed until January 1995.
[d] This is essentially a count of psychrotrophic bacteria.

standards laid down; these include both microbiological (see Table 13.2) and non-microbiological standards (e.g. phosphatase). Additional standards are imposed for UHT and sterilized milks; these milks, after pre-incubation at 30°C for 15 days should have a total viable count (30°C) of no more than 10 per 0·1 ml and, at the same time, there should be no observable deterioration in quality.

The Directive finally requires that pasteurized milk must be stored and transported at 6°C or below although doorstep deliveries may be exempted; interestingly, the legislation also permits the use of milk churns.

Recent legislation in the form of proposals for Regulations would amend and extend the legislation in Council Directive 85/397/EEC. Proposals Com (89) 667 and Com (89) 672 tend to overlap and it seems likely that the two will be combined into one Regulation. The former proposal covers raw milk and milk-based commodities produced from goats, sheep and buffaloes as well as from cows. It contains detailed microbiological and other standards covering the above foods. The sale of raw milk for human consumption is still permitted although most Member States prohibit such sales. The microbiological standards recommended now include a total viable count limit of 100 000 per ml raw milk whilst a *Staphylococcus aureus* maximum of 300 per ml is added where raw milk is consumed directly. The maximum storage temperature specified in the earlier Directive for raw milk not treated within 4 h is relaxed slightly from 5°C to 6°C.

Detailed microbiological standards for milk-based products are introduced in this proposal. The products covered are cheeses, butter, and liquid, powdered and frozen milk products. Amongst the most wide ranging standards are those proposed for soft cheeses. *Listeria monocytogenes, Salmonella* spp. and other pathogens must be absent from all 25 g samples analysed; limits are also prescribed for *S. aureus* (1000 per g maximum) and for 'index' and 'indicator' coliforms (see Section 4.3.7) although the standards are defined using the nomenclature described in Table 13.1.

Proposal Com (89) 672 is primarily a revision of Council Directive 85/397/EEC with the bacteriological standards and temperature requirements defined as in the proposal cited above. Still retained is the clause concerning raw milk not heat treated within 36 h, the only difference being that the total count standard for such milk is tightened from 600 000 to 200 000 per ml. The standards for pasteurized milk are essentially similar to those quoted in Table 13.2 but with the revised nomenclature (again see Table 13.1). Thus for coliforms (per ml): $M = 5$, $m = 0$, $n = 5$ and $c = 1$. Similarly the psychrotrophic count (per ml) now reads: $M = 10^5$, $m = 10^4$, $n = 5$ and $c = 1$. The standards for UHT and sterilized milk remain unchanged.

Commission Decision 91/180/EEC finally lays down techniques for the analysis of milk and milk-based products. It specifies sampling and microbial enumeration techniques in considerable detail; dilution methods, the pour plate counting technique and required media are all described.

13.2.2.6. Egg Products Legislation

Legislation dealing with egg products is to be found in Council Directive 89/437/EEC. This covers products only, made from the eggs of either chickens, ducks, geese, turkeys, guinea-fowls or quails. Member States must ensure that egg products have been prepared under suitable hygienic conditions in an approved establishment; the legislation excludes small concerns, shops and restaurants where foods are consumed within the premises or where eggs are used in the preparation of foods sold directly to the public.

The same general rules apply as before for the construction of premises, specialized equipment therein, the staff employed and inspection protocols. Special conditions include separate storage rooms for eggs and finished egg products, a separate room for egg breakage and facilities for the immediate removal and separate storage of empty shells, and eggs and egg products unfit for human consumption. Cracked eggs may be used for the preparation of egg products provided they are delivered directly from the packing centre or farm and are used as quickly as possible. Egg contents, generally, must be treated as soon as possible and if not immediately heat treated they must be stored at or below 4°C for a maximum of 48 h, or frozen immediately. Packaging, general storage and transportation requirements are also covered in the Directive. Sampling for laboratory analysis is also required and the establishment must either have an approved laboratory or secure the services of one for the analyses.

Microbiological standards are included for finished egg products which must satisfy the following criteria:

(a) salmonellas—absent in 25 g or ml of egg product;
(b) aerobic mesophilic bacteria—100 000 maximum in 1g or 1 ml;
(c) Enterobacteriaceae—100 maximum in 1 g or 1 ml;
(d) *Staphylococcus aureus*—absent in 1 g or 1 ml.

The consignment is regarded as unsatisfactory if even one sample is above the limits defined.

13.2.2.7. *Water and Mineral Water Legislation*

Council Directive 80/778/EEC deals exclusively with drinking water and it specifies numerous parameters to be taken as guidelines in fixing standards. Most are of little interest including, as they do, maximum levels of chemical elements, organoleptic criteria (e.g. taste, colour) and physico-chemical factors such as the pH range; also included are levels of nitrates and nitrites to be permitted. Microbiological standards are also laid down. These are:

Total viable count (37°C incubation)	— 10 per ml maximum
Total viable count (22°C)	— 100 per ml maximum
Coliforms, faecal coliforms and faecal streptococci	— all absent in 100 ml
Sulphite reducing clostridia (i.e. mainly *Clostridium perfringens*)	— 5 per 100 ml maximum

In addition no pathogens (e.g. salmonellas, enteroviruses) should be isolated where examination is deemed necessary; in addition, drinking water should not contain parasites or algae. (See also Section 11.2.4.)

Council Directive 80/777/EEC deals with natural mineral waters which are defined as waters extracted from the ground or from natural springs and recognized as such by a responsible authority of a Member State. The Directive allows for unstable elements to be extracted from the water, for free carbon dioxide to be partially or totally eliminated, and for carbon dioxide to be reintroduced. The legislation stresses the need for microbiological criteria to be met. The standards prescribed are:

After bottling, total viable count (20–22°C) — 100 per ml maximum
and, total viable count (37°C) — 20 per ml maximum
At source, counts are 20 and 5 per ml respectively.

In addition, natural mineral water must be free from *E. coli*, other coliforms, faecal streptococci and *Pseudomonas aeruginosa* in any 250 ml sample examined; sulphite reducing clostridia must also be absent from any 50 ml sample examined. Finally, waters must be free of parasites and pathogenic microorganisms.

13.2.2.8. General and Prospective Food Hygiene Legislation
The legislation discussed so far has originated almost entirely from Directorate General 6 (Agriculture) but, as mentioned earlier, legislation of interest can also come from Directorate General 3 (Internal Market Affairs). The latter was responsible for Council Directive 89/397/EEC which lays down general principles for the overall control of foodstuffs, food additives and packaging materials. In many ways it covers the same general ground as much of the detailed DG6 legislation and thus the need for inspection of premises and laboratory analyses of raw materials, semi-finished and finished products is stressed. The Directive aims to establish trading standards within the Community and it seeks to encourage an exchange of information amongst Member States to determine the frequency and type of inspections required. A draft Council Directive (III/8420/89) from DG3 aims to provide a framework of hygiene principles which will protect human health from food contaminated microbiologically or chemically. It again covers the production and distribution of foods and a guide Code of Hygiene Practice is expected in a later revision. However this DG3 initiative conflicts with a proposal (Com (89) 492)† from DG6 covering health rules for the production of foods (e.g. game and honey) not already covered by other legislation; this proposal calls for greater co-operation between manufacturers and controlling authorities, and amongst Member States, in order to control dangers to public health from contaminated foods of animal origin. A further proposal (amended as Com (90) 259) for a Council Directive on general product safety will probably include foods; the proposal aims to introduce a requirement on producers, importers and, in this case, retailers to

†Former Proposal, Com(89)492 has recently been combined with existing legislation to form a new Directive (Council Directive 92/5/EEC).

supply only safe products to the consumer. As will be obvious by now there are numerous Directives applying to different aspects of food safety and hygiene; it could be questioned as to whether much of this recent legislation is causing anything but confusion?

Finally there is a proposal (Com (91) 310) which aims to establish measures to control specified zoonoses in animals and animal products in order to prevent food-borne infections and food poisoning. Zoonoses are regarded as diseases and infections naturally transmitted between animals and humans. Particular emphasis is laid on *Salmonella* spp., *Listeria monocytogenes, Campylobacter* spp. and *Yersinia enterocolitica* although other organisms could be included. The proposal is far reaching and Member States will be obliged to specify the structure and distribution of the domestic animal population, and detail the number and distribution of establishments producing foods of animal origin. Member States will be expected to report on the incidence of *Salmonella* spp. in humans and animals, in foods of animal origin and in animal feedingstuffs. *L. monocytogenes* will be monitored in cheeses as will *S. enteritidis* in poultry and eggs. It is also recommended that detailed sampling plans are introduced to detect the sources of these different organisms in animals, animal feedingstuffs and products of animal origin. It is further aimed to introduce control measures to limit the spread of these bacteria.

13.3. INTERNATIONAL FOOD STANDARDS

Since the beginning of this century it has been suggested that international food standards should be considered. One of the earliest organizations attempting to draw up such standards was the International Dairy Federation; this body discussed standards covering milk and dairy products and was later (1956) involved in initiating the first international food standards negotiations with the support of the Food and Agriculture Organization (FAO) and the World Health Organization (WHO). Arising from these negotiations a 'Code of Principles concerning Milk and Milk Products' was prepared and this became widely accepted internationally.

Independently a Codex Alimentarius programme was initiated in the early 1960s under FAO/WHO control with the specific aim of getting international agreements on food standards and codes of practice which would safeguard the health of consumers and generally encourage good practices in the food trade. It was hoped to agree standards, including food hygiene standards, for all the main foods whether processed, semi-processed or even raw. In fact, at the present time, Codex standards are mainly concerned with food composition and quality, additives, pesticide residues and labelling. There is a Codex Committee on Food Hygiene which is able to recommend microbiological test methods in order to verify hygiene requirements. Codex Alimentarius can establish microbiological standards for foods, these standards being for relevant pathogenic and spoilage organisms. They stress that microbiological criteria should be adopted only where a definite need for them has been shown and they are effective and practical. This restrained approach is rather different from some of the recent EC legislation involving microbiological standards.

Some of the 160 Member Countries have incorporated Codex standards and Codes of Practice for many foodstuffs into their national food laws. This includes some of the Member States within the EC whereas others regard these criteria as targets for the future. Recent EC legislation (e.g. draft Council Directive III/8420/89; see p. 461) has been influenced by Codex thinking and there is little doubt it has a major role to play in the coordination of food standards world-wide.

13.4. FOOD HYGIENE AND SAFETY LEGISLATION IN THE UNITED KINGDOM

In the United Kingdom modern food legislation is based upon a relatively small number of Acts of Parliament. The primary legislation is now contained within the Food Safety Act (1990) which is based upon relevant parts of the earlier, and now largely repealed, Food and Drugs Act (1955) and Food Act (1984). All these Acts are examples of 'enabling' or 'empowering' Acts under which Regulations, and Codes of Practice for that matter, can be introduced by the relevant ministry, i.e. the Department of Health or the Ministry of Agriculture, Fisheries and Food. However it should always be borne in mind that any proposed national legislation that the United Kingdom, or any other Member State, wishes to introduce should be agreed tacitly by the European Commission.

13.4.1. The Food Safety Act (1990)

The Food Safety Act (1990) aims to control all aspects of food safety right through the so-called food chain from farms, and even animal feedingstuffs, to retail outlets. The Act thus aims to protect human health by ensuring the overall quality of food which must be produced and handled in a suitably hygienic environment. The Act therefore covers the construction, layout, maintenance and cleanliness of food premises, and the equipment therein, where food is prepared or handled in some other way. Hygiene requirements for staff are also covered and in this respect an important new provision has been introduced. It is now required that proprietors and employees involved in food businesses must have food hygiene training. Many personnel already receive some form of training through courses organized by the food industry and local authorities but standardization of courses is necessary; here local authorities will have a bigger part to play.

The Food Safety Act (1990) makes it an offence, either to sell food which is injurious to health, or is not of the nature, substance or quality demanded by the purchaser, or to fail to comply with food safety requirements, or to falsely label a food. The Act also requires that premises must be registered and that they are only used for certain prescribed purposes acceptable on hygiene grounds (e.g. as slaughterhouses or to manufacture ice-cream). The Act strengthens powers of entry to food premises by local authority officers; powers of enforcement are similarly tightened by the introduction of local emergency prohibition orders for premises where a health risk is deemed to exist. It is now also possible to use emergency control orders where urgent action is needed on more widespread problems. Food can be seized for failing to meet the aforesaid requirements and food can be taken for microbiological testing or other forms of analysis. Finally, the Act provides power to prosecute offenders for whom penalties have been substantially increased. For example, imprisonment for up to two years is now possible for certain offences and heavy fines can also be imposed; at the same time the Act allows the right of appeal to independent courts.

13.4.2. Codes of Practice

It is pertinent to reiterate that legislation has only been introduced in the United Kingdom where a need for it has been shown to exist and, when introduced, the legislation has tended to be fairly general. To supplement the legislation, guidelines for the food industry are made available in the form of Codes of Practice; these have no statutory powers but contain advice on how the food legislation aims may be achieved, whilst at the same time recognizing that other methods may achieve the same objectives. In the case of food hygiene some ten Codes have been published: these cover a wide range of topics and include, for instance, *The Hygienic Transport and Handling of Fish* (No. 4), *Poultry Dressing and Packing* (No. 5), *Hygiene in the Bakery Trade and Industry* (No. 6), *Hygiene in the Meat Trades* (No. 8) and *The Canning of Low Acid Foods* (No. 10). These Codes are prepared by government officials with advice from experts in the food industry, an arrangement which has helped to create and foster a close and harmonious relationship between the two groups.

13.4.3. Food Hygiene (General) Regulations (1970)

The principal regulations laid down under the Food and Drugs Act (1955) are to be found in the Food Hygiene (General) Regulations (1970). In essence the Regulations require a suitable standard of construction and cleanliness in food premises, the hygienic handling of food, cleanliness on the part of employees and suitable facilities for them; these Regulations therefore aim to ensure a suitable environment for food production. Flexibility and practicability are important features of these Regulations but it should be borne in mind that they specify the *minimum* standards required; it is assumed that in most instances far higher standards will be attained.

The following points from the Regulations are of particular interest:

1. **Premises.** No food business can be permitted in premises which are insanitary or where food is exposed to the risk of contamination as a result of the poor condition, construction or situation of the premises. The premises must be well lighted and ventilated. Every food room should be kept clean and in good repair to prevent, *as far as is reasonably practical*, any risk of infection. Adequate space, suitably sited, must be provided for waste materials prior to disposal. No refuse should be allowed to accumulate in a food room *except so far as may be unavoidable.*

2. **Equipment.** Articles or equipment in contact with food must be kept clean and in good repair. Construction should enable thorough cleaning to be performed and prevent, *so far as is reasonably practical*, any risk of food contamination. The construction materials used should prevent, *so far as is reasonably practical*, any matter being absorbed onto them.

3. **Facilities and amenities.** A clean and wholesome water supply must be available. Adequate wash-hand basins, with hot and cold

water supplied, must be readily accessible to food handlers; soap or other suitable detergents, nail brushes and clean towels or other suitable drying facilities must be provided. Sanitary conveniences of a suitable standard must be provided and must display notices requesting users to wash their hands. Suitable accommodation for outdoor clothing must also be provided as must appropriate first-aid materials including waterproof dressings.

4. **Food handlers.** Persons handling food or responsible for cleaning food equipment must take all steps, *as may be reasonably necessary*, to protect the food from the risk of contamination. Food handlers must keep themselves as clean as is *reasonably practical*, cover any open cut or abrasion with a waterproof dressing and refrain from spitting or smoking. Persons suffering from, or carrying, certain types of harmful microorganisms (e.g. salmonellas) must inform the employers who must, in turn, notify the Medical Officer of Environmental Health of that district. N.B. Such an infected person may be required to discontinue work until free from infection under the Public Health (Infectious Diseases) Regulations (1968).

5. **Holding temperatures for certain foods.** Whilst on catering premises certain foods, primarily comprising meat, fish, gravy and imitation cream, must be kept at either above 62·7°C or below 10°C. Food delivered to catering premises must be brought to these temperatures without any avoidable delay. A few deviations from these temperatures are allowed, where the preparation requires it, provided the requirements are observed later, and where sale conditions make them impracticable.

The above passages have not been quoted verbatim and the italics are those of the author—the tolerances allowed in the Regulations are thereby emphasized. The above Regulations relate to all foods except milk and cream (see Section 13.4.5), foods covered by the Slaughterhouses (Hygiene) Regulations and a small number of miscellaneous exceptions.

13.4.4. Food Hygiene (Amendment) Regulations (1990)

The Food Hygiene (Amendment) Regulations (1990) have made major changes in the requirements for the temperature control of foods. The new Regulations apply not only to catering premises, as in the Food

Hygiene (General) Regulations (1970), but also to all stages in the manufacture, storage, distribution and retailing of foods. In essence the new Regulations apply to food in which there is danger of food poisoning bacteria growing if temperature controls are not applied. The foods covered include: soft cheeses, e.g. Brie, Danish Blue, Camembert; cooked products (whether reheated or not) containing meat, fish, eggs, or their substitutes, cheese, cereals, pulses or vegetables; smoked or cured fish and meat (which has been cut/sliced after treatment); certain dairy based desserts which have a pH of 4·5 or above; prepared vegetable salads; uncooked or partly cooked pastry products containing meat or fish; sandwiches and rolls containing meat, fish, eggs, soft cheeses or vegetables and cream cakes. There are many foods which are exempt from the temperature controls and these include: bread, biscuits and cakes; icecream; milk; uncooked bacon and ham; fruit pies, and preserves and jams. Other foods are excluded on the basis of their being sold within a limited period after preparation.

Two temperatures are specified, 8°C as a maximum holding temperature for the specified foods and 63°C as a minimum for foods kept and served hot. These temperature requirements came into effect on 1 April 1991 but, from 1 April 1993, the maximum holding temperature for certain foods is further lowered to 5°C. These are the cooked products (listed above) intended to be eaten without reheating, smoked or cured fish and meat (sliced/cut after treatment), cut soft cheeses and relevant sandwiches and vegetable salads.

The new legislation has undoubtedly tightened the temperature requirements for a wide range of foods but it suffers from a serious and surprising omission; that is, no maximum time for holding foods at the required temperatures has been specified and this must weaken the value of the legislation. Time is mentioned only when dealing with unavoidable situations such as refrigerator defrost cycles or temporary equipment breakdown. Here a maximum period of 2 h is specified during which time a temperature rise of up to 2°C above the temperatures stipulated is permissible.

13.4.5. The Milk (Special Designation) Regulations (1977) and (1986)

Until recently milk was dealt with by different legislation from other foods with numerous special regulations for milk having been issued. Milk is now treated like other foods under general provisions, a legislative change that was introduced in the Food Safety Act (1990). The main regulations concerning milk are the Milk (Special Designation) Regulations of 1977 and 1986. These Regulations cover the treatment of 'untreated', 'pasteurized', 'sterilized' and 'ultra heat treated' (UHT) milks. The 1977 Regulations exemplified the attitude adopted in England and Wales by stipulating processing conditions and certain features required of the processing equipment rather than introducing microbiological standards. Thus with pasteurized milk temperatures are specified so that the milk should be held at a temperature of 'not less than 62·8°C and not more than 65·6°C for at least 30 min and be cooled immediately to a temperature of not more than 10°C' (i.e. in the 'holder' method) or, as is usual these days, the milk is held at '71·7°C for at least 15 s' before cooling (i.e. in the HTST method; see Section 3.7.1.2).

The Regulations also stipulate that the apparatus in which the milk is pasteurized 'shall be so constructed as to secure the protection of the milk from risk of atmospheric contamination'; and, again, the Regulations require that any apparatus in which the milk is heated and maintained at a temperature above 65·6°C 'shall be provided with a device which shall automatically divert the flow of any milk which is not raised to the authorised temperature'.

The 1986 Regulations retain all the above requirements but microbiological standards, to be used as guidelines only, are introduced to come into line with relevant EC legislation (i.e. Council Directive 85/397/EEC). Thus the standards for pasteurized milk are identical to those in Table 13.2 (step 2), whilst raw milk is allowed a total viable count maximum of 200 000 per ml (30°C incubation) and a coliform maximum of 100 per ml.

The two sets of Regulations both stipulate that raw milk must satisfy the methylene blue test (see Section 4.3.12.1) whilst pasteurized milk must pass the phosphatase test; this test is, in essence, a measure of the efficiency of pasteurization since the phosphatase enzyme is only destroyed when satisfactory heat treatments have been achieved.

For UHT and sterilized milks appropriate heat treatments are again specified (i.e. 'retained at a temperature of not less than 132·2°C for not less than one second' for UHT milk; 'heated to and maintained at such a temperature, not less than 100°C, for such a period as to ensure that it will comply with the turbidity test' for sterilized milk) as are requirements for the processing equipment. For these milks a microbiological standard is also stipulated (i.e. < 1000 organisms per ml using a 30–37°C incubation temperature for 2 days, for a sample pre-incubated at the same temperature for 24 h).

General hygiene legislation on milk is to be found principally in the Milk and Dairies (General Regulations) (1959). They require dairy farms and dairies to be registered, cattle to be in good health and inspected regularly, buildings to be suitably constructed, milk to be rapidly cooled after milking, milk to be protected against contamination, equipment to be thoroughly cleaned and milk to be distributed with suitable care.

13.4.6. Ice-Cream (Heat Treatment) Regulations (1959); Liquid Egg (Pasteurization) Regulations (1963)

These Regulations again illustrate the emphasis that has been placed on processing standards in the United Kingdom. The first set of Regulations specify that ice-cream mixture may not be kept for more than 1 h at a temperature above 7°C before heat treatment, after which the temperature must be reduced to 7°C or below within 1 h and so kept until frozen. Four alternative methods of heat treatment are permitted, viz. 65·6°C for 30 min, 71·1°C for 10 min, 79·4°C for 15 s and 149°C for 2 s. These requirements do not apply to mixtures with a pH of 4·5 or less, it being recognized that pathogenic bacteria could not multiply in such substrates.

The Regulations also include requirements for the heating apparatus; it must be capable of maintaining a constant flow of the mixture throughout the heating cycle and must be provided with a flow diversion system. Similar features are required for apparatus used in the pasteurization of liquid egg (Liquid Egg (Pasteurization) Regulations, 1963) but in this case the Regulations require that the liquid egg is held at 64·4°C for 2½ min and then immediately cooled to 3·3°C. The treated egg must pass the α-amylase test, this again being a check that pasteurization has been effective, the enzyme being destroyed if treatment has been adequate. Of course, a negative test result is no absolute guarantee that the egg is free of salmonellas; there have been a number of instances where the amylase has been destroyed but *Salmonella* spp. have been found in the liquid egg, presumably due to post-pasteurization contamination.

13.4.7. Enforcement of the Regulations

In the United Kingdom the principal responsibility for legislation enforcement is in the hands of elected local authority councils; such councils have a statutory obligation to employ officers concerned with the protection of health in the broadest sense. Until now Environmental Health Officers (EHOs) have been responsible for food hygiene inspection of premises and, where necessary, enforcement of the food hygiene legislation. Ideally their role is not concerned with enforcement but is primarily advisory and educational since there is a tendency to encourage transgressing food processors to correct their faults, unless of a serious nature, rather than prosecute them. Perhaps attitudes are now hardening and it will be interesting to see what effect the recent Food Safety Act (1990) and EC food hygiene legislation will have on the relationship between food manufacturers and caterers, and the enforcement authorities.

EHOs have been allowed a great deal of freedom in their actions but their powers are used with great discretion. They are empowered to confiscate and analyse foods (particularly important at ports and airports) but other duties can range from the investigation of food poisoning outbreaks to the running of courses in food hygiene. In order to perform their various duties effectively the EHOs must work in close conjunction with the local Medical Officer of Environmental Health, the local Public Analyst, Trading Standards Officers, the Public Health Laboratory Service, the DoH and MAFF; in the case of the last-named it will again be interesting to see how official veterinary officers and their qualified auxiliaries choose to interpret the legislation and how they liaise with EHOs.

13.5. FOOD LEGISLATION IN SOME EUROPEAN COUNTRIES

A detailed review of food legislation worldwide would obviously be well outside the scope of this book but, were such a review possible, it would soon require emendation. Hence the principal aim in the remainder of this chapter is to compare national food legislation in selected countries by taking examples of legislation related to microbiological or food hygiene principles. As we have seen, food legislation in the United Kingdom has been couched in general and practical terms and because of this enforcement is relatively easy. However, as already pointed out, national legislation of Member States must observe the intent of EC legislation so that there must be a common philosophy in those states for foods for which Directives have been issued. Of course, Member States have the right to impose additional legislation if thought necessary provided the intent of the legislation is respected and the free movement of goods between Member States remains unaffected.

In certain of the following countries legislation is much more comprehensive and perhaps because of this and the frequent lack of practicability, sanctions have been less vigorously applied; in the following examples attention is drawn to differences in the legislation rather than to the similarities.

13.5.1. France

The basic objective of French food legislation, operating through the much amended Act of the 1st August, 1905, was the repression of fraud rather than the protection of health although the latter was covered in 1983 by the Consumers' Safety Act. Food legislation in France is extremely detailed and different quality grades are applied to certain categories of foodstuffs (e.g. pasteurized milk) although the minimum standard acceptable is still specified.

The general hygiene regulations detail the processing conditions necessary for a wide range of food materials; these regulations embrace food premises, the equipment therein, raw materials, storage premises, food handlers, the transport of foods and conditions at the point of sale. Much of the more general food hygiene legislation was introduced in a recent Decree (1991) which must, of course, reflect EC legislation. In general, however, the regulations appear to be more stringent than those applying in the United Kingdom; for example, the list of harmful microorganisms which, if carried, requires food handlers to cease work is an extensive one and includes *Escherichia coli* and *Streptococcus haemolyticus* A.

The importance of temperature control is stressed at all stages. Maximum temperatures are given for the transport of frozen foods (e.g. frozen meat, $-12°C$; butter, $-14°C$). The regulations dealing with raw and the various types of processed milk are again extensive and here a maximum holding temperature of 4°C is stipulated for milk before pasteurization (cf. 10°C in the United Kingdom; Milk and Dairy (General) Regulations, 1959, as amended) although slight flexibility is allowed in that the temperature may rise to 8°C for up to 4 h. The importance of temperature control is again stressed for retailed foods; maximum temperatures permitted for perishable foods are: fish, 2°C; pre-packed cooked meats, 3°C; milk, cream and butter, 6°C; raw red meats, 7°C; eggs and soft cheeses, 8°C; hard cheeses, 15°C.

Table 13.3
Examples of Microbiological Standards Applied to Foods in France[a]

Commodity	Standard
Raw milk (at time of collection)	Total count: 500 000 per ml max
Pasteurized milk	Total viable count: 100 000 per ml max immediately after treatment, 200 000 per ml max at point of sale
	No microorganisms that would be destroyed by pasteurization at point of sale
Packaged pasteurized milk (4 days after packing)	Total aerobes: 30 000 per g max
	Coliforms: 10 per g max
	Faecal coliforms: absent in 1g
	S. aureus: 10 per g max
	Salmonellas: absent in 250 ml
Dried milk	Viable aerobic bacteria: <50 000 per g
	Coliforms: <5 per g
Raw cream	Faecal coliforms: 100 per g max
	S. aureus: 1000 per g max
	Salmonellas: absent in 25 g
Ices and ice-creams	Total aerobes: 300 000 per g max
	Coliforms: 100 per g max
	Faecal coliforms: 1 per g max
	S. aureus: 10 per g max
	Salmonellas: absent in 25 g
Pasteurized egg products	Total aerobes: 100 000 per g max
	Enterobacteria: 10 per g max
	S. aureus: 100 per g max
	Salmonellas: absent in 25 g
Sliced fish, crumbed or not; fresh fish fillets, refrigerated	Total aerobes: 50 000 per g max
	Faecal coliforms: 10 per g max
	S. aureus: 100 per g max
	Anaerobic sulphite reducers[b]: 2 per g max
	Salmonellas: absent in 25 g
Shellfish, bivalves, urchins live	Faecal coliforms: 300 per 100 ml max
	Faecal streptococci: 2500 per 100 ml max
	Salmonellas: absent in 25 g
Meat, minced in advance or on demand	Total aerobes: 500 000 per g max
	Faecal coliforms: <100 per g
	S. aureus: <100 per g
	Anaerobic sulphite reducers: <30 per g
	Salmonellas: absent in 25 g
Meat joints, packed under vacuum or not, chilled or frozen	Total aerobes: 50 000 per g max
	Faecal coliforms: 100 per g max
	Anaerobic sulphite reducers: 2 per g max
	Salmonellas: absent in 25 g

Table 13.3 — *contd.*

Commodity	Standard
Whole poultry refrigerated, chilled or frozen	Salmonellas: absent in 25 g of pectoral muscles
Roast meat, cooked, whole or sliced; stuffed meat slices, cooked or pre-cooked	Total aerobes: 300 000 per g max Faecal coliforms: 10 per g max *S. aureus*: 100 per g max Anaerobic sulphite reducers: 10 per g max Salmonellas: absent in 25 g
Cooked poultry meat, mechanically separated	As above but anaerobic sulphite reducers 30 per g max
Frogs' legs, shelled snails, chilled or frozen	*Clost. perfringens*: 1000 per g max Coliforms (frogs): 100 per g max *S. aureus* (frogs): 100 per g max Salmonellas (frogs): absent in 25 g Salmonellas (snails): absent in 1 g

[a] From Marshall (1986).
[b] Anaerobic sulphite reducers can be regarded as *Clostridium perfringens* together with a number of related species.

One of the areas where there is greatest detail is that relating to microbiological standards for foodstuffs. The full list of foods for which there are standards is an extensive one (Marshall, 1986) but a few examples are given in Table 13.3; general test methods are also included in the legislation. It is worth repeating that these strict microbiological standards can only be applied nationally and not on goods imported from another Member State unless the standards are in accord with EC legislation. However it would seem that the receiving Member State could utilize these stringent microbiological standards to block foods being imported, specifically on health or food safety grounds.

13.5.2. Italy

The main provisions of food legislation in Italy are to be found in the Act (No. 283) dated 30th April, 1962, and in a major later Amendment (Presidential Decree No. 327, 1980). The principal aim of Italian food legislation is to protect public health and the Act (1962) is concerned primarily with hygiene standards in the production of food and drink. It is necessary to obtain health authority licences for all food processing factories; applications for these licences must be accompanied by detailed information on the location of the premises, a description and plan of the premises, the types of food to be produced, the water supply, waste disposal arrangements and the cleaning systems to be employed. This legislation is similar in principle to more recent EC food hygiene legislation.

The regulations relating to the running of food factories and the health of employees are again relatively detailed and similar to EC legislation. Thus personnel employed in the production or handling of foodstuffs must possess a certificate of hygiene to show that they are neither suffering from an infectious or contagious disease nor are carriers of pathogenic microorganisms; employees are liable for work suspension should their health be suspect. Such discretionary clauses are frequent in Italian food legislation so the responsible authorities retain a degree of freedom in their judgements.

Food equipment must be designed for easy and efficient cleaning and the required characteristics for food contact surfaces are specified. Some attention is given to cleaning procedures and it is a requirement that, after the application of detergent and disinfectant to premises and equipment, residues must be removed by washing thoroughly with potable water. A similar cleaning protocol is also laid down for transport vehicles and these are also the subject of detailed legislation with regard to their construction; again these requirements have been incorporated in recent EC legislation. As in France, temperature control is again emphasized in Italian food legislation. There are standards for the transport of frozen foods (e.g. frozen meats, $-10°C$) and for a range of non-frozen foods (e.g. untreated or pasteurized milk in tankers, $0-4°C$; meats, $-1-4°C$; yoghurts, $0-4°C$).

There are relatively few microbiological standards for foods in Italy although the legislation allows for maximum microbial counts to be established where deemed necessary. At the present time dairy products are the principal foods covered (see Table 13.4) although, interestingly, there are no specific hygiene requirements for milk production. Microbiological standards are accompanied by details of the analytical methods to be used.

13.5.3. Germany

The German foodstuffs legislation is based on the Law on Trade in Foodstuffs, Tobacco Products, Cosmetics and Other Commodities which was fully revised in 1974 since when there have been further

Table 13.4
Examples of Microbiological Standards Applied to Foods in Italy[a]

Commodity	Standard
Raw milk	Total aerobes : 20 000 per ml max Coliforms: 100 per ml max
Pasteurized milk	Coliforms (in five samples): <0·3 per ml in four samples and <12 per ml in any one sample
UHT milk	Total mesophilic aerobes: 110 per ml max Thermophiles: 110 per ml max
Sterilized milk	Total mesophilic aerobes: 12 per ml max Thermophiles: 12 per ml max
Ice-cream (based on milk, cream or other milk derivatives)	Total aerobes (five samples): 100 000 per g max in four samples and 1×10^6 per g max in one sample Coliforms (five samples): 120 per g max in four samples and 1100 per g max in one sample
Frozen pasteurized liquid egg, white or yolk	Total aerobes (five samples): 100 000 per g max in three samples and 1×10^6 per g max in two samples Salmonellas: absent in 25 g
Pasta, stuffed, fresh (produced in shops)	Total aerobes (5 samples): 100 000 per g max in three samples and 1×10^6 per g in two samples S. aureus (5 samples): 1000 per g max in three samples and 10 000 per g in two samples Salmonellas: absent in 25 g
Pasta, stuffed, pre-cooked and frozen	Total aerobes (5 samples) : 100 000 per g max in three samples and 300 000 per g in two samples S. aureus: 100 per g max Clost. perfringens: 30 per g max E. coli: absent in 1 g Salmonellas: absent in 25 g

[a]From Marshall (1986).

minor revisions. Protection of human health is again the principal aim of the legislation but the position is complicated by the lack of detail in the federal legislation at the present time, particularly where food hygiene legislation is concerned. This possible weakness will be rectified when the proposed EC food hygiene legislation is agreed by Member States (see Section 13.2.2.8.). Nevertheless the General Provisions of the 1974 Law embrace the German Food Code which is a compilation of guiding principles for the production, condition and other characteristics of foodstuffs; these principles, which were formulated by the German Food Code Commission, take account of international food standards recognized by the federal government.

Much of the responsibility for food hygiene and safety legislation in Germany has been delegated to the ten separate states and city areas (i.e. Laender) and this same responsibility has been extended to the five new Laender that formally comprised East Germany. Each is able to formulate its own separate regulations and hygiene regulations have been issued by every Laender covering different foodstuffs such as ice-cream, and bakery and meat products; unfortunately these regulations are often different in the individual Laender and it may well prove difficult to reconcile these differences in the short term.

There is, of course, federal legislation in existence specific to foods where EC legislation requires it. Thus Council Directive 64/433/EEC and its amendments dealing with fresh red meat have been incorporated into German law as the 'Fleischhygienegesetz' (1987, 1991); likewise Council Directive 77/99/EEC covering meat products appears as the 'Fleischhygieneverordnung' (1986). The regulations remain essentially practicable as in the United Kingdom. For example, the legislation on red meats repeatedly stresses the importance of the easy cleaning and disinfection of factory walls and floors, and of the fittings and equipment therein. Hygiene requirements for personnel coming into contact with meats are not too demanding and amongst other regulations are: (1) clothing worn must be easily washable, clean and light; and (2) personnel infected with skin disorders and the usual intestinal complaints are not allowed to work in contact with meats. Maximum permitted storage and transportation temperatures are stipulated (e.g. fresh meat by-products, 3°C; deer and rabbit, 4°C; other fresh meats, 7°C; deep frozen meats, −18°C). Minimum processing temperatures/times for different categories of meats to be heated are also included (e.g. an internal temperature of at least 65° maintained for at least 10 min is required for good quality meats).

Other regulations detail bacteriological analyses that should be performed on core samples of meat from slaughtered animals; such samples should be examined for total numbers of viable bacteria, for faecal indicator bacteria and for salmonellas and clostridia.

The German view on microbiological standards is that they should only be introduced for health protection, that is when dealing with 'high risk' foods; this view is not unlike that held in the United Kingdom.

Thus microbiological standards for foods remain uncommon in Germany. The main foods covered by federal legislation have been raw milk and egg products although other foods have now been included through EC legislation. Four classes of raw milk, at the farm, have been delineated based on total aerobic counts. These are:

Class 1 — count should not exceed 100 000 per ml
Class 2 — count should not exceed 300 000 per ml
Class 3 — count should not exceed 800 000 per ml
Class 4 — count above 800 000 per ml

In addition, coliforms should be absent in 0·1 ml, *Streptococcus agalactiae* in 1 ml and salmonellas should also be absent (Marshall, 1986).

13.6. FOOD LEGISLATION IN THE USA

The principal Federal food legislation in the United States is the Food, Drug and Cosmetic Act (1938) through which the Federal Food and Drug Administration (FDA) is able to control the food industry. The Act, which has been much amended, covers all foods other than meats and poultry and meat and poultry products. These latter foods are covered by the Federal Meat Inspection Act, the Poultry Products Inspection Act and the Egg Products Inspection Act which are under the jurisdiction of the United States Department of Agriculture (USDA). All four Acts are fairly general in content and their intentions are implemented through the Code of Federal Regulations which is revised and published annually. The Code, which is subject to Federal jurisdiction, is divided into 50 Titles which represent very general areas of interest. With the Federal system of control each State has the power to pass its own laws although the intent of any relevant Federal legislation must always be respected. In practice, detailed State legislation is often not required and in this way the system is in contrast to that pertaining in Germany at the present time: nonetheless, States do vary considerably in their attitudes to legislation although many have agreements with the Federal agencies whereby individual States enforce Federal food legislation.

There are two types of Federal regulation in existence in the US these being (a) 'substantive' (enforceable) and (b) 'advisory' (for guidance only), but the regulations covering food hygiene are all of the former type. Much emphasis is placed on the Good Manufacturing Practice (GMP) regulations which embody basic hygiene principles. The GMP regulations outline the minimum acceptable standards for the processing and storage of foods in general.

13.6.1. The GMP Regulations†

The GMP regulations are divided into four subparts, viz. General Provisions, Buildings and Facilities, Equipment, Production and Process Control, and salient points from these are described in turn.

1. *General Provisions.* Included in this subpart are regulations concerning personnel. No person affected by disease or boils, sores, infected wounds, etc., may work in food premises where there is a reasonable possibility of food being contaminated or other personnel infected. Personal cleanliness must be maintained and hygienic practices observed. Personnel responsible for hygiene and for the production of wholesome food should have the necessary education or experience to provide a level of competence required to achieve these aims.

2. *Buildings and Facilities.* The need for clean surrounds to factories is stressed so that pests are not attracted and food not contaminated. Factories should be constructed so that cleaning is facilitated and adequate working and storage space should be provided. Floors, walls and ceilings should be constructed to allow for adequate cleaning whilst fixtures, ducts and pipes should not be suspended over working areas where drip or condensate may contaminate food materials or food contact surfaces. The minimum requirements for water, waste disposal, toilet and hand-wash facilities are described.

    In the section 'Sanitary Operations' criteria for the general maintenance and cleaning of buildings and equipment are described; also included are the general requirements for pest control. The need to minimize the danger of contamination of food and food contact surfaces is stressed repeatedly.

3. *Equipment.* This subpart includes a very brief description of desirable features in the design and construction of food processing

†Appears as Part 110. Current Good Manufacturing Practice in Manufacturing, Packing or Holding Human Food (Subparts A to E). In: Code of Federal Regulations, *Title 21—Food and Drugs* (see Bibliography).

equipment. Design and construction must ensure that contamination of food with lubricants, fuel, metal fragments, etc. is avoided; ease of cleaning is stressed. Food contact surfaces must be corrosion resistant, made of non-toxic materials and be resistant to the action of any food material, cleaning or sanitizing agent with which they may come in contact. Seams of such surfaces should be smoothly bonded or maintained so as to minimize the accumulation of any debris.

4. *Production and Process Controls.* This subpart includes regulations for the inspection, cleaning and storage of raw materials, restrictions on the use of food premises and equipment, and regulations on the cleaning and maintenance of equipment. The need to protect food from contamination is again stressed and conditions are given which affect the growth of microorganisms in food. Thus food manufacturing operations should be conducted under conditions that minimize the growth of such organisms. Careful monitoring is therefore required at all stages of production by use of chemical, microbiological and other testing methods to check food quality and hygiene failures. Raw materials and other ingredients that are susceptible to contamination with aflatoxin and other toxins must comply with current FDA regulations and guidelines. Processing equipment must be maintained in an acceptable condition through cleaning and disinfection when necessary. Also covered are packaging processes and the need to use materials which do not contaminate the product in any way; absence of contamination is also stressed for the storage and transport of finished products and in these products conditions should be such as to prevent the growth of harmful organisms and the deterioration of the product.

It is accepted that some foods may contain natural or unavoidable defects that are not hazardous to human health. Maximum levels for such defects in foods are established by the FDA, based on what is feasible under GMP. Mixing of food contaminated above acceptable levels with sound material to dilute the defect is not permissible.

As can be seen from the foregoing, these GMP regulations are of a general nature stating what should be possible in reasonable circumstances; the regulations are similar in many ways, therefore, to the food hygiene legislation found in the United Kingdom and much of that adopted by the EC.

13.6.2. Regulations Concerning Specific Foods

In addition to the 'umbrella' GMP regulations, specific regulations have
been adopted for a wide range of foodstuffs; amongst these are canned
foods, frozen foods, bakery products, fruit pies, eggs and egg products,
fish and shellfish, and milk and cream. Many detailed recommendations
are given for the processing of these foods and the principal aims are to
ensure their safety and wholesomeness. There is much laudable information detailing suitable storage conditions for foods, appropriate cleaning
and disinfection techniques, suitable equipment for specific processes
and methods of testing different foods; the keeping of records is also
covered.

The regulations dealing with meat and poultry, and their products,
and the regulations dealing with dairy products will now be considered
in greater depth in order to illustrate the detailed legislation which exists
in the United States.

13.6.2.1. Meat and Poultry

The United States Department of Agriculture (USDA) has legal control, by means of the three Inspection Acts mentioned earlier, over the
inspection, handling and processing of meat and poultry, and their
products. Part 304 of the relevant Regulation† requires that before permitting an establishment to become operational, detailed plans must be
submitted for inspection showing the locations of the principal items of
equipment, drains, handwash facilities, hose connections for cleaning,
other buildings on the land and the principal roads and railways serving
the factory; a schedule showing the finish of walls, floors and ceilings
must also be provided. Specifications must also be submitted covering
general drainage, water supplies, refrigeration and other equipment,
lighting and proposed cleaning and inspection programmes.

Once approval has been granted, Federal inspectors have the right of
access to meat and poultry premises at all times. Care in the inspection
of animals is stressed as is the need for strict control over the disposal
of condemned meat. Hygienic conditions must be maintained during all
stages of processing by strict adherence to the GMP regulations which
have been expanded in a specific way so as to apply to both meat and

---

†Appears as Part 304. Application for Inspection: Grant or Refusal of
Inspection. In: Code of Federal Regulations, *Title 9—Animals and Animal
Products* (see Bibliography).

poultry processing. Thus Part 308† draws attention to: (1) the need for good lighting and ventilation; (2) the use of potable water for nearly every operation including floor washing; (3) the availability of hot water (at least 82°C) for cleaning purposes; (4) the construction materials for floors, walls, etc. which must be readily cleanable; (5) the requirements for sanitary facilities and accommodation for personnel. Before food processing equipment can be approved plan drawings and details of the proposed construction materials must be submitted. In the light of experience and over a period of time a list of approved equipment has been compiled and can be obtained from Scientific Services, Meat and Poultry Inspection Program, Food Safety and Inspection Service, US Department of Agriculture, Washington, D.C. 20250, USA.

Part 318‡ represents the heart of the legislation covering meats and meat products and there is much of interest in it. For example, it is possible for a manufacturer to obtain what is known as 'Total Plant Quality Control', a system that confers official USDA approval of the premises and enables the manufacturer to include a logo (Fig. 13.1) in

Fig. 13.1. Labelling logo permitted on products prepared at USDA Total Plant Quality Control establishments.

†Appears as Part 308. Sanitation. In: Code of Federal Regulations, *Title 9— Animals and Animal Products* (see Bibliography).
‡Appears as Part 318. Entry into Official Establishments: Reinspection and Preparation of Products. In: Code of Federal Regulations, *Title 9—Animals and Animal Products* (see Bibliography).

Table 13.5

Examples of Temperature/Time Heating Profiles for Cooked and Roast Beef, and Cooked Corned Beef[a]

Minimum internal temperature (°C)	Minimum processing time (min) after minimum temperature is reached
55	97
57	37
60	12
62	5

[a]From Code of Federal Regulations (1988). *Title 9—Animals and Animal Products.*

any label. To obtain this approval the quality control system must submit information on methods of raw material control, critical control points in the process(es), the nature and frequency of tests to be carried out, and the remedial or corrective action which can be called on where necessary; in addition it is necessary to show that the person ultimately responsible for quality control is not linked to production.

Where cooked meat products are concerned processing requirements are given in considerable detail. Included are the minimum temperature/time combinations that are necessary when cooking meats and examples of these are given in Table 13.5; cooking profiles must be monitored, preferably with recording devices to assure accurate time, temperature and humidity control. A maximum delay of 2 h for precooking preparation of any meat product is permitted after which it must either be cooked or held at 4·4°C or lower. After cooking, cooling must commence within 90 min and products must be chilled from 49 to 13°C in no more than 6 h and be further chilled to 4·4°C. To ensure uniform heat treatments for different joints, each should not vary in weight by more than 900 g nor in thickness by 5 cm at the thickest part.

The importance of avoiding cross-contamination of cooked by raw products, either directly or indirectly, is stressed in the following way:

Cooked (meats) shall be so handled as to assure that the product is not recontaminated by direct contact with raw product. To prevent direct contamination of the cooked product, establishments shall (1) physically separate areas where raw product is handled from areas where exposed cooked product is handled, using a solid impervious floor to ceiling wall; or (2) handle raw and exposed cooked product at different times, with a cleaning of the entire area after the raw material handling is completed and prior to the handling of cooked product in that area; or (3) submit a written procedure for approval....

To prevent indirect contamination of cooked product, (1) any work surface, machine, or tool which contacts raw product shall be thoroughly cleaned and sanitized with a solution germicidally equivalent to 50 ppm chlorine before it contacts cooked product; (2) employees shall wash their hands and sanitize them with a solution germicidally equivalent to 50 ppm chlorine whenever they enter the heat processed product area or before preparing to handle cooked product, and as frequently as necessary during operations to avoid product contamination; and (3) outer garments including aprons, smocks, and gloves shall be especially identified as restricted for use on cooked product only, changed at least daily, and hung in a designated location when the employee leaves the area.

Cooked product shall not be stored in the same room as raw product unless it is first packaged in a sealed water-tight container or is otherwise protected by a covering that has been approved....

Operating procedures are dealt with in depth in Part 381† when poultry is considered. Here the need for rapid chilling after killing is exemplified by the requirement that a bird of some 2 kg weight must be chilled to 5°C within 6 h unless it is immediately frozen or cooked. Regulations on chilling techniques include the proviso that water in chill tanks should never exceed 18°C. As with EC legislation spin-chiller systems are permitted although again, as with the EC, detailed requirements concerning the volume and quality of water and ice to be used are included. Poultry carcasses and pieces may be held overnight in static chilling tanks provided the temperature does not exceed 4·5°C. Regulations for air-chilled ready-to-cook poultry are less stringent and the cooling period is here extended to 16 h.

13.6.2.2. *Dairy Products*‡
As with meat and poultry plants, USDA approval can be granted following a survey of the premises, storage facilities, equipment, the vol-

---

†Appears as Part 381. Poultry Products Inspection Regulations. In: Code of Federal Regulations, *Title 9—Animals and Animal Products* (see Bibliography).
‡Appears as Part 58. Grading and Inspection, General Specifications for Approved Plants and Standards for Grades of Dairy Products. In: Code of Federal Regulations, *Title 7—Agriculture* (see Bibliography).

ume of raw material processed daily and the product handling facilities at the plant; surveys must be performed biannually to ensure that standards are being maintained. Detailed information is supplied on premises and building requirements extending the general principles laid down in the GMP regulations. For example, driveways and adjacent traffic areas should be of cement, asphalt or similar material to keep dust and mud to a minimum. The immediate surroundings must be free of refuse, overgrown vegetation and waste materials to prevent harbourage of rodents and other pests. Buildings and all openings to the outside should provide an effective barrier against the ingress of pests and dirt. Internally, new buildings should have window sills which slope at an angle of 45°; processing and storage rooms should have walls, ceilings and partitions which are smoothly finished with a suitable material, light in colour, substantially impervious to moisture and readily cleanable. Floors should be constructed of tiles, concrete or other equally impervious materials; all jointing compounds should also be impervious. Wood floors are only permitted for rooms storing 'certain packaged finished products' and for miscellaneous materials, e.g. containers. The importance of good lighting, heating and ventilation is stressed and additional criteria are laid down for rooms used for certain specified purposes; included here is the laboratory and the following passage is pertinent:

> ... an adequately equipped laboratory shall be maintained and properly staffed with qualified and trained personnel for quality control and analytical testing. The laboratory should be located reasonably close to the processing activity and be of sufficient size to perform tests necessary in evaluating the quality of raw and finished products.

Equipment used for processing milk should be constructed in such a way that it is easily dismantled for cleaning. Product contact surfaces should be constructed of stainless steel or other materials that are equally corrosion resistant. Non-metallic parts in contact with the product should comply with the appropriate 3-A Sanitary Standards. All pipework and fittings should be easily accessible for cleaning and inspection, and should be kept in good repair; all CIP systems should comply with 3-A Accepted Practices. Requirements for individual items of equipment are also described with frequent reference to 3-A standards and practices.

Table 13.6
Minimum Temperature/Time Heating Profiles for Pasteurization of Fluid Products[a]

Temperature (°C)	Time
62·7 — vat pasteurization	30 min
71·6 — high temperature, short time	15 s
88·3 ⎫	1·0 s
90·0 ⎪	0·5 s
93·9 ⎬ higher heat, shorter time	0·1 s
95.5 ⎪	0·05 s
100 ⎭	0·01 s

[a]From Code of Federal Regulations (1991). *Title 7—Agriculture.*

The importance of accurate heat treatments and their effective monitoring is again stressed. This need for accuracy is of paramount importance when, as in this case, a number of different heat treatments are acceptable; these treatments are illustrated in Table 13.6.

In dealing with the protection and transport of raw milk it is somewhat surprising to find that milk may still be transported in cans although, in such cases, it is stipulated that the milk should either be delivered to the processing depot within 2 h of milking or should be cooled 'immediately after milking' to 10°C or lower. Bulk cooling tank systems should be capable of cooling milk to 40°F (4·4°C) within 2 h of milking and the milk should be held at or below 45°F (7·2 °C) until collected; a maximum pre-collection storage period is no longer specified for raw milk.

In the processing of liquid milk the importance of correct operating procedures, after reception at the depot, is underlined by the detailed nature of the regulations. For example, the conditions for the storage of raw milk are clearly defined (it must normally be maintained at 7°C or lower until processing commences) as are the precise details of cleaning treatments; the need to maintain adequate records, for periods of up to 1 year, is also stressed.

13.6.3. Microbiological Standards in the USA

In the USA there are relatively few Federal microbiological standards for foods other than for milk and milk products which are USDA ap-

Table 13.7
Examples of Microbiological Standards for Foods in the USA[a]

Commodity	Standard
Raw milk (no. 1 classification)	500 000 organisms per ml max (by direct microscopic or plate count)
Raw milk (no. 2 classification)	3×10^6 organisms per ml max (by direct microscopic or plate count)
Butter	Proteolytic count: 100 per g max
	Coliforms: 10 per g max
	Yeasts and moulds: 20 per g max
Frozen cream	Total count: 30 000 per g max
	Coliforms: 10 per g max
	Yeasts and moulds: 20 per g max
Cottage cheese	Coliforms (three out of five consecutive samples): 10 per g max
	Psychrotrophs (as above): 100 per g max
	Yeasts and moulds (as above): 10 per g max
Ice-cream	Total count: 50 000 per g max
	Coliforms: 10 per g max but 20 per g max for flavoured ice-cream (three out of five consecutive samples)
Spray dried milk (non-fat)	Total count: 50 000 per g max
Dried whole milk (Extra grade)	Total count: 50 000 per g max
	Coliforms: 10 per g max
Dried whole milk (Standard grade)	Total count: 100 000 per g max
	Coliforms: 10 per g max
Frozen fish cakes[b]	Viable aerobic bacteria: 100 000 per g max (in three or more of five samples) or 1×10^6 per g max in any one of five samples
	Coliforms: 150 per g max (in three or more of five samples) or 1500 per g max in any one of five samples
Crab meat, fresh or frozen[b]	*E. coli*: 4 per g max
Shrimps, raw, breaded[b]	Viable aerobic bacteria: 100 000 per g max
	E. coli: 4 per g max
	S. aureus: 100 per g max

[a]From Code of Federal Regulations (1991). *Title 7—Agriculture.*
[b]From Marshall (1986).

proved; the remaining standards are limited to various fish products (Marshall, 1986). Examples are given in Table 13.7. The raw milk standards also specify standard times for the resazurin reduction test; this test is similar in principle to the methylene blue reduction test mentioned earlier (see Section 4.3.12.1).

At the present time Federal microbiological standards or, more correctly, guidelines which are not strictly enforceable by law, are being developed for foods not yet included in the Code of Federal Regulations. The responsibility for these changes is in the hands of a National Advisory Group composed of several government and private agencies including, for example, the Food Safety and Inspection Service of the Department of Agriculture.

Many States have their own microbiological standards or, more usually, guidelines. In a survey performed by Wehr in 1978 it was found that 21 States had some microbiological criteria or guidelines, a figure that had increased to 25 States in a later survey (Wehr, 1982). The latter indicated that the most extensive lists of foods covered were for Ohio, Georgia and Rhode Island with about 20 guidelines each. Standards included those for total counts, coliforms, *E. coli*, salmonellas and *S. aureus* but Nebraska and Ohio also included yeast and mould counts for a number of foods. Generally an enormous but often different range of foods has been covered by State legislation with different types of shellfish, various raw and cooked meats, and salads being most commonly ascribed microbiological criteria. Few States have specific microbiological standards for foods as opposed to guidelines. The State of New Jersey has nine such standards covering salad foods (e.g. chicken salad: total count, 100 000; coliforms, 100: *S. aureus*, 100; salmonellas, absent—all figures quoted per gram). The majority of States would have preferred Federal guidelines to be introduced and hoped that they would be developed and issued in the future although some doubted whether it was possible to relate microbial counts to product quality.

13.7. FOOD LEGISLATION IN JAPAN

The principal legislation concerned with food hygiene in Japan is the Food Sanitation Law (1983) which is the latest revision of the earlier and similarly titled law introduced in 1947. The purpose of the law, which is enforced through Ordinances and Regulations, is to prevent health hazards arising from the consumption of food and in so doing to promote and extend public health in general.

The general food hygiene legislation is detailed but written in practical terms without stringent constraints. There are regulations dealing with the construction of buildings and equipment, in both instances the

Table 13.8
Examples of Microbiological Standards for Foods in Japan[a]

Commodity	Standard
Raw milk	Total count of bacteria: 4×10^6 per ml max
Pasteurized milk	Total count of bacteria: 50 000 per ml max
	Coliforms: absent in 1 ml
Cream, pasteurized	Total count of bacteria: 100 000 per ml max
	Coliforms: absent in 1 ml
Ice-cream	Total count of bacteria: 100 000 per g max
	Coliforms: absent in 1 ml
Frozen foods	
require only thawing	Total viable bacteria: <100 000 per g
before consumption;	Coliforms: absent in 1 g
heated before consumption	Total viable bacteria: $<3 \times 10^6$ per g
	E. coli: absent in 1 g
Meat and whale meat products	Coliforms: absent
Raw oysters	Total viable bacteria: 50 000 per g max
	E. coli: 230 per 100 g max

[a]From Marshall (1986).

requirement for easy cleaning being stressed. Equipment handling food should be washed at least once per day; surfaces coming into direct contact with food are to be cleaned 'especially carefully' and should be disinfected, after washing, with hot water at not less than 76·5°C or with a safe bactericide. The regulations concerned with the health of food handlers include clauses excluding them from work if suffering from infectious diseases or suppurating wounds or 'growths on any part of the body'; the importance of clean habits is also emphasized. Penalties for infringing the regulations are severe with heavy fines or imprisonment for up to 3 years being possible.

13.7.1. Microbiological Standards in Japan

In Japan, microbiological standards exist for a number of foods (Marshall, 1986) and examples of these are illustrated in Table 13.8. The tests used are limited almost exclusively to estimations of the total numbers of bacteria and the number of coliforms. On the other hand the methodology of the various tests is given in great detail and is very explicit and this must be regarded as one of the strengths of the Japanese legislation.

13.8. CONCLUDING REMARKS

This chapter has dealt with aspects of food hygiene and safety legislation and there are many similarities amongst the different countries which have been reviewed. This becomes increasingly inevitable for Member States within the EC who are aiming for harmonization by the end of 1992. However both the United States and Japan have much food legislation in a similar vein to that being introduced within the EC. It is primarily the depth and detail of the regulations which vary from country to country.

The United Kingdom has adopted a more lenient attitude, especially towards microbiological standards and temperature control of foods. But microbiological standards are being progressively introduced through EC legislation and the recent Food Hygiene (Amendment) Regulations (1990) have brought in stringent temperature controls even if holding times have not been specified. Both the United Kingdom and Germany have preferred to place the emphasis on processing standards to achieve the necessary food quality and safety. In contrast, France has an extensive list of foods covered by microbiological standards to support the food hygiene legislation.

A perusal of the microbiological standards shows that acceptable microbiological levels can vary significantly for a specific food in those countries where standards are applied. For example, ice-cream standards (expressed as total count of bacteria per gram) can vary from 50 000 (in the United States), through 300 000 (France) to 1 million (for one sample in five, Italy). For the same food, coliform levels can vary from 'absent in 1 ml' (Japan) to 1100 per g (Italy); in addition, France specifies standards for *S. aureus* and salmonellas. The United States is unusual in including yeast and mould counts as standards for certain foods and this format is repeated for a small number of individual States.

With such diverse approaches to the implementation of microbiological standards it is little wonder that the advent of truly international microbiological standards remains nearly as far away as it was some two to three decades ago when first mooted. Progress has not been helped by the difficulty of first agreeing on standardized testing techniques. Countries favour certain methods, sometimes for nationalistic rather than scientific or technical reasons but prejudices are slowly being broken down. The aim must always be to agree on methods giving the highest and most consistent recoveries of microorganisms provided that the constraints of finance and technical simplicity are satisfied.

In the final analysis, however, all the legislation in the world does not guarantee the quality and safety of the food that is consumed. Far more depends on the sincerity with which the legislation is applied and whether or not it is applied at all. A mass of food hygiene legislation and detailed microbiological standards are meaningless if the legislation is impossible to apply in practice.

Of greatest importance, however, is the care taken in the whole chain of food handling operations. This chain extends from the raw food materials right through to the finished product at the point of purchase; in reality it even extends to consumer treatment thereafter. The quality of raw materials, the environment within the food processing factory, the processing standards applied and the attitude of factory personnel are all of crucial importance. All these aspects relate to that vital ingredient, food hygiene, which must be taken so very seriously.

BIBLIOGRAPHY

Anon. (1982). *Hygiene Alimentaire. Textes Généraux*. Paris, Journal Officiel de la République Française.

Code of Federal Regulations. (1991). *Title 7—Agriculture*, Parts 53 to 209. Washington, Office of the Federal Register.

Code of Federal Regulations. (1991). *Title 21—Food and Drugs*, Parts 100 to 169. Washington, Office of the Federal Register.

Code of Federal Regulations. (1988). *Title 9—Animals and Animal Products*, Parts 200 to end. Washington, Office of the Federal Register.

Codex Alimentarius Commission. (1986). *Procedural Manual*, 6th edn. Rome, FAO/WHO.

Dehove, R. A. (1981). *La Reglementation des Produits Alimentaires et Autres. Qualité et Répression des Fraudes*, 10th edn. Paris, Commerce Editions.

Department of Health (1991). *EC Single Market Update. Food Safety and Hygiene*, 5th edn. Dept of Health, EHF3A, July.

Fallows, S. J. (1988). *Food Legislative System of the UK*. London, Butterworths.

Flowerdew, D. W. (1985). *A Guide to the Food Regulations in the United Kingdom*, 3rd edn. Leatherhead, British Food Manufacturing Research Association.

Food Sanitation Law. (1983). Tokyo, Japan Food Hygiene Association.

Howells, G. G., Bradgate, J. R. & Griffiths, S. M. (1990). *Blackstone's Guide to the Food Safety Act 1990*. London, Blackstone Press Ltd.

International Business Intelligence. (1989). *1992—Planning for the Food Industry*. London, Butterworths and Birmingham, Eurofi plc.

Japanese Legal Requirements for Food and Additives. (1981). Tokyo, Japanese External Trade Organization.

Jukes, D. J. (1987). *Food Legislation of the UK*, 2nd edn. London, Butterworths.

Katsuyama, A. M. & Strachan, J. P. (1980). *Principles of Food Processing Sanitation*. Washington, Food Processors Institute.

Olsen, J. C. (1978). Microbiological specifications for foods: international activities. *Food Technology*, **32**(1), 55–7.

REFERENCES

Commission Decision, 85/446/EEC. (1985). Concerning the on-the-spot inspections to be carried out in respect of the intra-Community trade in fresh meat. *Official Journal of the European Communities*, **L260**, 19–21

Commission Decision, 91/180/EEC. (1991). Laying down certain methods of analysis and testing of raw milk and heat-treated milk. *Official Journal of the European Communities*, **L93**, 1–48.

Council Directive, 64/433/EEC. (1964). On health problems affecting intra-Community trade in fresh meat. *Official Journal of the European Communities*, **L121**, 2012–64.

Council Directive, 69/349/EEC. (1969). On health problems affecting intra-Community trade in fresh meat. *Official Journal of the European Communities*, **L256**, 431–9.

Council Directive, 71/118/EEC. (1971). On health problems affecting trade in fresh poultry meat. *Official Journal of the European Communities*, **L55**, 23–39.

Council Directive, 72/462/EEC. (1972). On health and veterinary inspection problems upon importation of bovine animals and swine and fresh meat from third countries. *Official Journal of the European Communities*, **L302**, 7–31.

Council Directive, 75/431/EEC. (1975). Amending Directive 71/118/EEC on health problems affecting trade in fresh poultry meat. *Official Journal of the European Communities*, **L192**, 6–26.

Council Directive, 77/99/EEC. (1977). On health problems affecting intra-Community trade in meat products. *Official Journal of the European Communities*, **L26**, 85–100.

Council Directive, 78/50/EEC. (1978). Supplementing, as regards the chilling process, Directive 71/118/EEC on health problems affecting trade in fresh poultry meat. *Official Journal of the European Communities*, **L15**, 28–31.

Council Directive, 79/923/EEC. (1979). On the quality required for shellfish waters. *Official Journal of the European Communities*, **L281**, 47–52.

Council Directive, 80/777/EEC. (1980). On the approximation of the laws of the Member States relating to the exploitation and marketing of natural mineral waters. *Official Journal of the European Communities*, **L229**, 1–10.

Council Directive, 80/778/EEC. (1980). Relating to the quality of water intended for human consumption. *Official Journal of the European Communities*, **L229**, 11–29.

Council Directive, 83/90/EEC. (1983). Amending Directive 64/433/EEC on health problems affecting intra-Community trade in fresh meat. *Official Journal of the European Communities*, **L59**, 10–33.

Council Directive, 83/201/EEC. (1983). Establishing exceptions from Directive 77/99/EEC for certain products which contain other foodstuffs and only a small percentage of meat or meat product. *Official Journal of the European Communities*, **L112**, 28–30.

Council Directive, 85/323/EEC. (1985). Amending Directive 64/433/EEC on health problems affecting intra-Community trade in fresh meat. *Official Journal of the European Communities*, **L168**, 43–4.

Council Directive, 85/324/EEC. (1985). Amending Directive 71/118/EEC on health problems affecting trade in fresh poultry meat. *Official Journal of the European Communities*, **L168**, 45–6.

Council Directive, 85/397/EEC. (1985). On health and animal-health problems affecting intra-Community trade in heat-treated milk. *Official Journal of the European Communities*, **L226**, 13–29.

Council Directive, 88/288/EEC. (1988). Amending Directive 64/433/EEC on health problems affecting intra-Community trade in fresh meat. *Official Journal of the European Communities*, **L124**, 28–30.

Council Directive, 88/657/EEC. (1988). Laying down the requirements for the production of, and trade in, minced meat, meat in pieces of less than 100 grams and meat preparations and amending Directives 64/433/EEC, 71/118/EEC and 72/462/EEC. *Official Journal of the European Communities*, **L382**, 3–14.

Council Directive, 88/658/EEC. (1988). Amending Directive 77/99/EEC on health problems affecting intra-Community trade in meat products. *Official Journal of the European Communities*, **L382**, 15–31.

Council Directive, 89/227/EEC. (1989). Amending Directive 72/462/EEC and 77/99/EEC to take account of the introduction of public health and animal health rules which are to govern imports of meat products from third world countries. *Official Journal of the European Communities*, **L93**, 25–35.

Council Directive, 89/397/EEC. (1989). On the official control of foodstuffs. *Official Journal of the European Communities*, **L186**, 23–6.

Council Directive, 89/437/EEC. (1989). On hygiene and health problems affecting the production and the placing on the market of egg products. *Official Journal of the European Communities*, **L212**, 87–100.

Council Directive, 91/492/EEC. (1991). Laying down the health conditions for the production and the placing on the market of live bivalve molluscs. *Official Journal of the European Communities*, **L268**, 1–14.

Council Directive, 91/493/EEC. (1991). Laying down the health conditions for the production and the placing on the market of fishery products. *Official Journal of the European Communities*, **L268**, 15–34.

Council Directive, 91/497/EEC. (1991). Amending and consolidating Directive 64/433/EEC on health problems affecting intra-Community trade in fresh meat to extend it to the production and marketing of fresh meat. *Official Journal of the European Communities*, **L268**, 69–104.

Council Directive, 92/5/EEC. (1992). Amending and updating Directive 77/99/EEC on health problems affecting intra-Community trade in meat products and amending Directive 64/433/EEC. *Official Journal of the European Communities*, **L57**, 1–26.

Marshall, J. P. (1986). *Microbiological Standards for Foodstuffs*, 2nd edn. Food

Legislation Surveys, No.9. Leatherhead, Leatherhead Food Research Association.

Proposal, Com(89)492, for a Council Regulation (EEC). (1989). Laying down general health rules for the production and placing on the market of products of animal origin and specific health rules for certain products of animal origin. *Official Journal of the European Communities*, **C327**, 29–37.

Proposal, Com(89)667, for a Council Regulation (EEC). (1990). Laying down health rules for the production and placing on the market of raw milk, of milk for the manufacture of milk-based products and of milk-based products. *Official Journal of the European Communities*, **C84**, 112–120.

Proposal, Com(89)672, for a Council Regulation (EEC). (1990). Adopting health rules for the production and placing on the market of heat-treated drinking milk. *Official Journal of the European Communities*, **C84**, 130–41.

Proposal, Com(90)259 (amended), for a Council Directive (EEC). (1990). Concerning general product safety. *Official Journal of the European Communities*, **C156**, 8–14.

Proposal, Com(91)310, for a Council Regulation (EEC). (1991). Concerning measures for the prevention of specified zoonoses and of specified zoonotic agents in animals and products of animal origin in order to prevent outbreaks of food-borne infections and intoxications. *Official Journal of the European Communities*, **C253**, 2–12.

Wehr, H. M. (1978). Attitudes and policies of state governments. *Food Technology*, **32**(1), 63–7.

Wehr, H. M. (1982). Attitudes and policies of governmental agencies on microbial criteria for foods—an update. *Food Technology*, **36**(9), 45–54 and 92.

INDEX

Note: main references are indicated by **bold page numbers**;
text in Figures and Tables by *italic page numbers*

495